Xingni Zhou, Qiguang Miao and Lei Feng

Programming in C

Also of interest

Programming in C, vol. 2: Composite Data Structures and Modularization
Xingni Zhou, Qiguang Miao, Lei Feng 2020
ISBN 978-3-11-069229-7, e-ISBN (PDF) 978-3-11-069230-3,
e-ISBN (EPUB) 978-3-11-069250-1

C++ Programming
Li Zheng, Yuan Dong, Fang Yang, 2019
ISBN 978-3-11-046943-1, e-ISBN (PDF) 978-3-11-047197-7,
e-ISBN (EPUB) 978-3-11-047066-6

Elementary Synchronous Programming
Ali S Janfada, 2019
ISBN 978-3-11-061549-4, e-ISBN (PDF) 978-3-11-061648-4,
e-ISBN (EPUB) 978-3-11-061673-6

MATLAB® Programming
Dingyü Xue, 2020
ISBN 978-3-11-066356-3, e-ISBN (PDF) 978-3-11-066695-3,
e-ISBN (EPUB) 978-3-11-066370-9

Programming in C++
Laxmisha Rai, 2019
ISBN 978-3-11-059539-0, e-ISBN (PDF) 978-3-11-059384-6,
e-ISBN (EPUB) 978-3-11-059295-5

Xingni Zhou, Qiguang Miao and Lei Feng

Programming in C

Volume 1: Basic Data Structures and Program
Statements

DE GRUYTER

Author

Prof. Xingni Zhou
School of Telecommunication Engineering
Xidian University
Xi'an, Shaanxi Province
People's Republic of China
xnzhou@xidian.edu.cn

Qiguang Miao
School of Computer Science
Xidian University
Xi'an, Shaanxi Province
People's Republic of China
qgmiao@xidian.edu.cn

Lei Feng
School of Telecommunication Engineering
Xidian University
Xi'an, Shaanxi Province
People's Republic of China
fenglei@mail.xidian.edu.cn

ISBN 978-3-11-069117-7
e-ISBN (PDF) 978-3-11-069232-7
e-ISBN (EPUB) 978-3-11-069249-5

Library of Congress Control Number: 2020940233

Bibliographic information published by the Deutsche Nationalbibliothek
The Deutsche Nationalbibliothek lists this publication in the Deutsche Nationalbibliografie;
detailed bibliographic data are available on the Internet at http://dnb.dnb.de.

© 2020 Walter de Gruyter GmbH, Berlin/Boston
Cover image: roberthyrons/iStock/Getty Images Plus
Typesetting: Integra Software Services Pvt. Ltd.
Printing and Binding: CPI books GmbH, Leck

www.degruyter.com

Preface

Ideas of the book

This book is written based on years of teaching experience. To clear up students' confusion in programming learning, it is more focussed on introducing problems, analyzing problems, and discussing solutions to the problems. In the process of teaching programming courses (e.g., C programming, data structure) and discussing with students after class, we discovered that the common problems in programming learning could be generalized as four challenges: (1) concepts are too abstract to understand, (2) there are too many rules to remember, (3) there are few general principles to follow in programming, and (4) it is hard to debug programs.

Leonhard Euler, a well-known mathematician, once said that teaching mathematics would be meaningless if we do not show students the thinking process of solving problems. This holds for other subjects or courses as well. It is crucial in our learning to understand the theory and know the way of thinking when solving problems. While analyzing the difficulties that students face during learning to program, we find that the main issue is it is hard to build up the concept of programming in mind. Besides, it is hard to learn debugging techniques. We try to tackle these issues, in the order of difficulty, using the following four strategies: (1) focusing on thinking, (2) revealing the nature of problems, (3) putting more emphasis on debugging, and (4) less on syntax. We also devote more words to the introduction of problems and why some mechanisms are needed.

1 Focusing on thinking

Donald Knuth, recipient of ACM Turing Award and one of the pioneers of modern computer science, wrote in his masterpiece, *The Art of Computer Programming*, that programming is the process of translating solutions to problems into terms that computers can "understand," which is hard to grasp when we first try to use computers [1]. Among all high-level programming languages, C is generally acknowledged to be one of the hardest as it is tedious, abstruse, and has a large set of rules.

A computer is an automated tool. When we try to use computers to solve problems, limitations on what we can do exist due to their capabilities. In such systems, where rules are different from the ones we are used to, the experience we had may not be of any help. Many concepts in programming are unfamiliar to students who have only had exposure to subjects like mathematics, physics, or chemistry. We find that students struggle to understand when we simply follow a traditional textbook.

In *Presentation Zen*, Garr Reynolds wrote that "Stories can be used for good – for teaching, sharing, illuminating, and, of course, honest persuasion. . . Story is

https://doi.org/10.1515/9783110692327-202

an important way to engage the audience and appeal to people's need for logic and structure in addition to emotion" [2]. Many examples in this book start from interesting stories. We extract programming-related topics from the stories, raise a problem, and guide readers to think. Then we compare how people and computers solve the problem, analyze the similarities and differences, and eventually introduce the programming concept behind the problem. To provide our readers with immersive experiences, we engage Prof. Brown and his family into our discussions. Sometimes Prof. Brown raises a question from a beginner's perspective and tries to find a solution; at other times, he takes part in the discussions as an expert. Mrs. Brown, on the other hand, knows nothing about programming and sometimes says funny things regarding programming. Their son, Daniel, is still in elementary school and often asks naive questions as well. Students, colleagues, and relatives of Prof. Brown also make cameo appearances in our storytelling. This is also a practice of lessons from Prof. Takeo Kanade, who talked about his success in research – "Think like an amateur, do as an expert" [3].

> As we work with the same concept from slightly different angles and investigate questions surrounding it, we build even more and deeper connections. Collectively, this web of connections and associations comprises what we think of informally as *understanding* . . . "For a memory to persist, the incoming information must be thoroughly and deeply processed. This is accomplished by attending to the information and *associating it meaningfully and systematically with knowledge already well established in memory*" . . . Rather than memorizing individual bits of information, we are dealing with patterns and strands of logic that allow us to come closer to seeing something whole. (*The One World Schoolhouse: Education Reimagined*, Salman Khan) [4]

By analyzing similarities and differences between how humans and computers solve problems, this book explores methods of doing logical thinking based on characteristics of computers. With these characteristics in mind, we introduce the list method for reading programs, methods for designing algorithms, and classic description methods for algorithms. Using these methods, students can learn reading programs before writing programs, which enable them to grasp general approaches to programming at the macro level and to establish a mindset of programming.

2 Revealing the nature

> Concentrating on the nature of problems is particularly important for programmers. The amount of knowledge a programmer needs to know is enormous and is still increasing. Programmers often find themselves falling behind the trend and focusing on nature is the only solution. Many new technologies are based on concepts that have been established for dozens of years. Never will knowledge of low-level architecture become obsolete, nor will that of algorithms, data structures and programming theories. (*Dark Time*, Weipeng Liu) [5]

There is no formula for programming. Although one may find some patterns from studying numerous examples, it can be tricky to figure out why certain rules exist in programming languages. It is crucial to understand both how and why because one's understanding of syntax rules can be strengthened through exploring the theories behind them. By this, one becomes more familiar with rules and can eventually apply them in practice. Focusing on the nature helps students learn to solve problems with computers more efficiently. This book explains many concepts by introducing problems in practice. We compare correlated concepts from different perspectives and extract key elements from important or correlated concepts so that students can obtain a better understanding.

3 Emphasizing debugging

"No matter how well a program is designed or how self-explanatory its documents are, it is worthless if it outputs wrong results. (*Debugging C++: Troubleshooting for Programmers*, Chris H. Pappas & William H. Murray) [6]" Errors often exist in human-made devices or equipment, with software being an exception. Software is delivered in the form of binary code, which does not tolerate errors. However, the way we think, that is, fuzzy and error-prone, makes it difficult to write completely correct code in the first attempt.

Yinkui Zhang pointed out in his book *Debugging Software* that "debugging techniques are the most powerful tools to solve complicated software problems. If solving these problems were a battle, debugging techniques would be an unstoppable weapon that strikes critically. It is not hard to learn debugger commands, but it is tricky to use debuggers to find bugs" [7]. It takes effort to gain experience and master debugging skills, especially for beginners. This is also why many students become afraid of programming.

"You can draw an analogy between program debugging and solving a homicide. In virtually all murder mystery novels, the mystery is solved by careful analysis of the clues and by piecing together seemingly insignificant details. (*The Art of Software Testing*, Glenford J. Myers) [8]" "Debugging is somewhat like hunting or fishing: the same emotions, passions, and excitement. Lying long in ambush is in the long run rewarded by a victory invisible to the world. (Eugene Kotsuba) [9]" By mastering debugging skills, readers can find and fix bugs independently in their learning and developing, which in turn increases their interest and helps them gain confidence.

In addition to finding errors in programs, debugging also helps us understand many concepts in programming such as address, memory, assignment, passing arguments, and scope. Demonstrating the debugging process gives students a more intuitive explanation than describing the concepts using abstract words. "Not only do debuggers help us finding errors in programs, they also walk us through other software, the operating system, and underlying hardware. (*Debugging Software*, Yinkui Zhang) [10]" Debuggers share very similar, if not identical, ways of working,

"The first debugger in the MS-DOS world was Debug.com . . . New debuggers appear like mushrooms after a warm rain. However, most of them are not far in advance of the prototype, differing from it only in the interface. (*Hacker Debugging Uncovered*, Kris Kaspersky) [11]" Mastering debugging techniques helps learn other computer science subjects, so it should be an essential part of programming courses.

I had been a developer in industry for many years and spent over 4 years as a member of the development team of a State Science and Technology Prize-winning software. Furthermore, I engaged in other software engineering activities as well, including installing and setting up software for users and customer services. This allowed me to gain practical experience in testing and debugging. When my students ask me for help on their codes, I could quickly find errors by debugging and asking them if the results are as expected, even if I did not know the logic of their tasks. I have always insisted on demonstrating the debugging process in class. I would show the debugging process of example programs at students' request. However, I later discovered that students still failed to understand, even if I did this in class. The reason is that debugging is a complex process, and it is tricky to explain different data structures, code logic, and debugging skills in a few words. While students may manage to understand in class, there are few written resources they can refer to when reviewing later. Hence, a large amount of debugging processes of important examples and skills we used are "persisted" into this book, so that students can refer to when learning to debug. Debugging can be extremely different for programs and the number of skills used can be large, so this book will only cover the basics, yet there are few books of the same kind that cover as many skills like these. Not many books exist that specialize in debugging either.

Programming is a process that requires continuous changes. A program often needs to be tested and debugged multiple times. This book also covers how and when test cases are designed to make readers realize the importance of testing and grasp the concept of program robustness from the very beginning.

4 Less on syntax

"Putting less emphasis on syntax" does not mean ignoring it. Instead, we start from core syntax rules and let beginners remember after understanding them. Therefore, readers can master syntax step by step instead of feeling confused by being exposed to all the rules of the C language at once. For advanced, sophisticated, or uncommon syntax, it suffices to know how they are categorized and how to look up their usage in documents.

Introduction

This book explores the methodology of the entire process of solving problems with computers. Following the workflow of how computers solve problems, the book walks through how data are stored, processed, and the results are produced. This book analyzes concepts by introducing problems and drawing analogies. It describes the entire workflow in a top-down manner: from the description of algorithms, analysis of data and code implementation to testing, as well as debugging and validation of results. In this way, it is easier for beginners to understand and master programming thinking and methods. This book makes new concepts easy to learn by introducing real-life problems and discussing their solutions, leading to a less stressful and more exciting learning process. With the help of figures and tables, the contents of this book are straightforward for readers to understand easily.

https://doi.org/10.1515/9783110692327-203

Structure of content

This book studies the methodology of problem-solving with computers. Following the workflow of computers, we walk through how data are stored, processed, and how the result is produced. By introducing real-life examples and drawing analogies, we describe the whole process in a top-down manner: from the description of algorithms, analysis of data and code implementation to testing, as well as debugging and validation of results.

The introduction to data starts from their basic forms. As the complexity of problems increases, we gradually show how data are organized and stored in computers by discussing different methods of organizing data, such as arrays, memory addresses, compound data, and files. In addition, we also cover input and output methods of data.

Algorithms describe procedures and ways of solving problems. Computer algorithms should be designed following the traits of computers. Computer algorithms are implemented by program statements, which have their own syntaxes or usages. Programs have basic control flows and their development needs specific procedures and methods.

As the problems become more sophisticated, it is necessary to use multiple modules of code instead of one. This book demonstrates how to use functions by showing mechanisms we need for larger-scale problems.

When coding is completed, we need to test the code and debug if the results are not as expected. This book introduces the principles of designing test cases, runtime environment of programs, and techniques of debugging.

There are various exercises in this book, from relatively simple warm-up and basic exercises to normal and hard homework problems. This helps readers progress smoothly and stay motivated.

This textbook comprises two volumes. Volume I, *Basic Data and Programming Statements*, covers basic programming concepts such as introduction to algorithms, basic data, and programming statements; whereas Volume II, *Composite Data and Modularity*, concentrates on advanced concepts such as arrays, composite types, pointers, and functions.

https://doi.org/10.1515/9783110692327-204

Division of work

Among all chapters in the two volumes, the ones on preprocessing and files are written by Lei Feng. Xingni Zhou wrote the rest. Final compilation and editing was done by Qiguang Miao.

https://doi.org/10.1515/9783110692327-205

Notes

Created in 1972, C is "old" compared to many high-level programming languages. Starting from the American National Standards Institute C programming (ANSI C), there have been a series of standards after continuous revision. This book is based on ANSI C standard and includes syntaxes that have been modified in C99 or C11. As they do not interfere with the main contents of this book, we decided to follow our "Less on Syntax" principle and did not modify them according to the latest standard.

There are two types of examples in this book, namely "Example" and "Program reading exercise." An "Example" usually includes analysis of data structure, description of algorithms, code implementation, and debugging process. A "Program reading exercise," on the other hand, only describes the problem and demonstrates the sample code, along with the analysis for the readers as a practice of the list method.

To be typesetting-friendly, codes are formatted compactly. For instance, opening brackets do not have their own lines.

All sample programs have been tested under Visual C++ 6.0 environment. Despite being outdated, it has a smaller installation size and better compatibility. Moreover, the theory behind debugging is universal and is not limited to a specific language or runtime.

https://doi.org/10.1515/9783110692327-206

Acknowledgments

After working in the industry as a programmer for years, I came back to college to become a teacher. During the first few years of teaching, I held many discussions on methods and ideas in programming teaching with my father, who had been teaching further mathematics all his life. Sometimes, he found my ideas valuable and would encourage me to write them out.

After spending more time teaching C language and data structure courses, I gradually realized what was challenging for students to learn programming during my interaction with them. With this in mind, I tried to change my way of teaching so that students could gain computational thinking. I was then suggested by my students in the data structure class to write a book on data structure because they thought my methods were helpful and could make a unique book. As teaching data structure and teaching C language share the same ideas, I decided to write on both topics.

I would like to acknowledge my father for inspiring my dream and my students for making this dream come true. Friendship with my students is heartwarming and overwhelming. It is their support and help that makes this book possible. It is them who had encouraged me to complete this book aiming to help beginners enter the realm of programming. I wish this book can become a torch that lights up the road of exploring for every learner so that they gain more satisfaction instead of frustration and enjoy their learning process.

I would like to appreciate Xin Dong from Xi an Academy of Fine Arts for the beautiful illustrations in this book.

My appreciation also goes to my colleagues Zhiyuan Ren and Dechun Sun for their help on exercises in this book.

I am grateful to my students Yucheng Qu, Renlong Tu, Meng Sun, Shan Huang, Bin Yuan, Yu Ding, Liping Guo, Yunchong Song, and Jingzhe Fan for their help in completing this book. My thanks also go to colleagues and students that shared their opinions and suggestions. They made me to introspect about drawbacks in my past teaching and writing. Consequently, I started to think from psychological and cognitive perspectives and made improvements, such as reinforcing problem introduction and changing my way of storytelling. These improvements can be found in *Data Structures and Algorithms Analysis – New Perspectives*.

My thanks also go to Mr. Zhe Jiang for his work on localization of the manuscript.

Rewrite of this book after years (Chinese edition has already been reprinted) is like a rebirth. I would like to quote my 2019 spring appreciation to conclude this

https://doi.org/10.1515/9783110692327-207

acknowledgment: "Profusion of flowers, blossoming of lives; along with auspicious clouds, it is spring we celebrate."

Xingni Zhou
xnzhou@xidian.edu.cn
In Chang'an, midsummer 2020

Contents

1 Introduction to programs

Main contents
- Concept of flows
- Concept of programs
- Methods of program design
- A brief introduction to C programs

Learning objectives
- Know the concept of programs
- Understand the basic steps of program design
- Know the basic structure of C programs

1.1 Concept of flows

We use computers to help us work efficiently. How do computers work then? Before answering this question, let us take a look at how humans solve problems and then analyze how we think and what methods we use when solving problems.

1.1.1 About flows

Let us look at some flows in real life first.

There are several sessions in the opening ceremony at a college, as shown in Figure 1.1, where the order of operation is an order of time and is represented by the arrowed line. Arranging every session in the order of their time of completion, we obtain a stream of procedures, which we call a flow.

Many of us have traveled by train before and have experience of purchasing railway tickets. Figure 1.2 shows steps of buying tickets at the ticket office. As shown in the figure, the flow of buying tickets is a description of the entire process that starts from setting up a task and completes when achieving the goal by executing some actions.

Bread is a typical staple food. Baking bread is somewhat a complicated process. The main steps of its production process are shown in Figure 1.3. We process raw materials using specific devices in a particular order and eventually obtain finished goods. This is called the "production flow."

Many of us may travel by air for longer trips. The boarding flow shown in Figure 1.4 can clearly guide first-time flyers.

By observing these examples, it is clear that the purpose of flow, be it a work flow or a production flow, is to achieve a certain goal or to obtain a certain product.

https://doi.org/10.1515/9783110692327-001

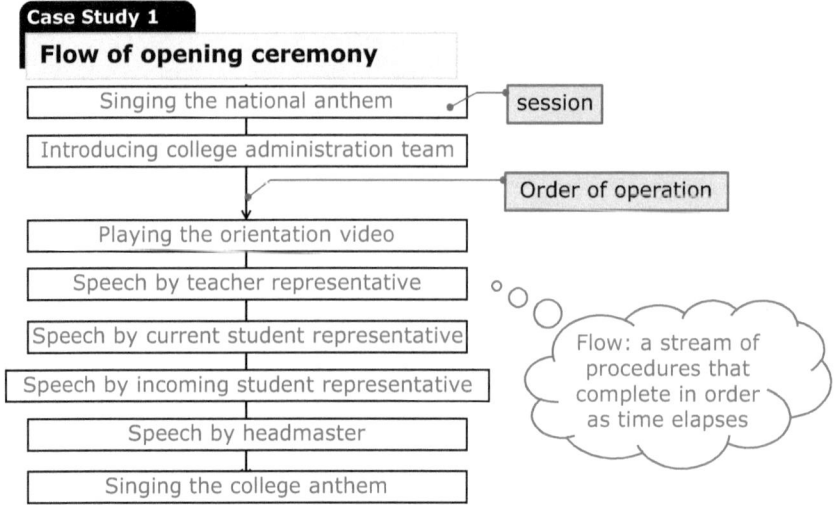

Case Study 1

Flow of opening ceremony

Singing the national anthem	session
Introducing college administration team	
	Order of operation
Playing the orientation video	
Speech by teacher representative	
Speech by current student representative	
Speech by incoming student representative	Flow: a stream of procedures that complete in order as time elapses
Speech by headmaster	
Singing the college anthem	

Figure 1.1: Flow of opening ceremony.

Case Study 2

Flow of purchasing tickets through ticket office

Step 1: The passenger provides information on trip date, destination, etc.
Step 2: The staff finds trains available on that day
Step 3: The passenger chooses a train and determines number of tickets to buy
Step 4: The passenger pays the fare and collects tickets

Figure 1.2: Flow of purchasing tickets at the ticket office.

Case Study 3

Flow of baking bread

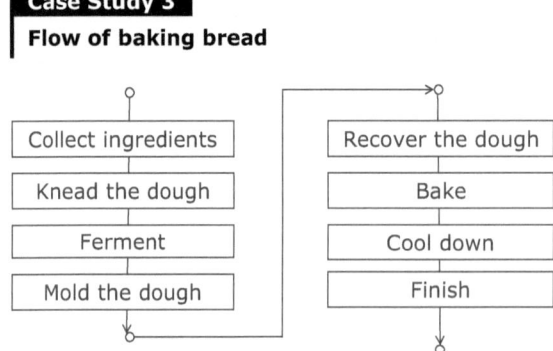

Collect ingredients	Recover the dough
Knead the dough	Bake
Ferment	Cool down
Mold the dough	Finish

Figure 1.3: Flow of baking bread.

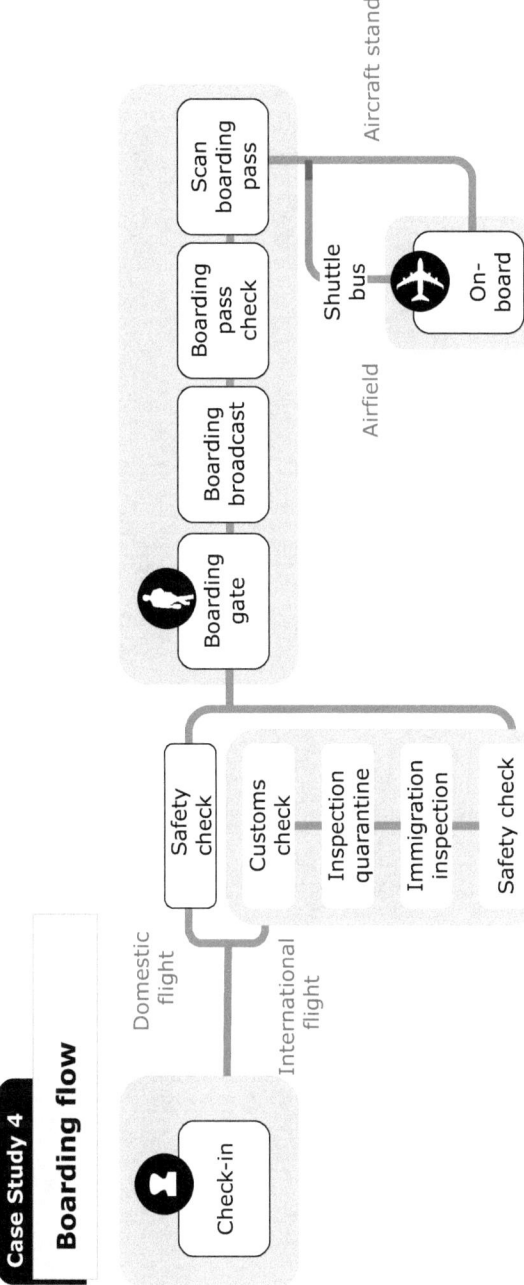

Figure 1.4: Boarding flow in airports.

It describes the entire process of completing a task, a job, or manufacturing a product. Every flow, regardless of what it describes, consists of a series of sessions and orders of operations as shown in Figure 1.5.

> **Flow**
>
> A flow describes the entire process of completing a task, a job or manufacturing a product. A flow consists of a series of sessions and orders of operations.
>
> Session: Phases or procedures during completion of a task, a job or manufacture of a product.
>
> Order of operation: Time order of sessions in a flow.

Figure 1.5: Concept of flows.

1.1.2 Expression of flows

Flows can be expressed in many ways, as seen in previous examples. The flow of an opening ceremony is shown in a flowchart. The flow of purchasing tickets is described in plain words. The boarding flow is shown in a figure. In addition, flows can be expressed as a table, model, video, etc. In programming, flowcharts and pseudo codes are often used to describe a flow as shown in Figure 1.6.

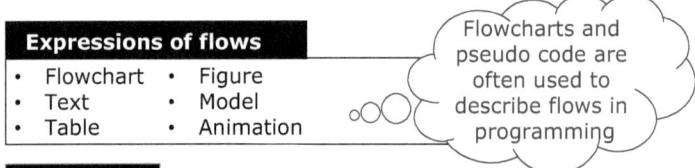

> **Expressions of flows**
>
> - Flowchart   • Figure
> - Text   • Model
> - Table   • Animation

Flowcharts and pseudo code are often used to describe flows in programming

> **Flowchart**
>
> A flowchart is one way to express steps in a process in figures. It consists of some shapes and flowlines , where shapes indicate type of operations, text and signs in each shape describe content of an operation and flowlines indicate order of operations.

> **Pseudo code**
>
> Pseudo code uses words and symbols that fall in between natural languages and computer languages to describe processing process of problems.

Figure 1.6: Common ways of expression of flows in programming.

A flowchart depicts a process as a figure, whereas the pseudo code uses words and symbols to describe a process. Pseudo code does not use graphical symbols, which makes it easier to write and understand, more compact, and more convenient to transform into programs.

This book uses pseudo code to describe program flows in most cases. Procedures human use to solve a problem is also known as "algorithms."

1.1.2.1 Flowchart

American National Standards Institute (ANSI) standardized some common flowchart symbols that have been adopted by programmers from all over the world. The most frequently used symbols can be found in Figure 1.7. Process symbol, decision symbol, and input/output symbol are used to represent sessions in different situations. The flowline symbol is an arrowed line, which is used to show the order of operation. Using graphical symbols to represent a flow is more intuitive and easier to comprehend.

Symbol	Name	Meaning	
⬭	Terminal	Indicates beginning and ending of a flow	
▭	Process	Represents normal operations	Session
◇	Decision	Determines whether a given condition is met and yields true or false. Result needs to be labeled at exits	Session
▱	Input/Output	Input and output of data	Session
⟶	Flowline	Connects process or decision symbol, shows path and direction of flows	Order of operation
◯	Connector	Connects flowlines drawn in different places	Order of operation

Figure 1.7: Common flowchart symbols.

1.1.2.2 Pseudo code

Pseudo code shows the execution process and algorithm of programs in the format of code. It does not rely on a certain programming language. It uses the structure and format of programming language to describe the execution process of a program. Hence, it cannot be compiled by compilers. Using pseudo code allows to show the execution process of programs in a way that is easier to understand and express.

Knowledge ABC Flowchart and pseudo code

From the late 1940s to the mid 1970s, flowchart has been the primary tool in process design. The main advantages of flowcharts are: it uses standard and straightforward symbols, is easy to draw, has a clear and logical structure, and is easier to describe and understand. Its intuitive depiction of control flows allows beginners to handle them painlessly. Moreover, flowchart is time honored and familiar to humans. Consequently, it is still widely used today although many people advocate obsoleting it due to its disadvantages. A flowchart is a description of methods, ideas, or algorithms people use to solve problems. However, avoiding flowcharts has been the latest trend. One of its major disadvantages is that it takes up more space. Besides, the use of flowlines has few restrictions, thus one can make a flow growing in an arbitrary direction. This results in challenges when reading or modifying a program. Secondly, flowchart is not helpful in the design of structured programs. It is not a tool that allows continuous improvement because it forces programmers to consider the control flow rather than the overall structure of programs in an early stage.

When implementing the same algorithm using different programming languages, people realize that these implementations (note that it is not functionality) are often different as well. It can be hard for a programmer who is proficient in one language to understand the functionality of a program written in another language, as the form of programming languages puts limitations on his/her understanding of the critical parts. Therefore, pseudo code was created.

We often use pseudo code when considering the functionality (instead of the implementation) of an algorithm. It is also used in computer science education so that all programmers can understand.

Pseudo code is written in the form of programming languages to indicate the functionality of an algorithm. It is similar to natural languages instead of programming languages (such as Java, C++, Delphi, etc). Using pseudo code, the execution process of an algorithm can be described in a way that is close to natural language. We may use any language in pseudo code, be it Chinese or English, but the key is to show what the program intends to do. Pseudo code helps us express algorithms without considering implementation.

1.1.3 Basic logical structure of flows

What logical features does the description of a flow that solves a problem have? We shall discuss the answer soon.

Let us consider a real-life flow first: setting up a washing machine.

1.1.3.1 Sequential structure

Mr. Brown is a computer science professor in a college who seldom does housework because he is usually busy working. However, his wife, Mrs. Brown, will be traveling for a few days, so he has to learn how to use the washing machine. She only told him the basic setup, as shown in Figure 1.8, as she was afraid that Mr. Brown could not remember all the functionalities the washing machine has if she did not do so.

The operations of the washing machine are executed in an order that is determined by logical relations between each operation. For example, soaking should happen before washing. Mr. Brown took notes carefully and made a flowchart of washing operations based on the execution order as illustrated in Figure 1.9. In this washing program, the preset operations are arranged sequentially. Thus, the structure of these operations is called a "sequential structure."

1.1.3.2 Branch structure

In his first attempt, he put many clothes into the washing machine and then configured it as how he was taught. However, he found that the washing machine was

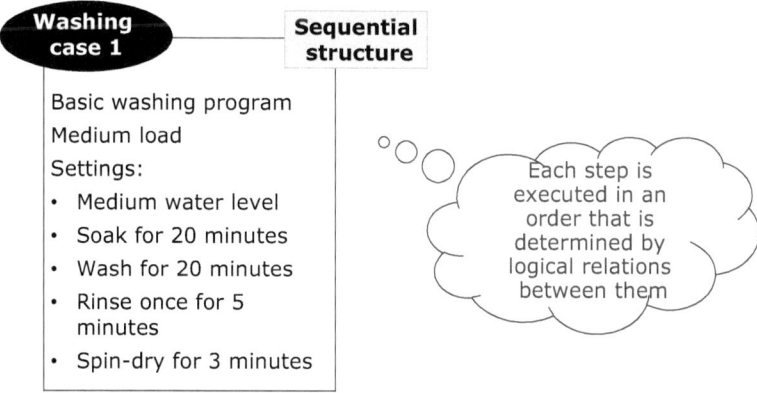

Figure 1.8: Washing case 1.

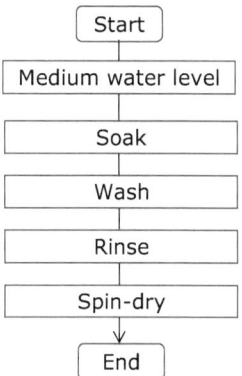

Figure 1.9: Flow of washing process with sequential structure.

not working smoothly, so he immediately called his wife for help. He was then told that this washing program was designed for a medium load of clothes and was not suitable for a large load.

The water level needs to be adjusted according to the clothes load. In this case, we need to setup water level after determining the amount of clothes as shown in Figure 1.10. We should choose a high water level for a large load. Otherwise, we use a medium level, after which we continue with our basic washing program.

Mr. Brown drew a new flowchart (Figure 1.11), which was more intuitive. He used different configurations for different clothes loads. Such flow where a decision needs to be made is called "branch structure." Note that the decision condition is put in a diamond symbol to make it more noticeable. The diamond symbol means a decision in the flowchart drawing standard.

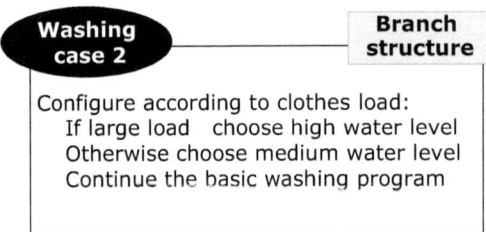

Figure 1.10: Washing case 2.

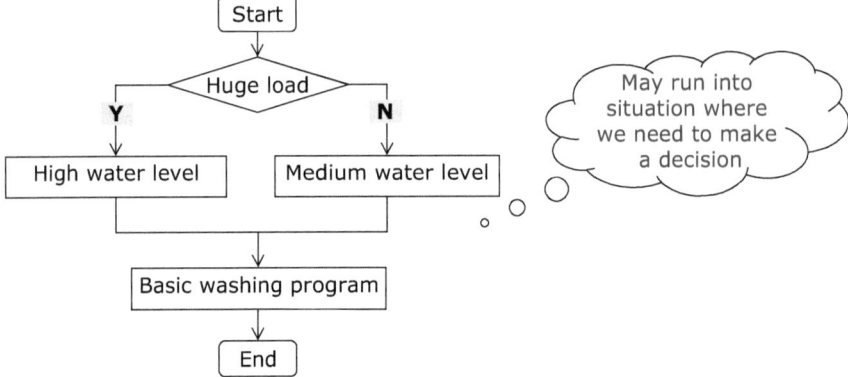

Figure 1.11: Flow of washing process with a branch structure.

Thinking about the problem caused by clothes load, Mr. Brown realized that there should be another water level option for a small load. He later found that it did exist after investigating the washing machine.

The complete flow of setting up water level is shown in Figure 1.12. If the clothes load is small, we choose the low water level; if medium, we choose the medium level; and if large, we choose the high level.

Mr. Brown then drew the multiple branch structure flowchart based on his wife's description as illustrated in Figure 1.13. This is a branch structure with three branches. Note that the clothes load is represented as a rectangle instead of a diamond as the diamond symbol is only used when there are two branches in a decision. More details are covered in the discussion of the selection structure.

1.1.3.3 Loop structure

Mr. Brown started to feel curious and started to investigate the washing machine as if he was experimenting. He noticed that the washing machine could not rinse properly with a huge load of clothes. This time he decided to solve the problem without calling

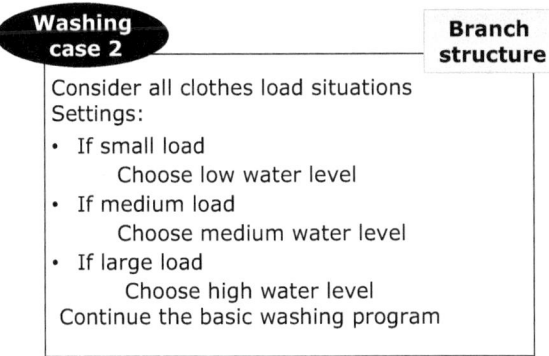

Figure 1.12: Washing case 2 with multiple branches.

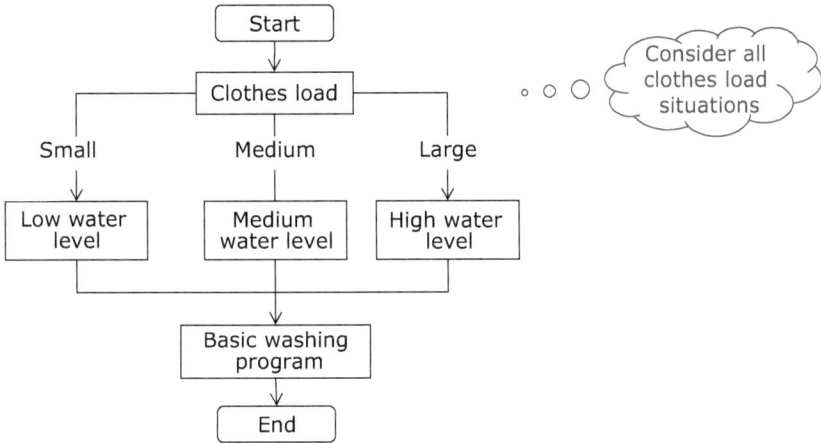

Figure 1.13: Flow of washing process with multiple branch structures.

Mrs. Brown for help. It seemed that he should increase the number of rinses in the washing program. He then checked the machine and pressed the "Rinse" button one more time. It turned out that the clothes were rinsed again. The case of multiple rinses is shown in Figure 1.14, where Mr. Brown increased the number of rinses to three without changing the rest of the washing program.

However, the way three rinses are expressed in the flow shown in Figure 1.15 seems cumbersome. It will be worse if we have even more rinses. Can we make any improvements?

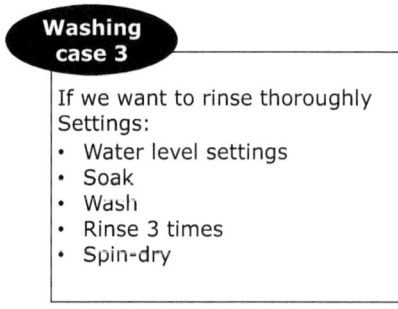

Figure 1.14: Washing case 3.

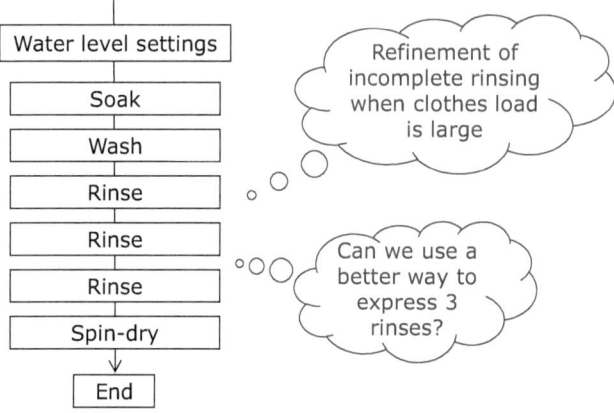

Figure 1.15: Flow of washing program with three rinses.

The refined flow is shown in Figure 1.16. The machine continues rinsing if it has not rinsed a preset number of times. Otherwise, it proceeds to execute other steps. A flow with steps that may be repeatedly executed is said to have a loop structure.

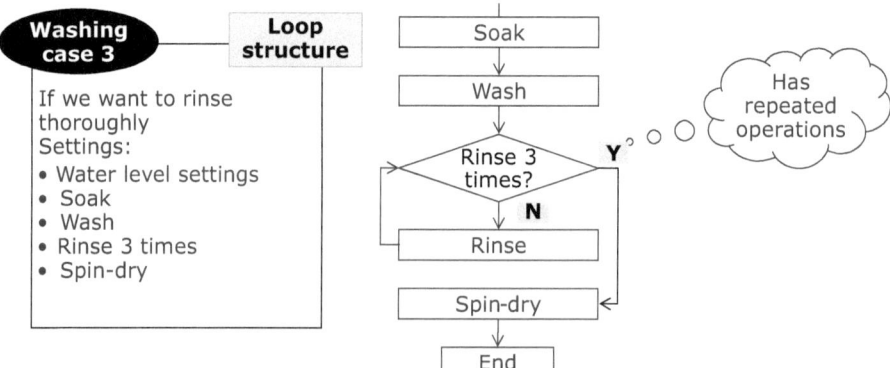

Figure 1.16: Flow of washing process with the loop structure.

1.1.3.4 Logical structure of flows

Based on these examples, the basic logical structure of flows is summarized in Figure 1.17. A flow structure is the logical structure of the execution of each step. If there is no decision to be made, it is a sequential structure; if there is a decision but no repetition, it is a branch structure; otherwise, it is a loop structure with both decision and repetition.

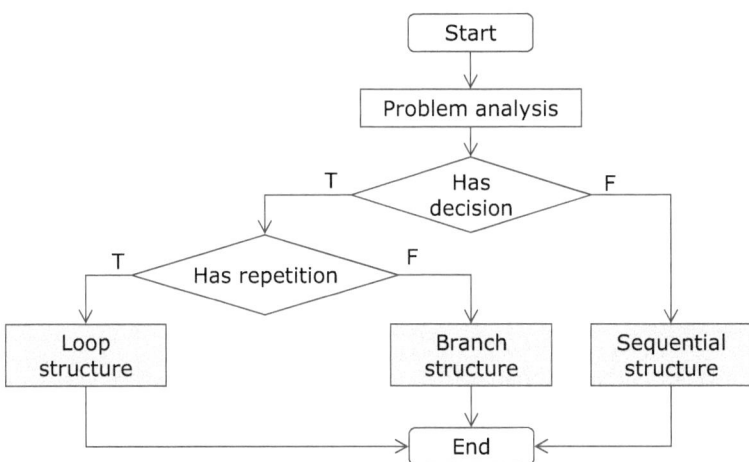

Figure 1.17: Basic logical structure of flows.

Sequential structure, branch structure, and loop structure are the three fundamental structures of programs. It has been proven in practice that any flow, no matter how complex it is, can be constructed by these three basic structures as shown in Figure 1.18. It is similar to how the three primary colors can be combined to produce a gamut of colors. They can also be nested to generate programs that are called "structured programs."

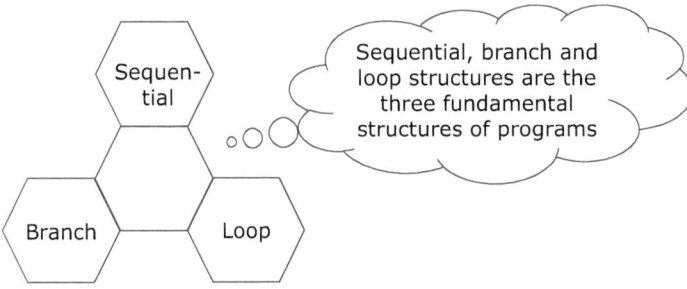

Figure 1.18: Fundamental structures of programs.

1.1.3.5 Expression of basic flow structures

Now we can summarize expressions of basic flow structures.

The sequential structure consists of several steps that are executed in order. It is the simplest and most fundamental structure that every algorithm has. To express a sequential structure, simply list the operations in time order. The rectangle symbol in flowcharts means "Process" and it represents operations as shown in Figure 1.19.

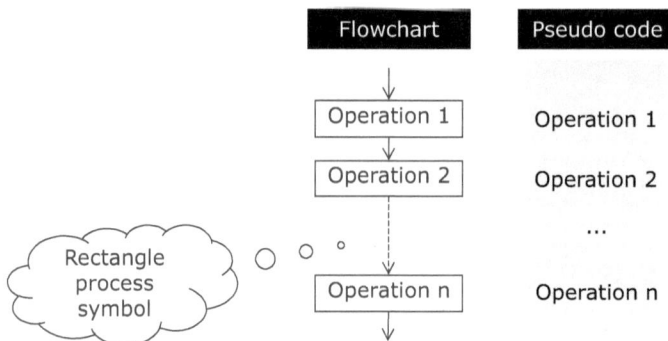

Figure 1.19: Expression of sequential structure.

The branch structure tests a preset condition and controls the flow based on the result as shown in Figure 1.20. In flowcharts, we put the condition into a diamond symbol. The flow enters different branches based on whether the condition is evaluated to be true or not. If true, the condition is met, the operation set A is executed (herein the operation set means a set or series of operations); otherwise, the condition is not met and operation set B is executed. We need to pay attention to the indentation of curly brackets when using pseudo code representation.

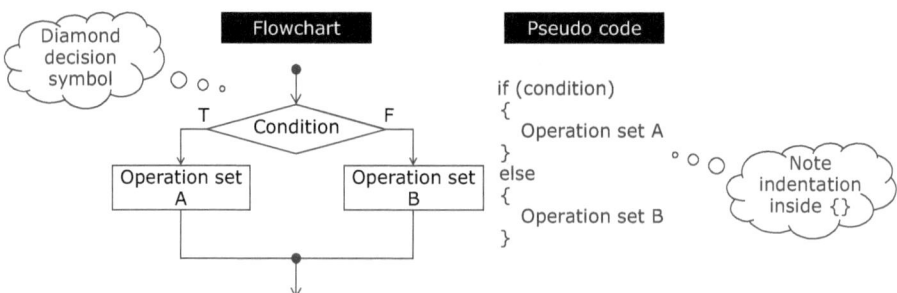

Figure 1.20: Expression of branch structure.

The loop structure contains steps that are executed repeatedly under certain conditions. There are two kinds of loop structures: while loops and do-while loops.

"While loops" test the loop condition first. If it is met, the operation set A is executed. Otherwise, the loop is terminated. In other words, it tests and executes as shown in Figure 1.21.

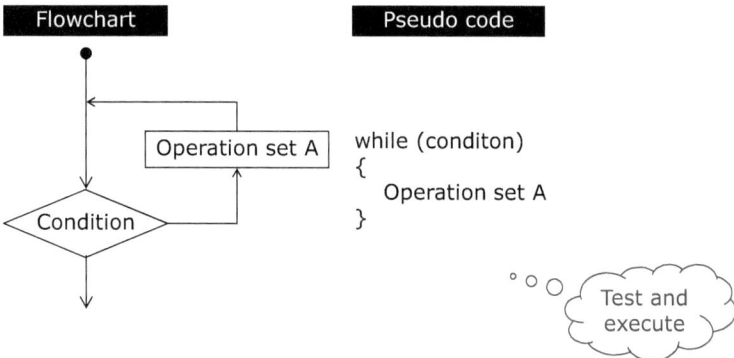

Figure 1.21: Expression of while loop structure.

Do-while loops execute operation set A first and test loop condition when the operations are completed. If the condition is met, the operation set A is executed again. The loop would n be terminated until the condition is no longer satisfied. In other words, it executes and tests as shown in Figure 1.22.

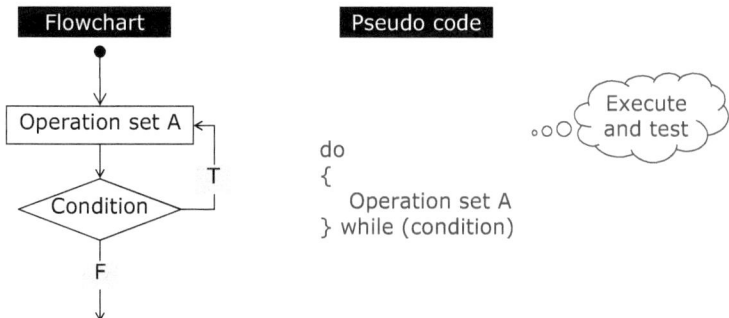

Figure 1.22: Expression of do-while loop structure.

1.2 Concept of programs

In the previous section, we discussed methods, procedures humans use to solve practical problems, and how they are represented. Herein we examine how computers solve problems.

1.2.1 Automatic flows

We have discussed how to configure the washing process in the introduction of the concept of flows. Herein we are going to focus on how the washing machine runs. To use a washing machine, we setup the washing program and start it. The machine automatically completes the necessary operations as shown in Figure 1.23. A program is a flow that machines can execute automatically after we setup the steps. In fact, it is the computer program installed inside that enables washing machines to run automatically.

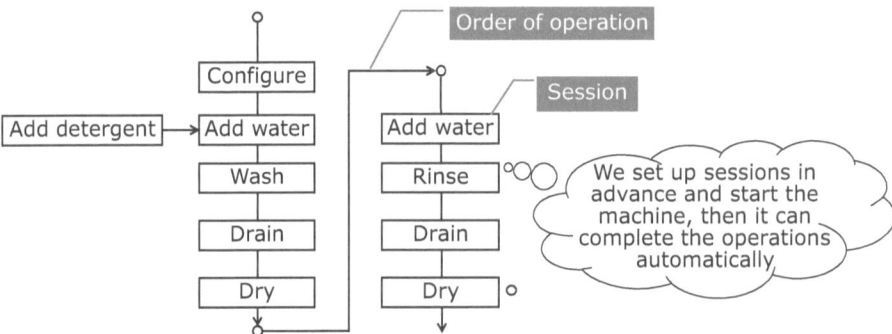

Figure 1.23: Automatic flow of washing machine.

Nowadays, it is quite convenient to buy tickets online. It is made possible by converting the process of purchasing through the ticket office into an automatic flow that computers can handle (Figure 1.24).

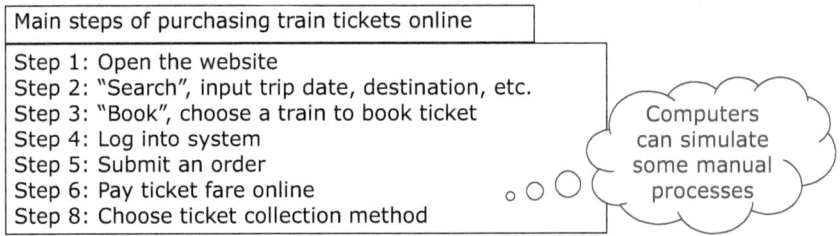

Figure 1.24: Flow of purchasing tickets online.

1.2.2 Concept of programs and programming languages

Based on the automatic flows mentioned, programs and programming languages can be formally defined as shown in Figure 1.25.

Program

A program converts the process of solving a practical problem into a sequence of instructions using programming languages. To be more specific, it is a flow designed to be executed by computers automatically, using data that can be accepted by computers and intending to produce expected result.

Programming language

A programming language is a computer-recognizable language used in communication between human and computers. Programming languages have fixed symbols and grammar rules.

Figure 1.25: Programs and programming languages.

1.2.2.1 Programming languages

Programming languages are languages used to write the computer programs. There are many kinds of programming languages, but generally, they can be categorized into machine languages, assembly languages, and high-level languages. Assembly and high-level languages are being widely used nowadays.

1.2.2.2 Machine languages

Computers are built from numerous electronic components. Videos we watch and music we listen to are merely variation and combination of high and low voltages. Hence commands that computers receive are simply variation of voltages, namely high and low voltages. Engineers and computer scientists use "0" and "1" to represent on and off. These 0's and 1's are called "binary codes." Computers can only recognize these codes.

Do we write programs using 0's and 1's then? The answer is yes. In the early days of programming, people wrote programs by punching on one-inch wide cards. There were eight holes on each line and each hole represented one binary bit. A punched hole represented 1 and an unpunched hole represented 0. Combinations of holes on each line represented commands. Code written using 0's and 1's in this way is written in machine language. Machine language can transfer commands to computers directly. Code written in machine languages is flexible and can be executed directly and quickly. However, it takes effort for developers to write programs and it is incredibly inefficient. Moreover, it is hard to comprehend a program in machine language. Reading and debugging are painful as well. Hence, machine languages are called "low-level languages." Machine languages are not universal either as they are machine-dependent. Specific machine languages can only be used in certain kinds of machines.

1.2.2.3 Assembly languages

It was not long before programmers found machine languages cumbersome, thus they started to look for other means to communicate with computers. Is there a way to write programs using symbols other than "0" and "1" and then translate them into machine languages? It turned out to be a great idea as assembly languages were created. In assembly languages, we use mnemonic symbols to replace machine commands. For example, ADD is used for adding numbers and SUB is used for subtraction. These symbols make assembly languages more readable. After writing a program, we need a translation program that converts the program into machine commands. Such a program is called an assembler. The process of writing a program in assembly language is shown in Figure 1.26.

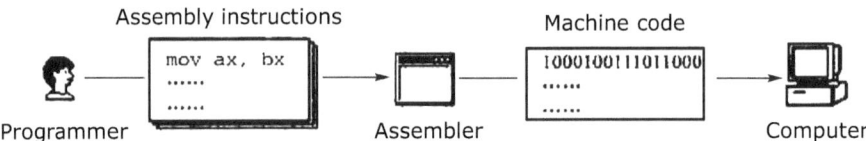

Figure 1.26: The process of writing program in assembly language.

The emergence of assembly languages allowed computers to be used in more areas. However, different computers often have different assembly languages, thus they were machine-dependent as well. Due to the weak universality, they were also categorized as low-level language. Nevertheless, they are still being used by people nowadays because they can be executed more quickly, save memory space, and manipulate hardware more efficiently. They can often be found in these areas: system (including embedded system) programming, such as operating systems, compilers, drivers, wireless communication, DSP, PDA, GPS, etc.; software development where resource, performance, speed, and efficiency are critical issues; as well as reverse engineering that aims at information security, software maintenance, and cracking software. Even if we are not going to work on system development or to become a hacker or a cracker, knowing assembly language is helpful for learning computer architecture, debugging software, and improving algorithms used in the critical part of programs.

1.2.2.4 High-level languages

To accelerate the developing process, humans created Fortran, the first high-level programming language, in 1954. This marked the beginning of a new era of programming.

High-level languages are similar to natural languages and mathematical expressions. Compared to assembly languages, they combine multiple correlated machine instructions into single instruction and eliminate hardware manipulation details that are unrelated to the functionalities of a program. Thus, instructions in a program are largely simplified and programmers no longer need a large amount of hardware knowledge.

The name "high-level language" means these languages are more advanced than assembly languages. It does not refer to a specific language. Instead, it refers to many programming languages, such as VB, C, C++, and Delphi, which vary in syntax and instruction format.

Programs written in high-level languages cannot be recognized by computer directly. They must be converted for computers to execute. Depending on the way of conversion, high-level languages can be categorized into two types:

- Interpreted languages: The conversion is similar to simultaneous interpretation in real life. The interpreter of a language executes a program directly while translating source code into target code (in machine language) at the same time. This process is inefficient and applications cannot work without the interpreter as there are no executable files generated. However, using an interpreter is more flexible as we can modify our applications dynamically.
- Compiled languages: Compilation is the process of applications being translated into object code (with .obj extensions) before being executed. Generated object programs can be executed without the language runtime, thus it is more convenient to use and more efficient. However, if any changes are needed, we have to modify the source code and recompile it into object code to execute. This can be inconvenient if we do not possess the source code. Many programming languages are compiled languages such as C, C++, and Delphi.

One of the advantages of high-level languages is better portability, which means programs can run on different types of computers. Compared to assembly languages, high-level languages are more comfortable to learn and master. Programs written in these languages are also easier to maintain. However, programmers cannot manipulate hardware and control operations of computers directly as high-level languages do not target specific computer systems. Besides, object programs are larger and run slowly compared to those generated from code written in assembly.

Knowledge ABC ANSI C and C standards

During the 1970s and 1980s, C was widely used on all types of computers, from mainframe computers to microcomputers. This led to different versions of the language. In 1989, the ANSI established an entire standard specification for C so that different companies could use the same set of syntax. This was the C89 standard (also known as ANSI C), which was the earliest standard of C. In 1990, the International Organization for Standardization (ISO) and the International Electronical Commission (IEC) adopted the C89 standard as the international standard of C language (known as C90 standard). After subsequent revisions, the C99 standard (published in 1999) became the second official standard of C. The C11 standard (published in 2011) is the third official standard and the latest one.

1.2.3 Execution characteristics of programs

Programs are flows of operations executed by machines. What are their characteristics?

A program is a process of setting up operations and executing them sequentially, which is similar to a domino show as illustrated in Figure 1.27. Many of us are familiar with domino shows. Their rules are simple: dominoes are aligned in sequence, each at a certain distance from the next. Builders can create patterns and images based on meticulous design. Once the first domino has been toppled, a chain reaction happens and the rest are toppled in order.

Figure 1.27: Characteristics of automatically executed flows.

Another one of such systems that feature meticulous design and chain effect is the Rube Goldberg machine. It is an extremely well-designed and complicated system that completes a task in an indirect and overly complicated way. Designers of a Rube Goldberg machine have to make sure everything is correctly calculated so that each device in the system can achieve a stated goal at the perfect time. The Rube Goldberg machine also creates visual effects similar to those of a domino show.

1.2.4 Workflow of computers

Programs are executed automatically in computers. Hence, we need to know how computers work.

In the mid-1930s, John von Neumann, a Hungarian-American mathematician, proposed to use the binary numeral system in digital computers and to write computing programs in advance so that computers could follow the computation sequence to complete numerical calculations. Electronic computer systems designed based on his concepts and theories are now called "von Neumann architecture" computers. A computer should have data input, data processing, and result output functionalities. To achieve this, computers must have five basic components as shown in Figure 1.28.

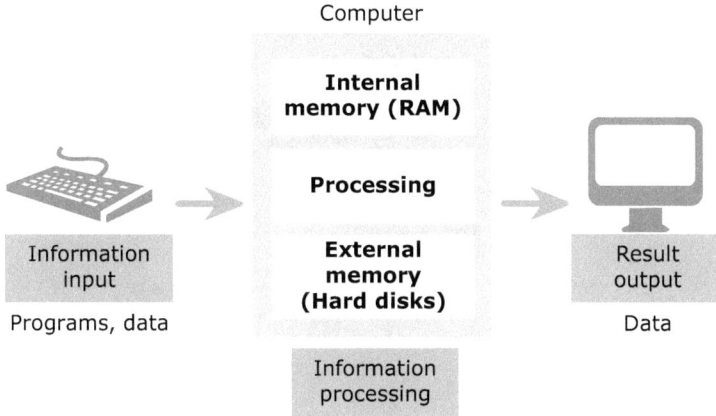

Figure 1.28: Basic components of computers.

They are:
- Input devices for data and programs: keyboards are the most common input devices.
- Output devices for results of processing: monitors are the most common output devices.
- Memory units for storing programs and data: there is nonvolatile memory (hard disks) that can retrieve stored information even after having been power-cycled and volatile memory such as random-access memory (RAM) that requires power to maintain the stored information.
- Central processing unit (CPU) processes data and controls the execution of programs.

CPU is the most critical component of computers. It consists of a processor and control units. The processor is used to complete arithmetic and logical operations. The decision-making process we have seen in flowcharts is categorized as a logic operation in computers. The control unit performs instructions fetched from memory units and provides control signals to other components.

The workflow of a von Neumann architecture computer is shown in Figure 1.29. The main steps are:
- Input programs and data: Programmers need to store programs to be executed and related data into random access memory (RAM).
- Fetch instruction: The control unit fetches the first instruction from RAM.
- Fetch data: Based on the instruction in step 2, data are retrieved from memory units and sent to processors.
- Process: Specified arithmetic and logical operations are carried out in processors.
- Store intermediate results: Intermediate results are sent to a specific address in RAM. This is done continuously until all instructions are completed.
- Output final result to output devices.

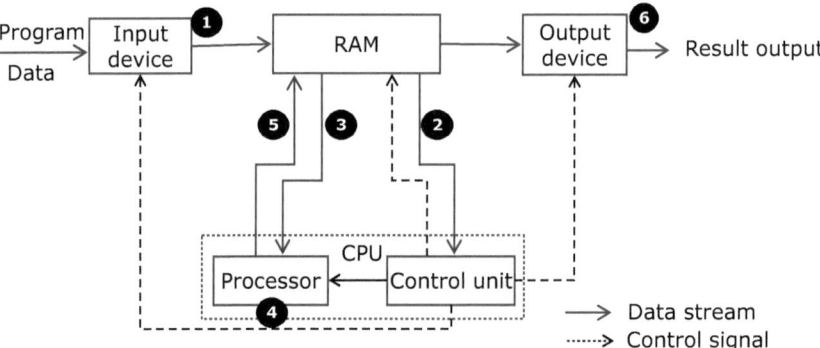

Figure 1.29: Workflow of computers.

1.3 Components of programs

In the workflow of computers, data are sent to memory units through input devices. The processors perform logical and arithmetic operations under the control of control units. The final result is presented to users through output devices. The flow that computers use to solve problems can be divided into three parts: data input, data processing, and result output as shown in Figure 1.30.

Figure 1.30: Flow of problem-solving with computer.

1.3.1 Problem-solving with computer: data

Data in programs are units of information that can be stored in and processed by computers. If we were computers, as tools for information processing, what issues we need to handle between us and the data to be processed? Considering the workflow we saw earlier, the main issues include how information is input, how data are stored and processed, and how results are output as shown in Figure 1.31.

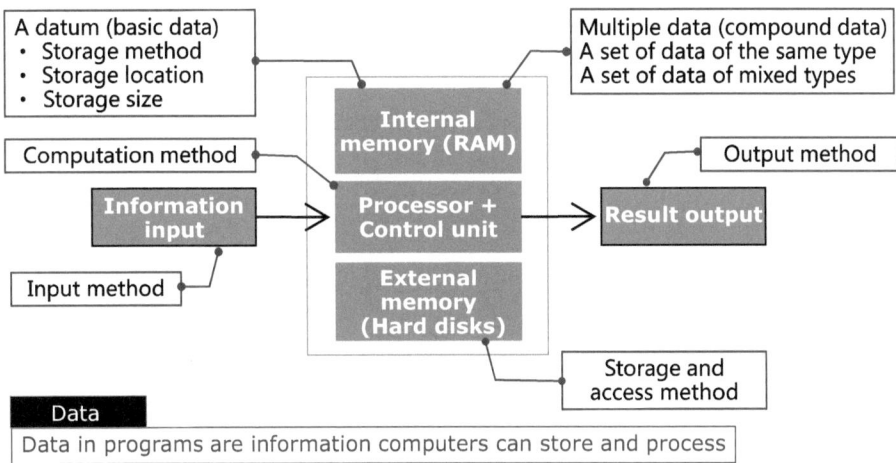

Figure 1.31: Relation between computers and data processing.

With hardware architecture and theories of computers in mind, we study the following data-related issues: how data are input/output; how data are stored in internal (and external) memory units, which address they are stored at and how much space they take up; and how related data are combined and stored according to their characteristics and types, which is a combinatorial data problem in programming.

Issues of data processing include how they are represented in programs and their computation rules.

In summary, issues related to data in programs can be categorized into memory-related and input/output devices related as shown in Figure 1.32.

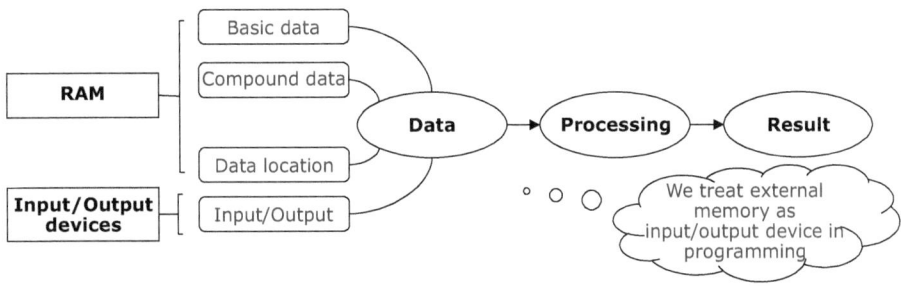

Figure 1.32: Problem-solving with computers: Data.

Memory-related issues include:
- Basic data issues that include how data are stored and calculated.
- Combinatorial data issues that include how data of same and different types are combined and processed.
- Address issues that determine which location in RAM data should be stored in.

Input/output devices-related issues include how data are input and output. Note that external memory is deemed input/output device in programming.

1.3.2 Problem-solving with computer: processing

Data processing consists of the description and implementation of the processing process as shown in Figure 1.33.

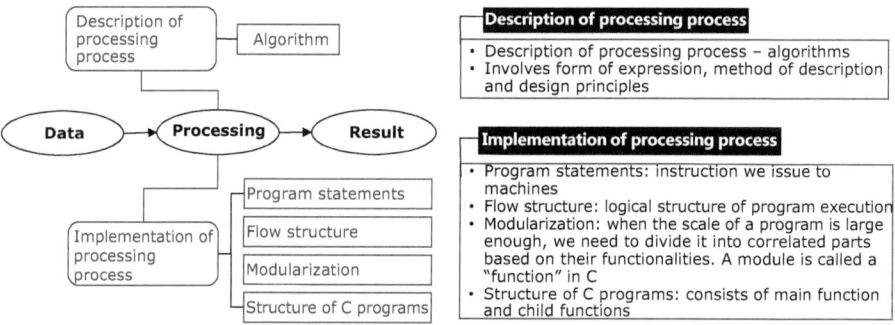

Figure 1.33: Problem-solving with computers: processing.

The description of the processing process is nothing but an algorithm. We need to consider the form of expression, method of description, and design principles when talking about algorithms.

The implementation of the processing process comprises four topics. The actual implementation is done by code. Flow structure is the logical structure of program execution. Basic logical structures of flows include sequential structure, branch structure, and loop structure. When the scale of a program is large enough, we need to divide it into correlated parts based on their functionalities. This division is called modularization. In our daily life, a complex task can be divided into multiple easy tasks for different people to complete. Each task can be regarded as a module. Modules are called "functions" in C programs. The structure of C programs consists of the main function and child functions.

1.3.3 Problem-solving with computers: results

To obtain the results of data processing, we need to do some work before execution as well as testing and debugging as shown in Figure 1.34. A program needs to be compiled into machine code and linked with necessary resources before generating executable instructions. Preprocessing is a series of code organizing tasks done before compilation. These tasks can include programs written by others (resource linking mentioned above), character replacement, or conditional compilation. More details of preprocessing can be found in the chapter "Execution of Programs."

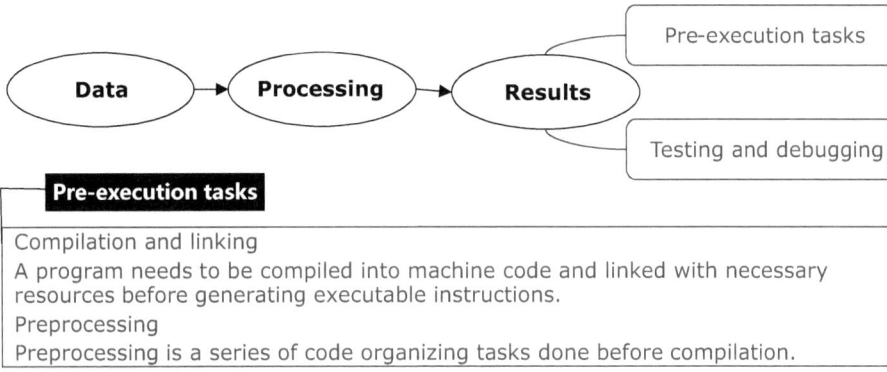

Pre-execution tasks

Compilation and linking
A program needs to be compiled into machine code and linked with necessary resources before generating executable instructions.
Preprocessing
Preprocessing is a series of code organizing tasks done before compilation.

Testing and debugging

Testing: input preset data, run program to obtain result, compare with expected result
Debugging: techniques of finding errors in programs

Figure 1.34: Problem-solving with computers: results.

Knowledge ABC Preprocessing
Preprocessor directives are instructions that begin with a # sign. They are invoked before compilation to complete some support tasks for compilers.
There are three kinds of preprocessor directives: macrodefinition directives, file inclusion directives, and conditional compilation directives.
Preprocessing is a series of work done by the preprocessor before the first scan (lexical scan and syntax analysis) of compilation. When compiling a source file, the system automatically invokes the preprocessor to complete preprocessing before the actual compilation. More details can be found in the chapter "Preprocessing."

We can compare the result returned by the program with our expected result. If the result is correct, our programming task is completed. Nonetheless, it is not rare that a sophisticated program produces wrong results at first. We need to find errors in the program and fix them by testing and debugging.

Computers solve problems with programs. Programs contain three components: data, processing, and results. Figure 1.35 is the knowledge map of C language. These concepts will be introduced in corresponding chapters.

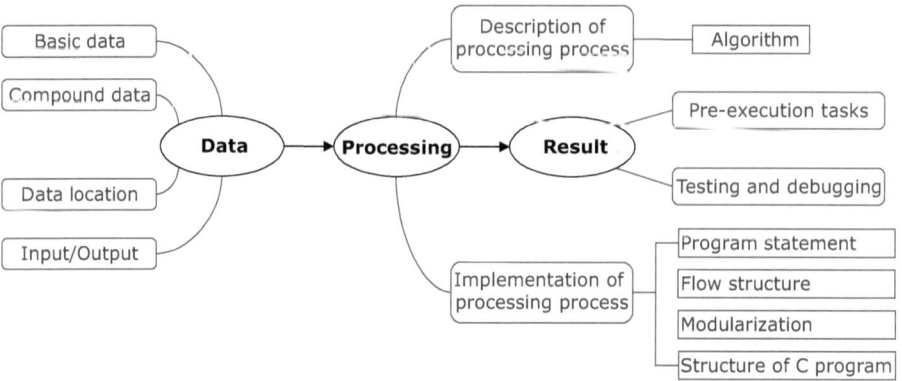

Figure 1.35: Knowledge map of C language.

If we compare data and code to raw materials, flow logical structure, and algorithms to manufacturing methods and requirements, the program will be the final product. This process can be generalized as a formula as shown in Figure 1.36.

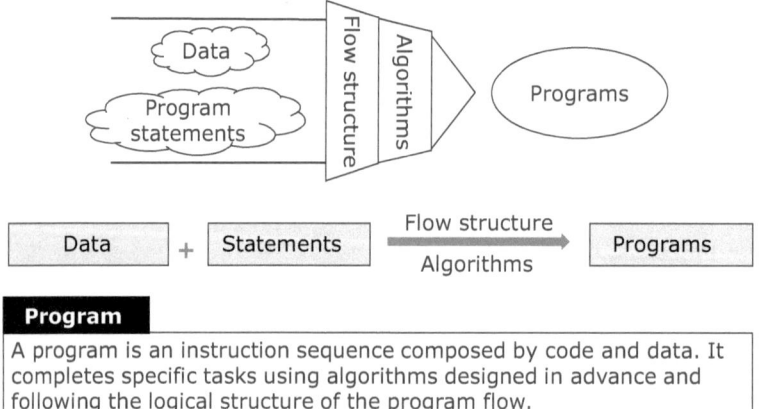

Program

A program is an instruction sequence composed by code and data. It completes specific tasks using algorithms designed in advance and following the logical structure of the program flow.

Figure 1.36: Components of programs.

A program is an instruction sequence composed of code and data. It completes specific tasks using algorithms designed in advance and following the logical structure of the program flow.

Niklaus Wirth, who is a Swiss computer scientist and recipient of Turing Award, summarized the above description of components of programs as the famous "Wirth's Equation" as shown in Figure 1.37. An algorithm is a strategy used to solve the problems. A data structure is the model of data used to describe a problem, which consists of inherent logical relation of data, storage methods of data, and operations allowed to be applied to data.

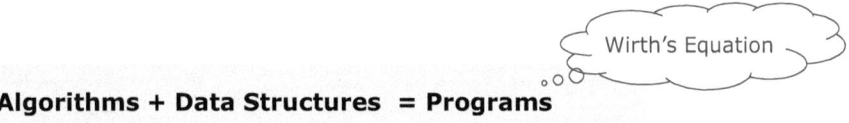

Algorithms + Data Structures = Programs

Figure 1.37: Construction of programs.

1.4 Development process of programs

Through discussion in previous sections, we roughly know how computers solve problems, but how about details behind the process? Let us examine an example first.

1.4.1 Case study

1.4.1.1 Using a calculator

We all know how to use a calculator to complete common calculations conveniently. The simplest arithmetic operations are addition, subtraction, multiplication, and division. Their corresponding operators are "+", "–", "×," and "/". Let "a" and "b" denote the operands. To complete a calculation, we need to press buttons to input necessary information, including operands and operators, and the calculator performs corresponding calculation automatically and displays the result on the screen. As shown in Figure 1.38, the key elements in this process are input, output, and processing.

1.4.1.2 Using a computer

Suppose the age of two brothers is "a" and "b," respectively and we need to use a computer to determine who is elder. Although the flow is simple, where the only difference from the calculator flow is that arithmetic operations are replaced with comparison operation as shown in Figure 1.39, the comparison operation cannot be done by a simple calculator. Compared with calculators, computers can complete complicated logical operations. The key elements in this process are input, output, and processing.

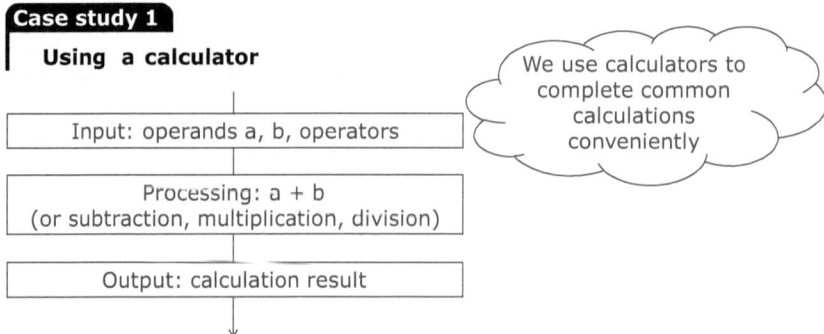

Figure 1.38: Using a calculator.

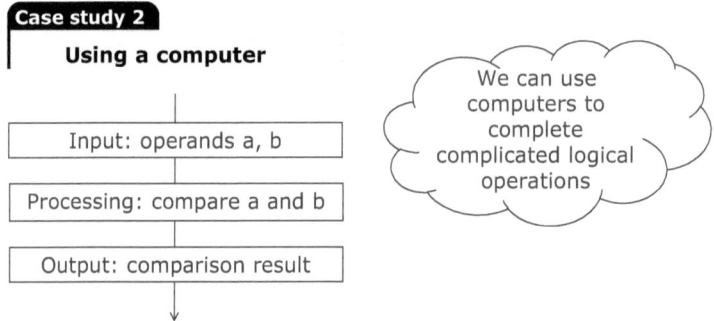

Figure 1.39: Using a computer.

1.4.2 Basic steps of program development

After years of practice, people have found that four main steps are needed to effectively solve complicated practical problems with computers as shown in Figure 1.40. The three key elements can also be found in these steps.

The first step is building a model that is done in the analysis phase. We extract functionalities and data from the problem, summarize objects involved, which are also information to be processed, and look for relations between them.

The second step is designing, which includes data structure design and algorithm design. Data structure design tries to find a way to organize and store data, whereas algorithm design tries to find a solution to the problem that satisfies requirements on functionalities.

The third step is coding. We write code in a certain programming language to implement the algorithm designed in the previous step.

Step	Tasks
Build a model	• Extract required functionalities • Extract data objects and analyze relations between them
Design	Data structure design and algorithm design
Coding	Write codes
Testing	Software testing and debugging

Figure 1.40: Steps of problem-solving with computers.

The fourth step is testing, which is also known as software testing. We test the code from step 3 and then try to debug if errors exist.

1.4.3 Example of problem-solving with computers

Example 1.1 Incentive system for employees

A company builds an incentive system to encourage its employees. The system works as illustrated in Figure 1.41. If the sales number is larger than or equal to five units, then the employee who achieved this number will be rewarded ¥ 1000. If the sales number is greater than 10, but not more than 50 units, the employee will be rewarded ¥ 200 times his/her sales number. If the sales number is greater than 50 units, the reward will be ¥ 250 times the sales number.

Sales number	Reward (¥)
sales number >=5	1000
10<=sales number<50	200*sales number
sales number >=50	250*sales number

Figure 1.41: The incentive system for employees.

Please design a program that outputs the amount of reward when given a sales number.

[Analysis]

1 Model building

Based on the information provided, let sales number be x and reward be y. We can write a piecewise function and buildup the mathematical model as shown in Figure 1.42.

Sales number x	Reward y
x>=5	1000
10<=x<50	200x
x>=50	250x

$$y = \begin{cases} 1000, & 5 \le x < 10 \\ 200\,x, & 10 \le x < 50 \\ 250x, & x \ge 50 \end{cases}$$

Mathematical mvodel

Figure 1.42: Model of the incentive system problem.

i **Knowledge ABC** Mathematical model
Mathematical models use mathematical language to describe characteristics or numerical relations of a system approximately. They are mathematical relation structures that reflect certain
problems or certain systems. We may also consider them to be mathematical expressions of
relations between variables in a system.

2 Algorithm design

We can derive the value of y piecewise based on the domain of x in the mathematical model. The process is represented in the flowchart (Figure 1.43).

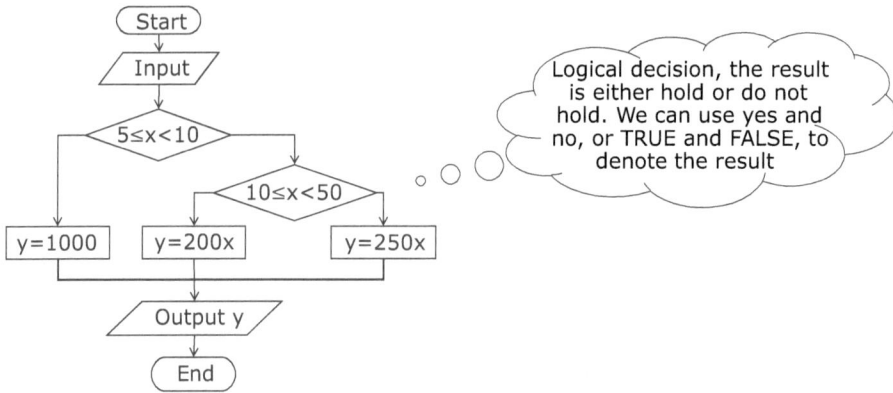

Figure 1.43: Flowchart of the incentive system problem.

Note that determining which interval x falls in is a conditional statement, which is
a logical operation in computers that outputs either true or false. Following the
flowchart, we can describe the execution process as follows:
- Input x, if $5 \le x < 10$, then y = 1000;
- otherwise, go to the no branch. If $10 \le x < 50$, then y = 200x; otherwise y = 250x;
  and
- output y.

We can also write pseudo code based on the mathematical model as shown in Figure 1.44. Note that we use indentation and alignment in pseudo code to indicate different levels of logical operations. In this case, the entire no branch is indented.

Pseudo code description	Program statements description
input x	scanf("%d",&x);
if then 5<=x<10 则 y=1000	if (x>=5 && x<10) y=1000;
else	else
if then 5<=x<10 则 y=200x	if (x>=10 && x<50) y=200*x
else y=250x	else y=250*x;
output y	printf("y=%f",y);

Figure 1.44: Pseudo code of the incentive system.

With the pseudo code, we can write the corresponding code. The syntax is introduced in the chapter "Program Statements," readers may take them for granted at this stage. In programming, each step of algorithms needs to be implemented by the corresponding code. By comparing flowchart with pseudo code, it is observed that flowcharts are more intuitive and easier to understand. However, it is hard to draw a flowchart. If we draw flowlines at our will, it will be difficult for others to read and modify. On the other hand, pseudo code avoids the use of graphical symbols and has a more compact form. It is easier to write and understand pseudo code. Converting pseudo code into real code is also effortless. Hence, it is now used worldwide to describe algorithms.

3 Code implementation

Now we may present the full C code following C grammar and program structure. Readers can run this program in their own environment.

```c
#include <stdio.h>
int main(void)
{
    int x;
    int y;
    printf("Please input sales number: ");
    scanf("%d",&x);
    if (x>=5 && x<10) y=1000;
    else
```

```
{
    if (x>=10 && x<50) y=200*x;
    else y=250*x;
}
printf("Reward is ¥%d\n",y);
return 0;
}
```

If we type 12 when prompted to input a sales number, we will see "Reward is ¥ 2400" is displayed. This is consistent with our expectations, so the program is correct.

4 Testing

Although the program produces a correct result, we still need to test it thoroughly.

Data for testing need to be designed in advance. This is also called the "test case design." Except for input data, we also need to know the expected results. More details on testing are covered in the chapter "Execution of Programs." When testing a program, we type in the input data and compare the output with expected results. If they are not identical, there are errors in our program. Test cases for this problem are shown in Figure 1.45. It is clear that the program outputs 2500 when $x = 2$ while we expect to see 0, so apparently, something is wrong in our program.

Test cases			
Input data		Expected result	Test result
Sales number x	Range of x	Reward y	Comparison result
2	$x<5$	0	Wrong
5	$5 \le x < 10$	1000	Correct
10	$10 \le x < 50$	200*10	Correct
20	$10 \le x < 50$	200*20	Correct
50	$x \ge 50$	250*50	Correct

Figure 1.45: Test cases of incentive system program.

Further investigation suggests that we forgot to handle the case where $x < 5$ when designing the algorithm. After finding the error, we need to modify the algorithm before we change the code. This example shows that algorithm designs can be optimized if we consider test cases in advance.

Example 1.2 Looking up telephone numbers

Telecommunication companies use a phone number record table to record information of their customers. Please write a program to inquiry about private phone numbers in a certain city or company. Given a name, the program should output the corresponding number if it exists in the record. Otherwise, the program should output "No such number" sign.

[Analysis]

The phone number record table is shown in Figure 1.46. To keep our description of data simple, we call each row a data element, which consists of multiple data items such as customer name, phone number, address, etc. A key is the data item that can identify data elements. In other words, a key is a data item that can uniquely identify data elements. For example, a phone number is unique, but an address may not be so.

Customer name	Cellphone number	ID card number	Address
Zhang1	138*****	6101131980***	***
Li2	152*****	6101131981***	***
Wang1	139*****	6101131990***	***
Zhang2	139*****	6101131972***	***
Li1	188*****	6101131976***	***
...			

Data element

Key

Figure 1.46: Phone number record table.

Solution 1: Sequential structure and sequential search

The model, design, and implementation are shown in Figure 1.47.

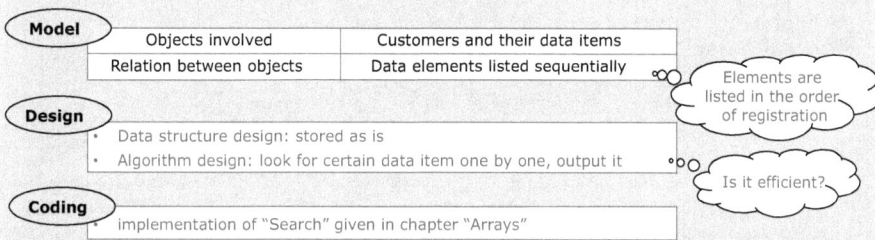

Figure 1.47: Sequential search solution for phone numbers.

When building the model, we notice that objects in this problem are customers and their data items. Objects are arranged sequentially, which means they are arranged in the order of registration and not sorted.

The design step consists of data structure design and algorithm design. Data structure design controls how the table is stored and accessed in computers. Algorithm design determines how to look up phone numbers. In this case, we query the "customer name" data item sequentially. If a value hits, the corresponding "phone number" data item value is returned; otherwise, "No such number" sign is returned.

Readers may refer to "Search" programs in chapter "Arrays" for the implementation of this algorithm.

A search program that looks for a target record one by one in the table is called "sequential search." The number of comparisons needed is related to where the record locates in the table. When the table is large, using a sequential search is time-consuming. To improve efficiency, we can rearrange the table so that data elements are stored in alphabetical order of customer family names.

Solution 2: Ordered structure and binary search
In an ordered structure, the customer names are in alphabetical order. We can perform a binary search in such a table as shown in Figure 1.48.

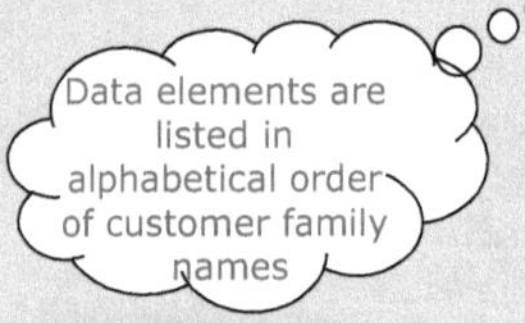

Customer name	Cellphone number	ID card number	Address
Li1	188*****	6101131976***	***
Li2	152*****	6101131981***	***
Wang1	139*****	6101131990***	***
Wang2	138*****	6101131986***	***
Zhang1	138*****	6101131980***	***
Zhang2	139*****	6101131972***	***
...	......		

Data elements are listed in alphabetical order of customer family names

Binary search
Compares the middle value in a sorted sequence with a given key,
If the key is larger, then it must be in the second half of the sorted sequence;
Otherwise it is in the first half.
The length of the sorted sequence is thus cut to half, we then compare the new middle value with the given key again and repeat this process.

Figure 1.48: Ordered structure and binary search.

The entire solution is shown in Fig. 1.49. In the modeling step, objects are still customers and their data items; objects are now arranged in the order of a data item since the table has been sorted alphabetically. We use the same data structure as before, while we opt in binary search to find the record. Implementation can be found in the chapter "Arrays" where we will cover code for binary search.

If the table is huge, we can also use indexed structure and layered search for higher efficiency. The table of contents in a book is the most common indexed structure.

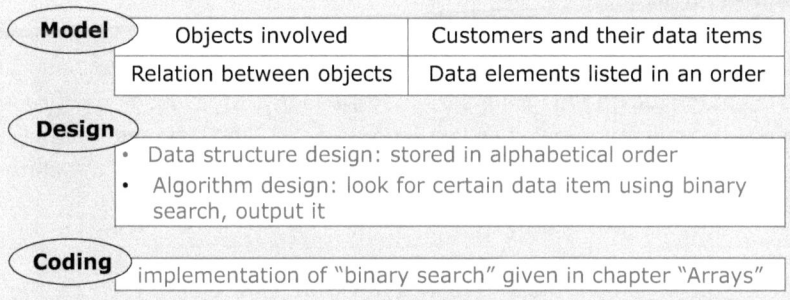

Model	Objects involved	Customers and their data items
	Relation between objects	Data elements listed in an order

Design
- Data structure design: stored in alphabetical order
- Algorithm design: look for certain data item using binary search, output it

Coding implementation of "binary search" given in chapter "Arrays"

Figure 1.49: Binary search solution of looking up phone numbers.

Solution 3: Indexed structure and layered search

We can put records with the same family name together and create an index table of family names, and the index table needs to be associated with the original table as shown in Figure 1.50.

Data table

Index Table

Family name	Address in data table	Quantity
Li	0x2000	***
Zhang	0x4000	***
Wang	0x6000	***
...	......	

Customer name	Cellphone number	ID card number	Address
Li1	188*****	6101131976***	0x2000
Li2	152*****	6101131981***	***
...	......		
Zhang1	138*****	6101131980***	0x4000
Zhang2	139*****	6101131972***	***
...	......		
Wang1	139*****	6101131990***	0x6000
Wang2	138*****	6101131986***	***
...	......		

Figure 1.50: Index table.

The entire solution is shown in Figure 1.51. In the modeling step, we buildup the index table and rearrange the original table accordingly. In the data structure design step, we need to find a suitable way of data organization so that we can store these two tables into computers. In the algorithm design step, our algorithm should look up a family name in index table first and then look up full name based on the address and length we obtain from the index table.

Model			
	Objects involved	Index table	Customer family names and corresponding addresses in data table
		Data table	Customers and their data item
	Relation between objects	Index table	Data elements listed in an order
		Data table	Data elements listed sequentially

Design
- Data structure design: index table storage
- Algorithm design: look up in index table first, then in data table

Figure 1.51: Indexed search solution of phone numbers.

1.4.4 Flow of program development

We should now have some general knowledge about solving practical problems by programming after studying the examples above. Usually, the main steps of problem-solving with computers are as shown in Figure 1.52. The rectangle symbol in the figure is a phase or result in the problem solving process, whereas the ellipse symbol represents a step.

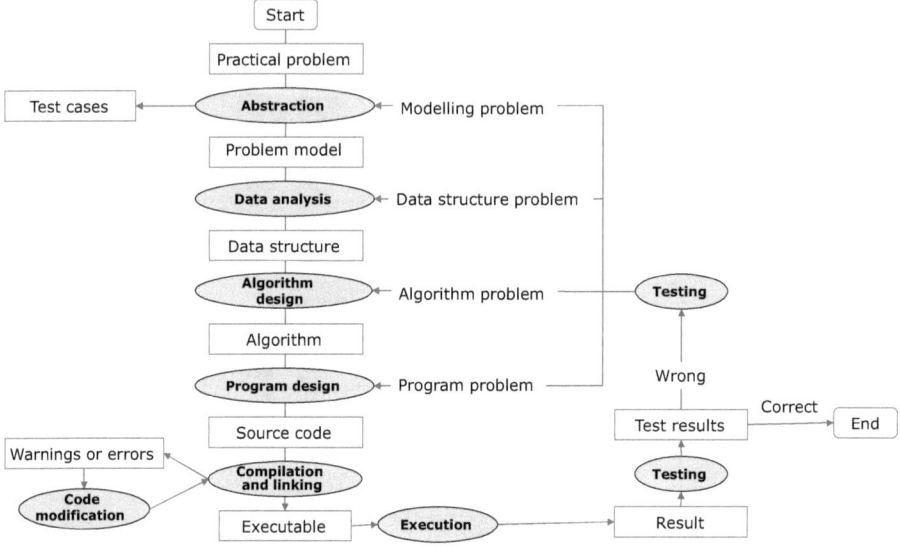

Figure 1.52: Flow of program development.

The full flow contains:
- Practical problem: It needs to provide known conditions and descriptions of required functionalities.
- Abstraction: First, extract functionalities and information from the problem description. Then, look for relations between these data and buildup a mathematical model. Test cases are also designed in this step. By building a model, the problem is converted into a form that computers can "understand" and "accept."
- Data analysis: We analyze what data are contained in the information provided, how they are correlated, and how they can be stored in computers. Based on our analysis, we designed proper data structures. While this book focuses on fundamental topics in programming, such as algorithms and programming concepts and methods, readers can refer to "Data Structure" courses for sophisticated data structure designing.

- Algorithm design: We formalize a solution to the problem based on the functionalities required.
- Program design: We "translate" the algorithm we designed into code to obtain
  source files.
- Compilation and linking: Programmers use compilers to translate code into machine code that computers can execute. If errors or warnings exist, we need to
  modify the code until it is successfully compiled and linked.
- Execution: Executable programs will produce results after executed in a runtime environment.
- Testing: We compare the results of our program with test cases. If they are
  identical, we have successfully solved the problem; otherwise, we need to
  debug to find where the defect lies. Once the mistake is found, we return to the
  corresponding phase or step to revise our solution and start over from there
  until we obtain the correct results.

1.5 Introduction to C programs

We are going to introduce C programs in this section briefly.

Computers solve problems using programs, which consist of data, processing,
and results. The processing part is done through programming. Programming is
similar to writing, where we carefully select words to use, arrange them into paragraphs, and organize contents into chapters. If we compare program statements to
words, flow structure would be paragraph structure. Consequently, functions and C
program templates would be chapter structure as shown in Figure 1.53.

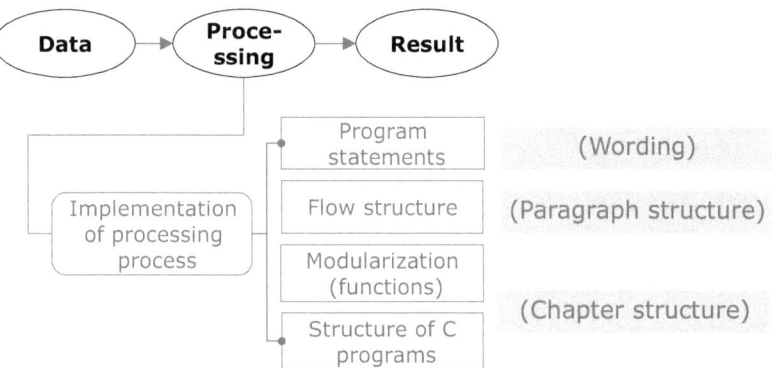

Figure 1.53: Implementation architecture of processing process.

1.5.1 Sample C programs

Example 1.3 Sample program 1
The program shown in Figure 1.54 displays "hello world!" on screen.

```
01  // Display hello world on screen
02  #Include <stdio.h>
03
04  // Program starts from main function
05  int main(void)
06  {
07    printf( "hello world!\n" );
08
09    return 0;
10  } // main function ends
```

Line numbers are used to read programs conveniently, they are not parts of the program

Functionality of program: display hello world on screen

Figure 1.54: Sample program 1.

1 Functions

A segment of code that has its own functionality is called a function in C. Code between lines 5 and 10 in Figure 1.55 is a function, which we call the main function. The structure of C programs contains the main function and child functions although there is no child function in this example. The curly brackets on lines 6 and 10 are used to wrap the body of the main function. They mark the beginning and the end of the main function. In programs, a line that ends with a semicolon is called a statement.

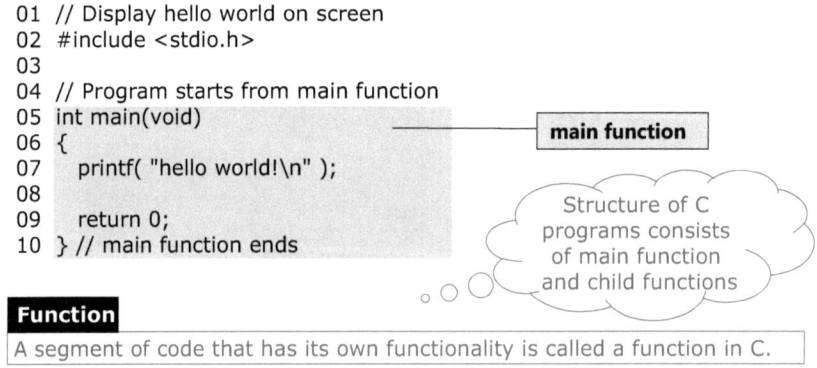

```
01  // Display hello world on screen
02  #include <stdio.h>
03
04  // Program starts from main function
05  int main(void)                          main function
06  {
07    printf( "hello world!\n" );
08
09    return 0;                   Structure of C
10  } // main function ends        programs consists
                                   of main function
                                   and child functions
```

Function
A segment of code that has its own functionality is called a function in C.

Figure 1.55: Functions in sample program 1.

2 Comments

In Figure 1.56, the sentence that starts with // on line 1 is a "comment." A comment is not a program statement. Note that it is not terminated by semicolon either. During compilation, comments will be ignored and not be translated into machine code. Comments are used to provide extra information about the code. As program statements are abstract expressions, other programmers may not fully understand them without explanations. Even the author may become confused if the code was written a long time ago.

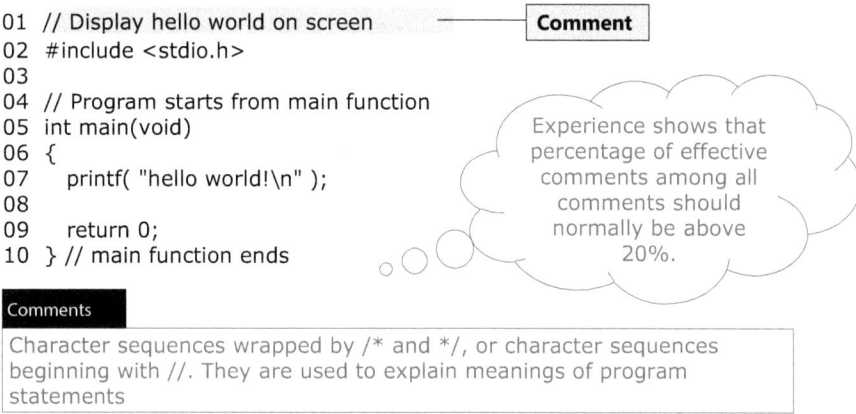

```
01  // Display hello world on screen          ———  Comment
02  #include <stdio.h>
03
04  // Program starts from main function
05  int main(void)
06  {
07     printf( "hello world!\n" );
08
09     return 0;
10  } // main function ends
```

Experience shows that percentage of effective comments among all comments should normally be above 20%.

Comments

Character sequences wrapped by /* and */, or character sequences beginning with //. They are used to explain meanings of program statements

Figure 1.56: Comments in sample program 1.

Good habits in programming
Normally, at least 20% of all comments written should be effective comments. The general rule of thumb is that effective comments should help us understand the program. Comments must be accurate, simple, and easy to comprehend.

3 Library functions and file inclusion

In Figure 1.57, printf() on line 7 is a printing function. In this case, it prints out "hello world!" onto the screen. The implementation of printf() is rather complicated, but programmers use it a lot. Hence, programs like printf() are implemented in advance and collected in a system library. They are called "library functions."

At line 2 is a preprocessing directive in C, which denotes "file inclusion." We can include an entire source file in our source code so that we can use it in our program. stdio is short for standard input and output.

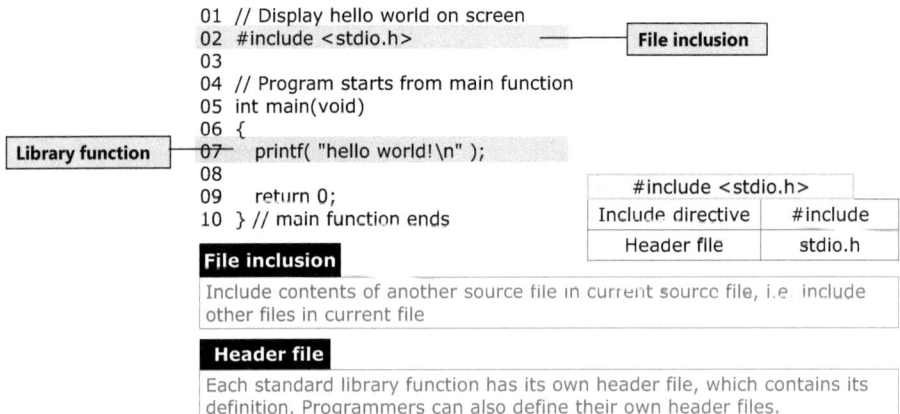

Figure 1.57: Library functions and files inclusion in sample program 1.

Definition of printf() is done in header stdio.h. Programmers can use printf() by using #include. printf() prints contents between quotation marks inside parentheses onto the screen. Programmers can fill in characters as needed.

> **Knowledge ABC** Library functions, header files, and file inclusion
> – Library functions: Library functions are not part of the C language. It is compilers that collect a series of programs that implement frequently used functionalities and put them into a system library. Users can include corresponding definition files (header files) to use these programs. In other words, we can use library functions by using "file inclusion" commands.
> – Building libraries provides a collection of reusable functions that can be shared by other programmers and programs. For example, multiple programs may need the same helper functions. We may put them into a library and link them into our programs with a linker, without copying and pasting their source code into each program. This simplifies code writing and maintenance.
> – Header files: The purpose of using header files is to put code shared by several C programs into one file so that the overall code size is reduced. Header files have extension ".h." Each library function needs a corresponding header file that contains the prototype of the function. To use a library function in programs, one must include the header file with the prototype of that function. Header files in C contain prototypes of every function in the standard library.
> – Refer to Appendix C for common library functions in C.
> – File inclusion: File inclusion replaces include statements with the included file so that it is linked with the current file to form a single source file.
> – Usage: #include <filename> (or #include "filename")
>
> **Notes:**
> (1) The included file can be provided by the system or written by programmers.
> (2) One include statement can only include one file. Use multiple include statements if multiple files are needed.
> (3) If the file name is surrounded by angle brackets, compilers will look for the file in a system specified directory. If it is in quotation marks, compilers will first look for it in the directory of the current source file. If no such file is found, compilers proceed to search it in the system directory.

Example 1.4 Sample program 2

This is a program with multiple functions. The main function asks for two integers from keyboard input, calls child function to calculate the maximum of them, and displays results onto screen.

The program is shown in Figure 1.58. main() receives two integers a and b from the input, calls child function max() to compute the maximum, stores result in c, and outputs to screen.

```
01 #include <stdio.h>        //File inclusion
02 int max(int x, int y);    //Function declaration or function prototype
03 int main(void)
04 {
05    int a,b,c;              //Variable definition                     ── Main function
06
07    scanf("%d,%d",&a,&b);  //Input integer a, b from keyboard
08    c=max(a,b);            //Call max() to compute the larger between a and b and store in c
09    printf("max=%d",c);    //Output value of c to screen
10    return 0;
11 }
12
13 int max(int x, int y)     //Information function max needs to handle is two integers x, y
14 {
15    int z;                                                            ── Child function
16
17    if (x>y)  z=x;         //Compare x and y, use z to store the larger one
18    else z=y;
19    return(z);            //Tell caller the value of z
20 }
```

Figure 1.58: Sample program 2.

Except for the main function, there is also a child function max() in this program. Similar to main(), the function body is also surrounded by curly brackets. The main function lies between lines 3 and 11, whereas child function max() is located between lines 13 and 20.

On line 7, inputs from the keyboard are stored into variables a and b. scanf() is a library function used to receive input from the keyboard and is defined in stdio.h as well. On line 8, child function max() is called to calculate the maximum of a and b. The result is stored in c and output to screen on line 9.

The only thing that child function max() does is to distinguish the larger number in its inputs. The input is obtained from x and y in the parentheses on line 13. This is the convention of how child functions get input data. The calculation result z is returned to main() through return the statement on line 19.

Main and child functions complete tasks that are relatively independent of each other. They work together through function calls to perform complex functionalities.

1.5.2 Structure of C programs

The normal structure of C programs is shown in Figure 1.59. Preprocessing statements are at the beginning of the program. The include directive seen earlier is also a preprocessing statement. There must be a main function in C programs. It is the

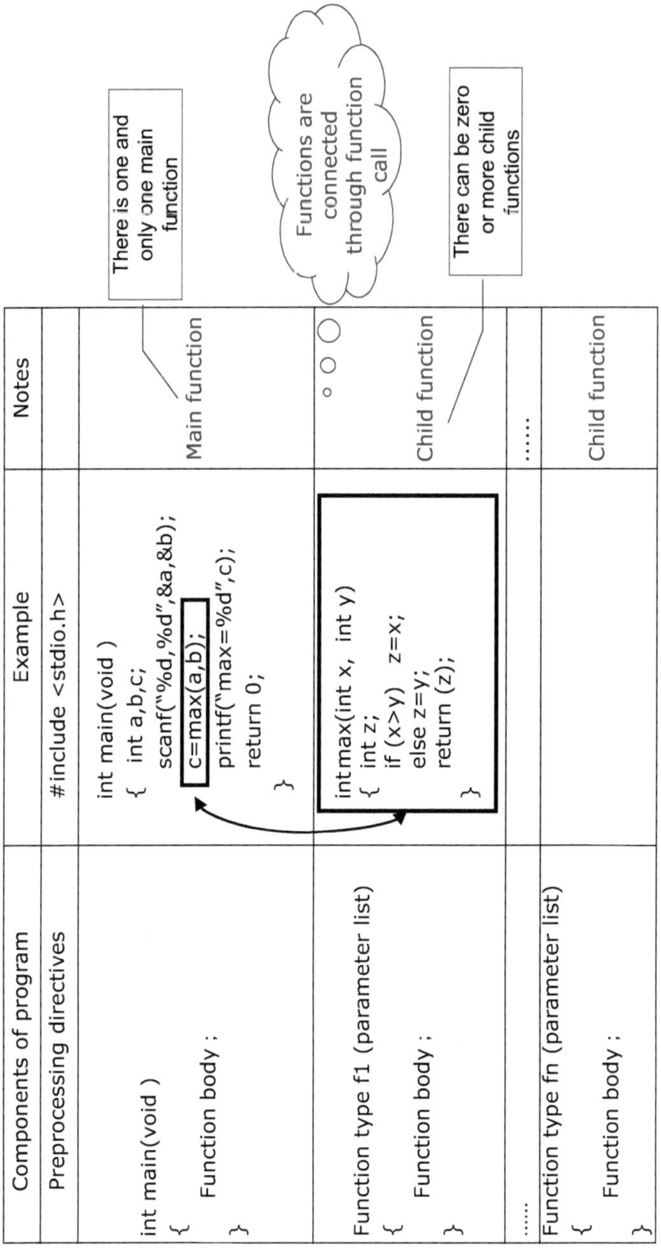

Components of program	Example	Notes
Preprocessing directives	#include <stdio.h>	

int main(void)
{
    Function body ;
	int main(void)
	{ int a,b,c;
	    scanf("%d,%d",&a,&b);
	    c=max(a,b);
	    printf("max=%d",c);
	    return 0;
	}
		Main function

		○ ○ ○

Function type f1 (parameter list)
{
    Function body ;
	int max(int x, int y)
	{ int z;
	    if (x>y) z=x;
	    else z=y;
	    return (z);
	}
		Child function

......		......
Function type fn (parameter list) { Function body ;		Child function

There is one and only one main function

Functions are connected through function call

There can be zero or more child functions

Figure 1.59: Structure of C Programs.

entry point of execution. There can be zero or more child functions. Functions are connected through function calls.

More details of functions are covered in the chapter "Functions."

1.5.3 Coding style requirements

As we need to follow format conventions when writing articles, we need to follow some guidelines when writing programs. The purpose is to make our code less confusing and more readable. Detailed requirements are shown in Figure 1.60.

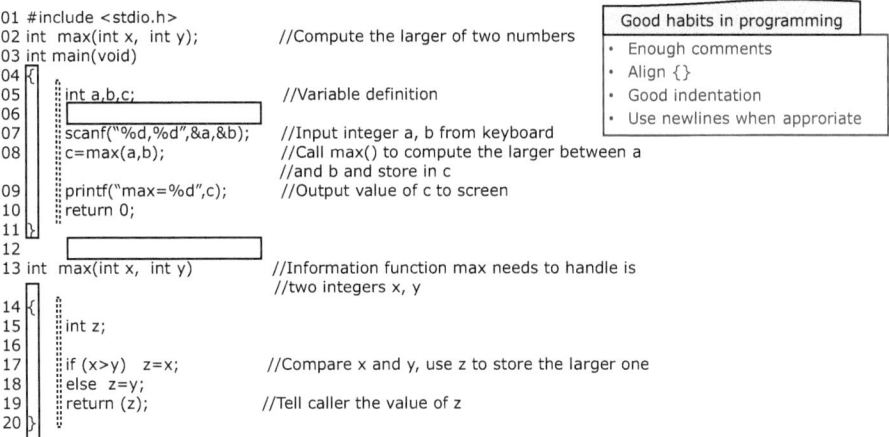

Figure 1.60: Coding style requirements.

C programs use spaces and newlines to split lexical terms. Special characters are used to identify syntax. For instance, semicolon represents the end of a statement. The flexible format makes a flexible coding style possible.

When writing programs, we should follow these guidelines on coding style:

– Enough comments: We need to provide adequate and precise comments to explain the functionalities of programs and the meaning of statements. At the beginning of a program, we should briefly introduce the functionality of this program. Critical variables should be commented with their usages. Our sole purpose in doing these is to improve the "readability of program." Programs with good readability allow users to understand them easily. Readable code also allows authors to recall how their programs work when they review them after a while.

- Indentation: Users should choose an indentation style at their will and use it consistently in their programs. Tab key can be used for this purpose. Tab key is the "tabulator key" on keyboards. Pressing tab usually inserts four spaces in programs. The main difference between the tab key and the space key is that it indents more efficiently. However, the size of tab may vary in different editors, therefore we need to be careful when using it.
- Alignment of {}: There could be multiple pairs of curly brackets in a function. Each pair should be aligned vertically. Together with indentation, it makes our programs easier to read as the structure is clearer and the scope of statements is more obvious.
- Appropriate use of newlines: We can use newlines to separate different functionality blocks, such as variable declaration, variable assignment, and execution statements. This creates a better visual effect and highlights the structure of the program.

> **ⓘ** **Knowledge ABC** Programming trivia
> The relation between software and program:
>   It is quite common to hear people say "programming is writing software." However, software and program are two different concepts. Software should contain the following elements:
> – Collection of commands or programs that can satisfy specific requirements when executed.
> – Data structures that allow the programs to process information reasonably.
> – Documents that describe requirements and usage of the programs.
>
> Hence, we may conclude that software = program + data + document.

1.6 Summary

Concepts related to programs and their expressions are shown in Figure 1.61.
    We can use flows to describe a series of operations in order.
    Flowcharts are intuitive, whereas pseudo codes are convenient.
    Sequential, branch, and loop are the three basic logical structures of flows,
    And they are the building blocks of programs.

    To solve a problem with computers, we need to extract data and functionalities.
    Data are processed to obtain the solution.
    Data can be received, stored, computed, and output, each operation has its own methods.
    Description of the problem-solving process is called an algorithm.
    We need testing and debugging to verify our results.

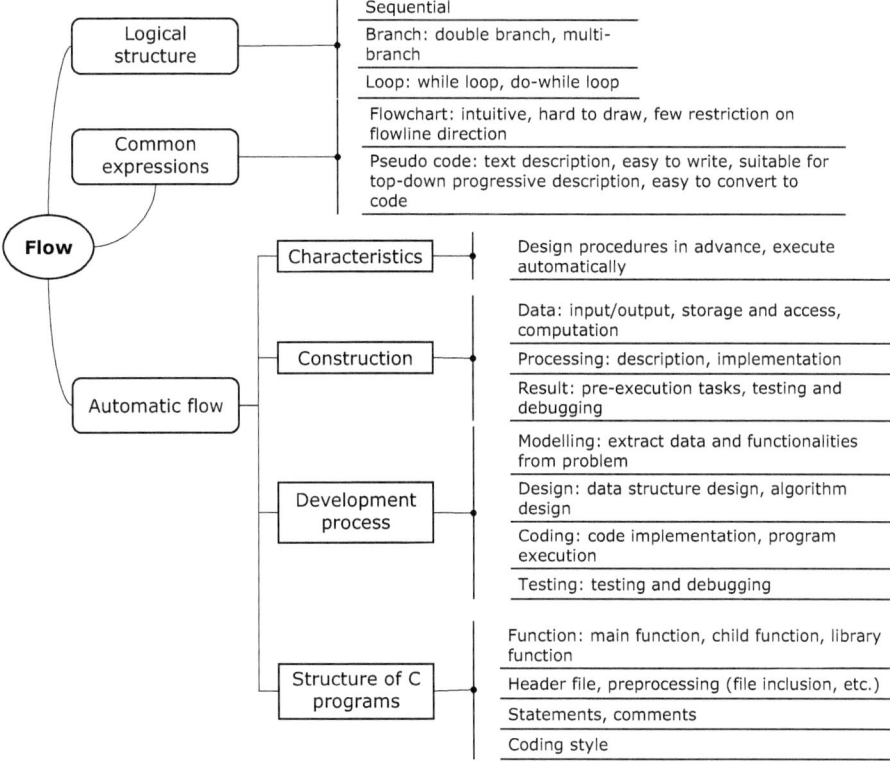

Figure 1.61: Concepts and expression of programs.

Programs are flows that can be executed automatically by machines, and we need to write code for them.

Writing code is like writing an essay, and we need to be deliberate on every detail.

Program statements are like words and comments are explanations.

Function modules are paragraphs and we build them up to form programs.

Referencing others' work, appellation, and whether we include a paragraph is done by preprocessing.

Sometimes spaces, sometimes newlines, we need them to achieve a good coding style.

1.7 Exercises

1.7.1 Multiple-choice questions

(1) [Concept of programs]
A command sequence that is designed to solve a particular problem is called a ( ).
A) Document   B) Language   C) System   D) Program

(2) [Concept of programs]
Which of the following statements is correct? ( )
A) Algorithm + data structure = program   B) An algorithm is a program
C) A data structure is a program   D) An algorithm consists of data structures

(3) [Concept of programs]
Which of the following statements is wrong? ( )
A) Computers can recognize programs in hexadecimal code.
B) A collection of commands that can be executed sequentially is called a "program."
C) A "program" is a language we use to "communicate" with computers.
D) Computers can recognize machine language code in 0's and 1's.

(4) [Concept of software]
A piece of software consists of ( )
A) Algorithm and data   B) Program and data
C) Program and documents   D) Program, data, and documents

(5) [Concept of debugging]
What is the purpose of debugging? ( )
A) Diagnosing and correcting errors in programs
B) Finding as many errors as possible in programs
C) Finding and correcting all errors in programs
D) Determining the nature of errors in programs

(6) [Concept of data structure]
A data structure consists of three components, namely ( ).
A) Storage structure of data, relations between data, and their representations.
B) Logic structure of data, relations between data, and their representations.
C) Logic structure of data, relations between data, and their storage structure.
D) Logic structure of data, storage structure of data, and operations on data.

(7) [Steps of problem-solving with computers]
There are four major steps (①~④) in the process of problem-solving with computers. Please determine the correct order of them. ( )
① Debugging   ② Problem analysis   ③ Algorithm design   ④ Coding
A) ①②③④   B) ②③①④   C) ②③④①   D) ③②④①

(8) [Flowcharts]

A flowchart is a tool to describe algorithms. Standard flowcharts are constructed by a few basic shapes. Which of the following shape stands for input/output? ( )

A) Parallelogram   B) Rectangle   C) Ellipse   D) Diamond

(9) [Programming languages]

Which of the following statements about programming languages is correct? ( )

A) High-level language is a natural language.

B) C is independent of platforms. C programs are platform-independent.

C) Machine languages are closely related to computer hardware. Programs in machine languages have better portability.

D) Programs must be translated before being executed on computers, regardless of which language they are written in.

(10) [Programming languages]

Which of the following statements is wrong about the characteristics of C? ( )

A) C has both merits of high-level languages and low-level languages and is highly efficient.

B) We can use C to write applications and system software.

C) Portability of C is poor.

D) C is a structured programming language.

(11) [Program structure]

Which of the following statements is correct? ( )

A) C program must use main as the name of its main function, starting from which the program is executed.

B) Users can use any function as the main function, starting from which the program is executed.

C) A C program is executed starting from the first function in the source file.

D) We can use different spelling forms of main (such as MAIN and Main) in the main function

(12) [Control structure]

Which of the following statements is wrong? ( )

A) Programs using the three fundamental structures can only solve simple problems.

B) A structured program is constructed by the three fundamental structures, namely sequential structure, branch structure, and loop structure.

C) C is a structured programming language.

D) The idea of modularization is recommended in structured programming.

(13) [Compilation and linking]

Which of the following statements is wrong?

A) Every executable statement and nonexecutable statement in a C program is converted to a binary machine instruction.

B) A C program must be compiled and linked to generate an executable binary machine instruction file.

C) A program written in C is called a source program. It is stored in a text file as ASC II code.

D) A source program in C is converted into an object program after compilation.

(14) [Grammar rules]

Which of the following statements is correct? ( )

A) C program must be stored in a single program file.

B) We can only write one statement in a line.

C) Comments in a C program must be at the beginning of the file or after statements.

D) The syntax of C is flexible. We can write a statement across multiple lines.

(15) [Grammar rules]

Which of the following statements is correct about comments in programs? ( )

A) Comments must be written between /* and */ or after //.

B) A comment must be in the front of or after the statement it explains.

C) We can write a comment in another comment.

D) Errors in comments lead to compilation errors.

(16) [Grammar rules]

The termination mark of statements in C is ( ).

A) ，　　　　　　B) ；　　　　　　C) 。　　　　　　D) 、

2 Algorithms

Learning objectives
- Know concepts related to algorithms
- Know the characteristics of computer algorithms
- Know representations of algorithms
- Know general procedures of algorithm design
- Can design algorithms in top-down stepwise refinement method

Donald Ervin Knuth once said, *"Programs are like blue poems."* If that is true, then algorithms are the soul of these poems.

2.1 Concept of algorithms

2.1.1 Algorithms in practice

Let us take a look at a real-life algorithm: ticket purchasing at the ticket office. When there was no internet, Mr. Brown had to buy train tickets at ticket offices when he needed to take business trips. The main procedures of purchasing tickets are shown in Figure 2.1, in which the key information includes the date, destination, train number, price, and fare. In fact, everything is done under certain conditions and through performing a series of operations. Algorithms are methods and procedures used to solve the problems.

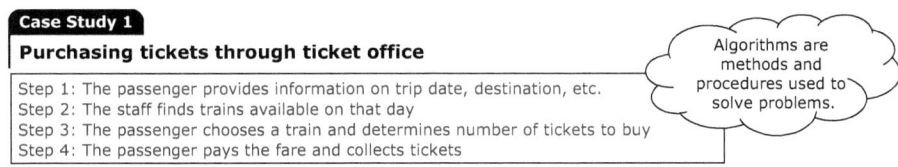

Case Study 1

Purchasing tickets through ticket office

Step 1: The passenger provides information on trip date, destination, etc.
Step 2: The staff finds trains available on that day
Step 3: The passenger chooses a train and determines number of tickets to buy
Step 4: The passenger pays the fare and collects tickets

Algorithms are methods and procedures used to solve problems.

Figure 2.1: Flow of purchasing tickets at the ticket office.

Nowadays, Mr. Brown can purchase tickets easily online. The main steps of online purchasing are shown in Figure 2.2. Compared with offline purchasing, online purchasing uses the same set of crucial information, except that we are dealing with computer systems instead of a human.

https://doi.org/10.1515/9783110692327-002

Case Study 2
Purchasing ticket online

Step 1: Open the website
Step 2: Click "Search" in search page, input trip date, destination, etc.
Step 3: Click "Book" in ticket booking page, choose a train to book ticket
Step 4: Log into system (if you have an account)
Step 5: Confirm passenger information, seat class, and submit the order
Step 6: Pay ticket fare online
Step 7: Choose ticket collection method

> Computers can partially simulate human minds. They are faster and more accurate.

Figure 2.2: Flow of purchasing tickets online.

Computers can partially simulate human minds. They can complete some tasks for our brains in a faster and more accurate manner. They liberate us from dull mental work. Nonetheless, they cannot, at least for now, solve problems on their own. Operations that computers can complete are determined by humans in advance.

Another example is setting up a washing machine. We set up a washing program and the machine executes corresponding operations after being started as shown in Figure 2.3. These operations that can be executed automatically by machines are programs. Note that programs are operations executed by programs, but algorithms are procedures humans or machines use to solve problems.

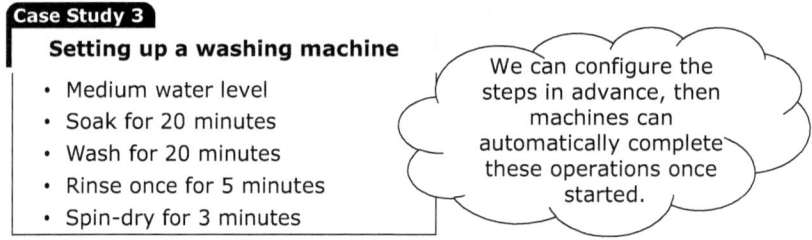

Case Study 3
Setting up a washing machine

- Medium water level
- Soak for 20 minutes
- Wash for 20 minutes
- Rinse once for 5 minutes
- Spin-dry for 3 minutes

> We can configure the steps in advance, then machines can automatically complete these operations once started.

Figure 2.3: Setting up a washing machine.

2.1.2 Definition of algorithms

Through these examples, we should roughly understand what algorithms are by now. Algorithms can be considered as an entire solution consisting of basic operations and order of operations. We may also deem algorithms as finite, definite computation sequences designed under specific requirements to solve a certain type of problem. Definition of algorithms is given in Figure 2.4, in which computer algorithms mean algorithms that can be executed by computers. In other words, steps in computer algorithms are intentionally designed to conform to computer characteristics: operations should be simple to be executed repeatedly. Problem-solving

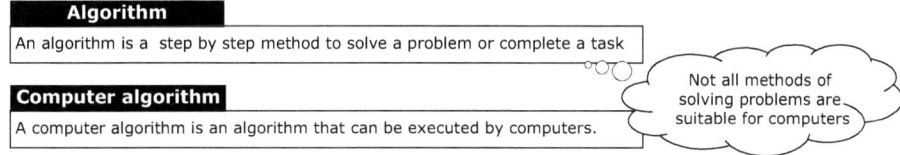

Figure 2.4: Definition of algorithms.

with computers has limitations as well. A solution feasible in daily life may not work on computers, that is, not all the methods we use to solve problems can be used by computers. We will elaborate on this topic in Section 2.3.

Characteristics of computer algorithms are summarized from the perspective of computers. Computers receive input data, execute certain operations, and produce results that fulfil functionality requirements. Hence, we consider input, functionality, and result to be the three key elements of algorithms as shown in Figure 2.5.

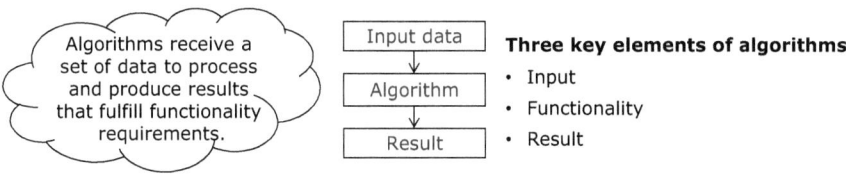

Figure 2.5: Key elements of computer algorithms.

Example 2.1 Example of computer algorithms: price guessing game
The TV show *Number Guessing* has the following rule: participants can obtain a product if they can correctly guess its price in the given time. The host will give hints to participants after a guess, telling them whether it is "too high" or "too low." Suppose the price of a product lies in the interval 0–2000 (the price is an integer), what strategy should we use to guess the answer in the shortest time on average?

[Analysis]
There is a classic solution to this problem called the "binary search." To use binary search, the search interval has to be sequentially increasing. The data in this game are prices that sequentially increase in the given interval so that we may apply binary search.

To be more specific, we should use a strategy called "guess the middle value" in this game as shown in Figure 2.6. We can see that there are inputs and outputs in this algorithm. Also, each step is feasible and we can find out a solution in finite steps.

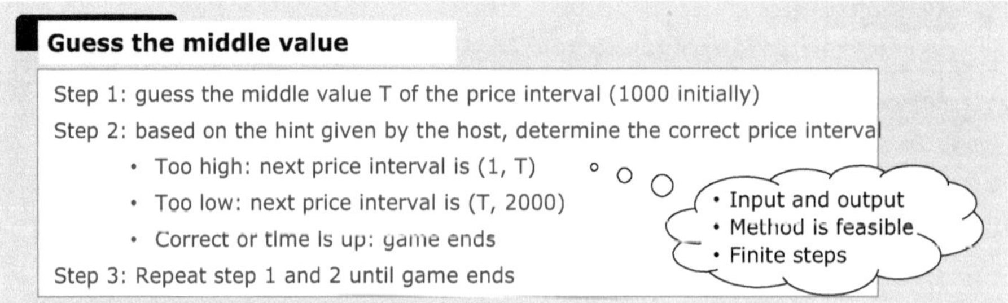

Guess the middle value

Step 1: guess the middle value T of the price interval (1000 initially)

Step 2: based on the hint given by the host, determine the correct price interval

- Too high: next price interval is (1, T)
- Too low: next price interval is (T, 2000)
- Correct or time is up: game ends

Step 3: Repeat step 1 and 2 until game ends

- Input and output
- Method is feasible
- Finite steps

Figure 2.6: Using binary search.

2.1.3 Characteristics of algorithms

We can now summarize the characteristics of algorithms as shown in Figure 2.7. Note that an algorithm may accept zero input in some cases. For instance, when solving an equation, an algorithm can find out the solution following each step of the algorithm and given conditions without any input data.

Characteristics of algorithms

- Input: have zero or more input
- Output: produce result that fulfills functionality requirement
- Finiteness: an algorithm should contain finite steps
- Definiteness: each step of an algorithm should be precisely defined without ambiguity
- Effectiveness: each step of an algorithm can be done effectively and generate certain result

Figure 2.7: Characteristics of algorithms.

2.2 Representation of algorithms

2.2.1 Top-down stepwise refinement method

Algorithms describe the steps and methods used to solve problems. Given a complex problem, we can use a global than local strategy to describe it in a top-down manner. For example, some daily activities of Mr. Brown are illustrated in Figure 2.8. In different time periods, each activity can be divided into multiple steps or subactivities.

When describing a problem, designers need a comprehensive understanding of the system to be designed. With this in mind, they can divide the system into several parts, do the same to each subsystem, and repeat this process. This is the standard way of program development, "top-down stepwise refinement" method as shown in Figure 2.9. We should not try to complete a program in one attempt. Instead, we do this in multiple steps, each implementation being more detailed than the last.

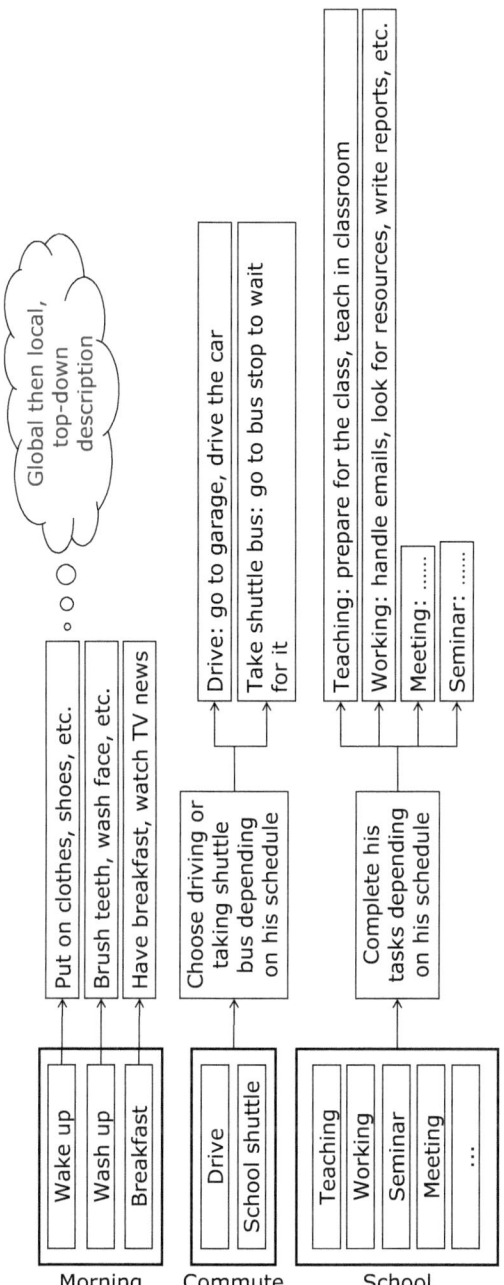

Figure 2.8: Daily life of Mr. Brown.

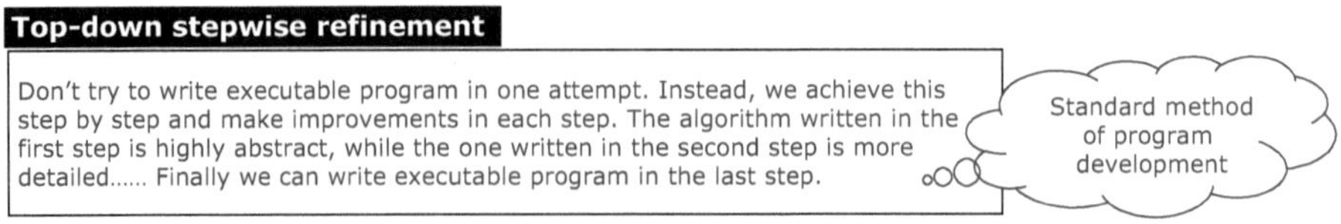

Top-down stepwise refinement

Don't try to write executable program in one attempt. Instead, we achieve this step by step and make improvements in each step. The algorithm written in the first step is highly abstract, while the one written in the second step is more detailed...... Finally we can write executable program in the last step.

Standard method of program development

Figure 2.9: Method of program development.

Knowledge ABC Methods of program development

1. Structured programming

Structured programming is a design principle that concentrates on module functionalities and process design. The concept, being a milestone in the history of software, was first proposed by E. W. Dijkstra in 1965. It advocates the use of a top-down stepwise refinement method of programming. Every program should be constructed using the three basic control structures, namely sequential, decision, and loop structure. Structured programming aims at improving the readability of programs.

2. Top-down and stepwise refinement

When programming, we should consider outline before details, global objectives before local objectives. Instead of bothering with numerous details, we should design our system starting from a global goal and proceed gradually. For complex problems, we should also set some subobjectives to achieve a smooth transition.

It seems tedious to program in this way, but it has many merits. It makes programs more comfortable to read, write, debug, maintain, validate, and verify. It brought a revolution to programming and soon became standard practice, especially in software engineering, which quickly developed later.

3. Modularized programming

A complex problem can usually be divided into several simpler problems. Modularization split the objective of a program into subobjectives that are easier to achieve and repeat the split process. Each subobjective is called a module.

2.2.2 Example of representation of algorithms

Let us consider an example of describing algorithms using pseudo code.

Example 2.2 Scoring in competitions

In TV karaoke shows or diving competitions, scores are computed by discarding the highest and lowest scores from the judges and computing the average of the rest. The main steps are shown in Figure 2.10, where some steps need further processing, for example, "discard the highest" and "compute the sum." We shall cover the details now.

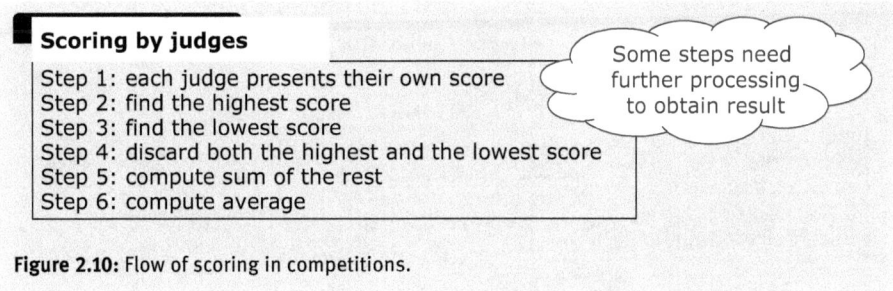

Figure 2.10: Flow of scoring in competitions.

1 Finding maximum

Given n numbers, find out and display the maximum.

[Analysis]

We will focus on finding the maximum. Without loss of generality, we suppose n = 10.

Unlike how it appears, this problem takes several steps to solve. The step-by-step solution is given in pseudo code as shown in Figure 2.11. Top-level pseudo code is a brief description of the problem. The first refinement indicates input, processing steps, and output. As it is already trivial to write actual code after the second refinement, we no longer need further refinement.

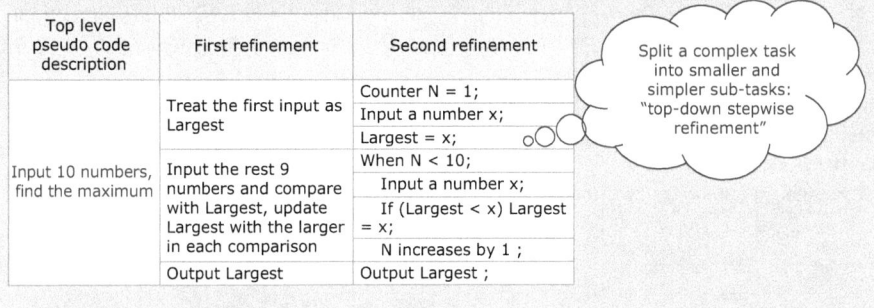

Top level pseudo code description	First refinement	Second refinement
	Treat the first input as Largest	Counter N = 1;
		Input a number x;
		Largest = x;
Input 10 numbers, find the maximum	Input the rest 9 numbers and compare with Largest, update Largest with the larger in each comparison	When N < 10;
		Input a number x;
		If (Largest < x) Largest = x;
		N increases by 1 ;
	Output Largest	Output Largest ;

Figure 2.11: Finding maximum.

We solved the problem by splitting a complex task into simpler tasks. This is what we call the "top-down stepwise refinement" approach.

2 Computing sum of scores

Given keyboard inputs of some positive integers, calculate and display the sum of them. Suppose that the user uses "–1" to mark the end of the input data.

[Analysis]

Pseudo code of the solution is given in Figure 2.12. Note how we make the solution more detailed with each refinement until we can quickly write out code.

Top level pseudo code description	First refinement	Second refinement
Input a series of positive integers, computer their sum. "-1" is used to mark the end of all input	Input a number	Input a number x;
	If the input number is not "-1" 1. Add the input number to existing sum 2. Continue inputting data	let sum = 0
		While (x not equal to -1)
		sum = sum + x;
		Input x;
	Output result	Output sum

Figure 2.12: Computing sum of scores.

The input of this problem is x. The required functionality is to add x to sum, whereas x is not the ending mark. The output is the sum.

The actual code of these two problems can be easily written after learning the syntax of C in chapter "Program Statements."

Example 2.3 Sorting poker cards

When playing poker, we want the cards to be sorted in order after we have drawn all cards. We may sort our cards as follows. Suppose we draw a 9 followed by a 3. As 3 is smaller than 9, we put it in front of 9. We then draw a 4 and put it behind 3 because 4 is larger. We then draw a 2 and put it in front of 3 and so on.

This sorting method is called a straight insertion sort in classic sorting algorithms. The basic idea behind it is to split the list to be sorted into two lists: one is the sorted list and the other is the list to be sorted. We then insert every single element from the list to be sorted into the sorted list until there is nothing left. Figure 2.13 shows an example of straight insertion sort. In step 3 of this example, the sorted list from step 2 contains 3 and 9 and the element to be inserted is 4. Because 4 is between 3 and 9, the new sorted list should be "3, 4, 9." There are eight elements in total, so we need seven such operations.

Step	Sorted list								List to be sorted							
①								9	9	3	4	2	6	7	5	1
②							3	9	3	4	2	6	7	5	1	
③						3	4	9	4	2	6	7	5	1		
④					2	3	4	9	2	6	7	5	1			
⑤				2	3	4	6	9	6	7	5	1				
⑥			2	3	4	6	7	9	7	5	1					
⑦		2	3	4	5	6	7	9	5	1						
⑧	1	2	3	4	5	6	7	9	1							

This insertion is done in the original memory space

Data to be inserted

Figure 2.13: Straight insertion sort.

When we start sorting, we may consider the first element as the first sorted list. Note that the implementation operates on the original memory space of the number list. In other words, we do not need temporary space for the lists. Instead, it suffices to use extra space of only one element.

Next we analyze the process of inserting an element carefully. Let us take steps 3 and 4 from Figure 2.13 as an example as illustrated in Figure 2.14. The sorted list now has "3, 4, 9" and the element to be inserted is 2. As previous element 9 is larger than 2, we need to move 9 to its right. However, it will overwrite the element that is already in that position, namely 2. To avoid this, we need a "sentry" that records the element to be inserted. After moving 9, we also need to move 4 and 3. Finally, we may put 2 into the first position.

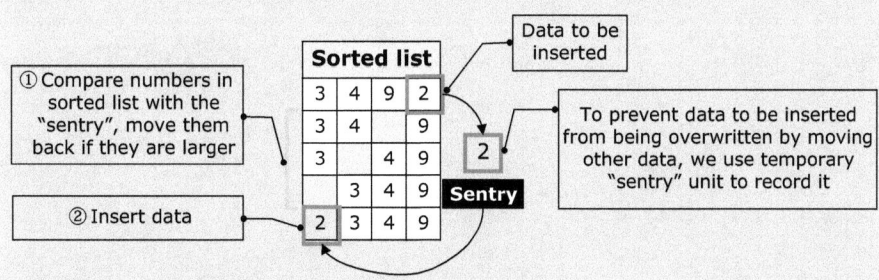

Figure 2.14: Insertion process of one element.

We can write out pseudo code for the algorithm based on the description above as shown in Figure 2.15.

Except for straight insertion sort, there are many other sorting algorithms as well. We will cover some of them such as bubble sort and selection sort, in the chapter "Arrays." Related concepts are usually introduced in data structure courses as well.

2.3 Effectiveness of algorithms

We have seen algorithms for some problems by now. Essentially, what are the differences between solving problems by humans and by computers?

In fact, the strategies humans use to solve problems can often be used with computers as well. However, exceptions do exist. To design an algorithm suitable for computers, we need to think from the perspective of computers. In other words, we need to have a computer mindset. Let us examine some examples of algorithm design.

Top level pseudo code	First refinement	Second refinement
Process recurrently from the second number	Process recurrently from the i-th number	Suppose there are N numbers, data[i] represents the i-th number
Compare numbers in sorted list with the "sentry" one by one, move back if larger	If the i-th number is smaller than the (i-1)-th	if (data[i] < data[i-1])
	Use the i-th number as "sentry" temp	temp=data[i]
	In the sorted list j = i-1 ~ 0 Move number in front of j back if it is larger than temp	j = i-1 while (j >= 0 and data[j] > temp) a[j+1]=a[j]; j=j-1
Insert "sentry" value into right position	Insert temp into the empty position	a[j+1]=temp
		i=i+1

Figure 2.15: Top-down stepwise refinement pseudo code description of straight insertion sort.

2.3.1 Example of algorithms

Example 2.4 Sample algorithm 1: data swapping
Swap data stored in two storage units without losing any after the swap.

[Analysis]
We usually swap data with the help of a third empty storage unit as shown in Figure 2.16. Double slashes in each step mark the beginning of comments in C.

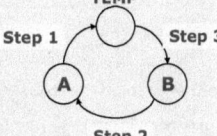

■ **Data swap**

Suppose we have data A = 10, B = 12, and a temporary unit TEMP
Step 1: TEMP = A; // TEMP = 10, content of A is put into TEMP
Step 2: A =B; // A = 12, content of B is put into A
Step 3: B = TEMP; // B = 10, content of TEMP is put into B

Figure 2.16: Flow of data swap.

In addition, we can use some "unconventional" methods as follows.

- Throw into the sky: suppose A and B are objects, we can throw them to the target location like how people in the circus would do.
- Transfer with tubes: we "connect" two storage units using two tubes and transfer A and B through them.
- Move trivially: if storage units of A and B are large enough to store both of them, we can simply move B to the unit of A, then move A to that of B.

In the real world, we can use all the conventional and unconventional methods. However, only the conventional swap and trivial move methods under certain conditions will work in computers.

Example 2.5 Sample algorithm 2: simulating elementary school students solving division problems
To simplify the problem, we suppose that all divisions are two-digit numbers divided by single-digit numbers and quotients have one digit, for example, $23 \div 3 = 7\,r\,2$.

[Analysis]
Elementary school teachers may tell students to do divisions by trying quotients. As an example, the process of solving $23 \div 3$ is shown in Figure 2.17. However, this method is limited in a way that we need to follow different rules for different divisions. For instance, when computing $37 \div 6$, we need to figure out a number that yields a result between 30 and 40 when multiplied by 6. It is not an easy and efficient method for computers to use as processing rules vary with data.

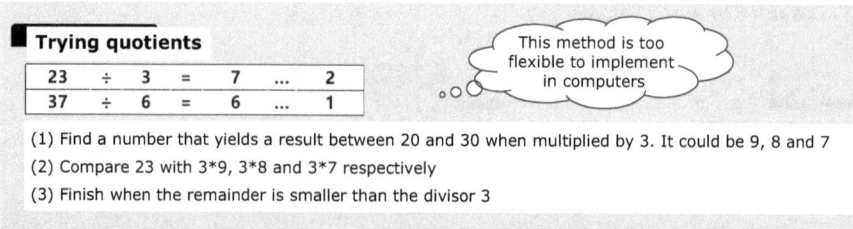

Figure 2.17: Solving divisions by trying quotients.

There are two methods suitable for computers as shown in Figure 2.18. The first one is trying quotients as well, except that we always choose 9 as our first attempt. This makes the rule simple and unified, so programmers can write code efficiently. The second method is to use "chunking," which means repeated subtraction. By definition of division, we can repeatedly subtract the divisor from the dividend and obtain quotient eventually. In fact, this is how division is done in computers internally.

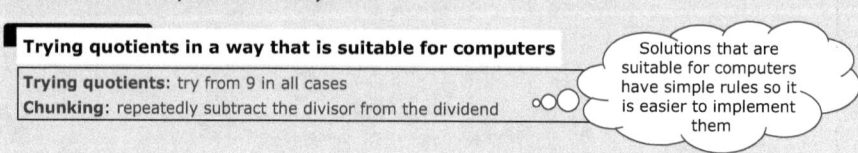

Figure 2.18: Trying quotients in a way that is suitable for computers.

Solutions suitable for computers should have a simple rule that is easy to follow. Being able to complete simple tasks tirelessly is considered by some people the "fundamental quality of computers."

Example 2.6 Sample algorithm 3: evaluate expressions
In C language, we call statements formed by connecting operators and operands expressions. For example, $1+(5-6/2)*3$ is an arithmetic expression. Note that division and multiplication are represented by slash and asterisk, respectively. All symbols in C should appear on keyboards as well so that we can insert them easily.
  Figure 2.19 shows how we are taught to evaluate expressions in elementary school and how computers evaluate expressions. Computers use a method called "Polish notation." Polish mathematician, Jan Łukasiewicz, first proposed it in 1920. It simplifies evaluation through two steps.

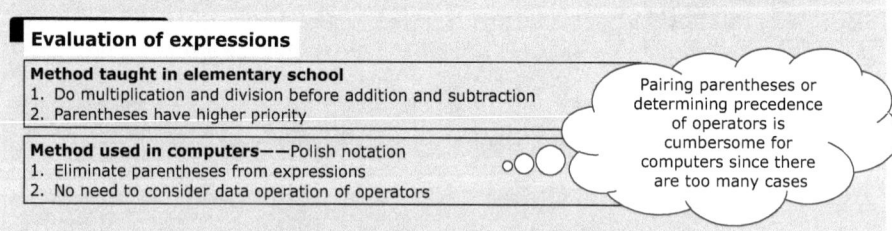

Figure 2.19: Evaluation of expressions.

The expression 1 +(5–6/2) *3 will become 1562/–3*+ after conversion. Details of such conversions are covered in data structure courses. Interested readers may refer to other resources for them. We shall take a look at how an expression in Polish notation is evaluated.

The # sign in Figure 2.20 is used to mark the end of expressions, while cells in which numbers are stored represent memory units. Herein we show changes in the stored data in each step. The figure also shows possible cases in expression scanning and corresponding operations computer take. After six steps of processing, the value left in memory is the final result.

$$1 \ \ 5 \ \ 6 \ \ 2 \ / \ - \ 3 \ * \ + \ \#$$

Scan Expression	Store if a number is scanned
	Withdraw number twice, compute and store result if an operator is scanned
	Withdraw result and end if # is scanned

Processing steps
(1) See number 1,5,6,2, store them
(2) See operator "/", compute 6 / 2 = 3, store 3
(3) See operator "-", computer 5 –3 = 2, store 2
(4) See number 3, store it
(5) See operator "*", compute 2 * 3 = 6, store 6
(6) See operator "+", compute 6 + 1 = 7, store 7
(7) See #, withdraw 7, end

	(1)	(2)	(3)	(4)	(5)	(6)
	2					
	6	3		3		
	5	5	2	2	6	
	1	1	1	1	1	7

Figure 2.20: Evaluation of expression in Polish notation.

2.3.2 Computational thinking

We have seen that methods we usually use may not work in computers in the last section. To design algorithms suitable for computers to execute, we need to know the characteristics of how computers solve problems, called "computational thinking." Professor Jeannette Wing from Carnegie Mellon University wrote that "Computational thinking builds on the power and limits of computing processes ... Computational thinking involves solving problems, designing systems, and understanding human behavior by drawing on the concepts fundamental to computer science." The essence of computational thinking is abstraction and automation. In procedure-oriented programming, we may also describe it as "program thinking."

> **i** **Knowledge ABC** Program thinking
> Abstraction in programming uses identifiers, constants, variables, arrays, and structures to describe and record information and relations between information. Automation is the process of operating information using statements and operators to achieve a particular goal. Functions are formed by organizing statements based on functionality. Functions are used to decompose a larger problem into multiple subproblems that are independent of but related to each other. Algorithms describe the steps and procedures of solutions to problems. To fit how humans think, they are expressed in a top-down stepwise refinement manner. Together, these concepts construct procedure-oriented programming and procedure-oriented languages.

2.4 Universality of algorithms

Universality requires that data in problems of the same type are handled consistently. Let us look at solutions to some classic problems first.

2.4.1 Solutions to classic problems

Example 2.7 Things whose number is unknown
There is a well-known problem in *Sunzi Suanjing*, a mathematical treatise in ancient China, called "things whose number is unknown." The problem is as follows: there are certain things whose number is unknown. If we count them by threes, we have two left over; by fives, we have three left over; and by sevens, two are left over. How many things are there?

[Analysis]
The problem does not restrict the number of solutions and there might be multiple solutions. As shown in Figure 2.21, we may apply exclusive induction. We look for three sets of numbers that satisfy the three conditions, respectively and seek common numbers among them. Readers may quickly notice from the figure that the minimal number of things that satisfy all three conditions is 23.

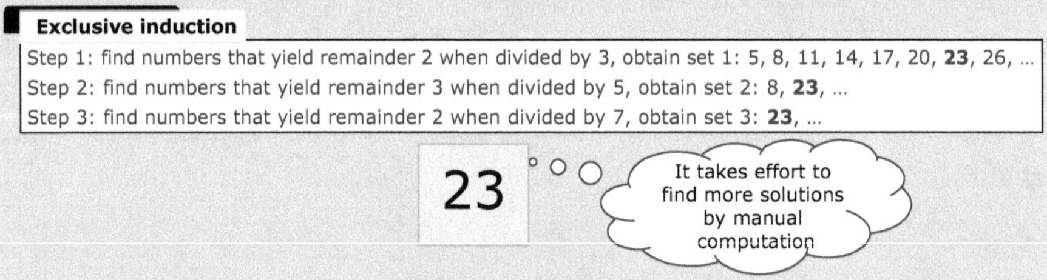

Exclusive induction
Step 1: find numbers that yield remainder 2 when divided by 3, obtain set 1: 5, 8, 11, 14, 17, 20, **23**, 26, ...
Step 2: find numbers that yield remainder 3 when divided by 5, obtain set 2: 8, **23**, ...
Step 3: find numbers that yield remainder 2 when divided by 7, obtain set 3: **23**, ...

23 ° ○ ○ It takes effort to find more solutions by manual computation

Figure 2.21: The first solution to "things whose number is unknown".

Now we examine how computers solve this problem. In C language, whether a number x has remainder 2 when divided by 3 is represented as x% 3 = = 2, where % is the remainder operator and = = (note that there are two equal signs) checks whether its operands are equal. Pseudo code of the algorithm is shown in Figure 2.22. The loop condition in the second refinement is "always true," which means the solution-finding process can run forever because we do not know the exact number.

Top level pseudo code description	First refinement
x start from 1	let x = 1

find result that satisfies all conditions	repeat following operations
	if x satisfies following conditions at the same time
	   "2 remaining if divided by 3, 3 remaining if
	   divided by 5, 2 remaining if divided by 7"

output result	then output value of x x increases by 1

Second refinement
x=1

while (loop condition is always true)
   if (x%3==2 and x%5==3 and x%7==2)
     output x
x increases by 1

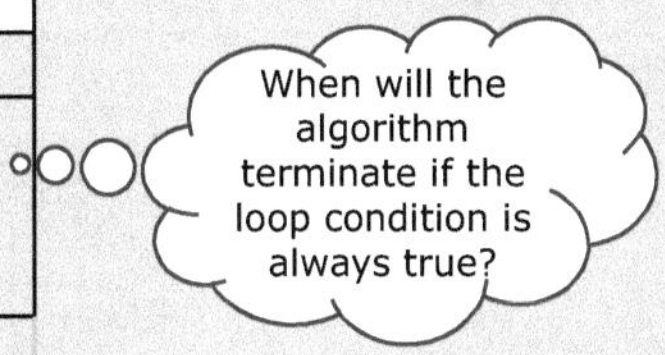

When will the algorithm terminate if the loop condition is always true?

Figure 2.22: Second solution to "things whose number is unknown".

But when should the algorithm terminate if the loop condition is "always true"? If we do not need all solutions, we can add a terminating statement, which, for instance, terminates the loop when x > 2000. More details about "always true" loops are covered in the section of loop statements.

Guessing solutions one by one, like we just saw in this example, is another demonstration of how computers can complete simple tasks repeatedly and tirelessly.

Example 2.8 Chickens and rabbits in the same cage
This is another classic problem from *Sunzi Suanjing*. Suppose there are several chickens and rabbits in the same cage. The total number of heads is 35 and the total number of legs is 94. How many chickens and rabbits are there in the cage?

[Analysis]
We can solve this problem using linear equations as shown in Figure 2.23. However, determining the positivity of coefficients of variables or doing substitution is tricky and cumbersome to implement on computers. Besides, different systems require different methods to solve.

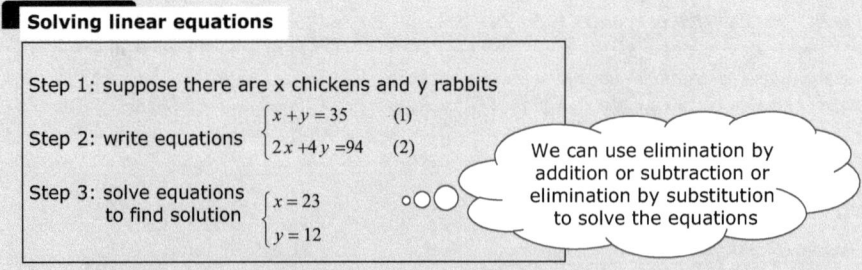

Solving linear equations

Step 1: suppose there are x chickens and y rabbits

Step 2: write equations $\begin{cases} x+y=35 & (1) \\ 2x+4y=94 & (2) \end{cases}$

Step 3: solve equations to find solution $\begin{cases} x=23 \\ y=12 \end{cases}$

We can use elimination by addition or subtraction or elimination by substitution to solve the equations

Figure 2.23: The first solution of chickens and rabbits in the same cage problem.

The universal solution is shown in Figure 2.24. Similar to the previous problem, we try every possible value of each variable to find the solution that satisfies all conditions.

Top level pseudo code description	First refinement	Second refinement
both x and y start from 1	x=1, y=1	x=1, y=1

find results that satisfy the equations	while x < 35, repeat following operations
	  if there is a value of y between 1 and 35 that satisfies the equations
	  output results
		while (x<35)
		  while (y<35)
		    if x+y=35 and 2x+4y=94
		      output x and y
		    y increases by 1
		  x increases by 1

output results	x increases by 1	

Figure 2.24: The second solution of chickens and rabbits in the same cage problem.

In the first refinement, both x and y have a value of 1 and terminating condition "less than 35," as there are 35 heads in total. When x = 1, we substitute every possible value of y, from 1 to 35 into the equations to test whether they are the solution. If none of these combinations work, we increase x by 1, and test possible values of y again. We repeat this process until x = 35. Note that this is a nested loop of two layers.

In the second refinement, the condition is further specified as "if x + y = 35 and 2x + 4y = 94" and we find a solution if it is met.

2.4.2 Three phases of problem-solving with computers

Based on the above-mentioned examples, we may conclude that there are three phases when solving problems with computers, namely, the beginning phase, the processing phase, and the ending phase. Each phase contains a set of operations that should be done as shown in Figure 2.25.

Three phases of problem solving with computers

Beginning phase	determine initial conditions of program execution
Processing phase	complete data processing based on requirements of problem, fulfill functionality requirements
Ending phase	determine terminating conditions of program, obtain final results

Figure 2.25: Three phases of problem-solving with computers.

2.4.3 Characteristics of computer algorithms

Having seen these examples, we can now summarize the characteristics of problem-solving with computers as shown in Figure 2.26. Each operation done by a computer should be simple, yet these simple operations can be combined to provide complex functionalities. Given problems of the same type, there should be a universal set of rules to process corresponding data.

Characteristics of problem solving with computers

Rules are simple: each step in data processing is simple
Rules are universal: operation rules for corresponding data in problems of the same type should be consistent

Figure 2.26: Characteristics of problem-solving with computers.

2.5 Comprehensiveness of algorithms

After midterm exams, Mr. Brown's son, Daniel, asked his mom whether they could go for a weekend trip. He then started to talk about what he wanted to do and bring. However, Mrs. Brown smiled and asked, "What if it rains?" Daniel was caught off guard. He thought for a while and suggested, "Then we can go to movies, as long as we don't just stay at home."

Like Mrs. Brown, experienced people try to consider every situation when solving practical problems so that they would not make mistakes or lose anything. Similarly, when solving problems with computers, we hope our algorithms are comprehensive so that they can handle all types of input data. In other words, our algorithms should:
- Correctly process normal data
- Correctly handle abnormal data

2.5.1 Algorithm analysis: Starting from normal cases

We shall explain the comprehensiveness of algorithms by analyzing the algorithm for n! problem.

2.5.1.1 Problem analysis

The mathematical definition of n! is shown in Figure 2.27. There is only an abstract variable n and we do not know its value. How should we compute it? For generalized problems like this where we do not know the exact value of variables, we can find a value that satisfies the condition and is easy to compute, for instance, $n = 5$, to analyze the pattern and characteristics of the problem.

$$n! = \begin{cases} 1, & \text{When } n = 0 \\ n * (n - 1)!, & \text{When } n \geq 1 \end{cases}$$

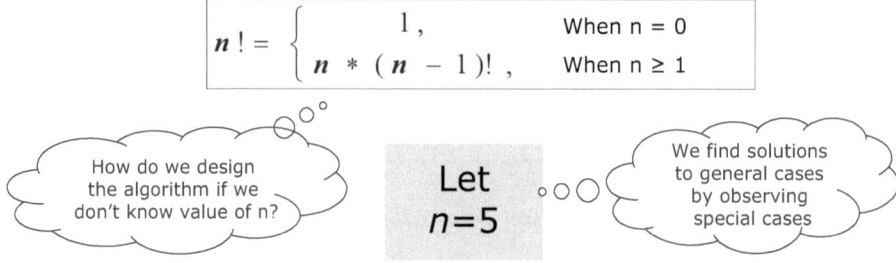

How do we design the algorithm if we don't know value of n?

Let $n=5$

We find solutions to general cases by observing special cases

Figure 2.27: n! problem.

Before designing a computer algorithm, we observe the universal method to compute 5! manually. Herein universal means that the method works for values other than 5 as well.

2.5.1.2 Manual method

To simplify the description, let variable S be the product of repeated multiplication. We may consider the variable as a box. Data can be put into or taken out of the box. Data stored in the box can also be modified. Detailed procedures of computing fac torial are shown in Figure 2.28, where the arrow symbol in representation column means "put into."

■ Manual computation

Step	Operation	Representation
1	1 times 2 yields 2, store into S	1*2→S
2	Multiply value of S, 2, with 3, obtain 6, store into S	S*3→S
3	Multiply value of S, 6, with 4, obtain 24, store into S	S*4→S
4	Multiply value of S, 5, with 5, obtain 30, store into S	S*5→S

Let variable S be product of repeated multiplication

Characteristics of variables
- Can be stored
- Can be accessed
- Can be updated

Variables can be regarded as boxes

Figure 2.28: The first algorithm of computing n!.

2.5.1.3 Analysis of computer solutions

Now we examine the universal method for computers to compute n!. We can think of computers as more advanced calculators, but they would not do anything unless we have provided the necessary data and algorithms "bit by bit." The necessary information in this problem is given in Figure 2.29. They can be obtained by computers using the following methods.

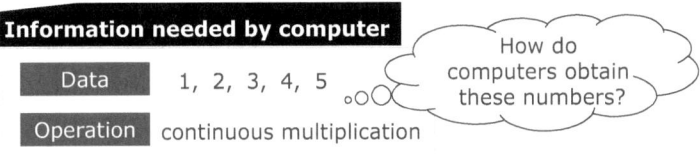

Information needed by computer

Data	1, 2, 3, 4, 5
Operation	continuous multiplication

How do computers obtain these numbers?

Figure 2.29: Information needed by computer.

Method 1: Type each number into the computer using a keyboard. Type one number after computer completes a calculation.

Unfortunately, this method becomes tedious when n is large, thus we need an alternative one. Considering the characteristics of factorial, we notice that each multiplier can be obtained by adding one to the previous one, except the very first multiplier, 1. Hence, we get our second method.

Method 2: Each multiplier is obtained by adding one to the previous multiplier or the multiplier from the previous iteration, except 1.

It is not hard to see that the second method is easier; therefore, we shall use it in further discussion as well. We will represent the multiplier as T, which is a variable as well, for easier referencing.

2.5.1.4 Comparison of manual method and computer method

We have analyzed both the manual and the computer methods. We may learn characteristics of problem-solving with computers by listing differences between these two methods as given in Figure 2.30.

	Computation model	Major steps in solution	Characteristics of problem-solving process
Manual		• Use factorial formula • Use multiplier directly • Store product into S	Directly use known information given in the problem
Computer	5 ! = 1*2*3*4*5	• Use factorial formula • Multiplier obtained by iteration:T+1->T • Store product into S: S*T->S	Data and operations need to be "provided" in advance

Let variable S denote the cumulative product and variable T denote the multiplier

Figure 2.30: Comparison of manual and computer method.

Although these two methods use the same model of computation and share common steps, they vary in the way of handling data: human use known information directly, whereas computers need to be "provided" with data and operations.

Computers are tools as well. We need to respect their limitations when using tools. We also need to keep their features in mind when designing methods and steps in solutions. The most significant thing in programming learning is to know the characteristics of problem-solving with computers, namely "computational thinking."

2.5.1.5 Algorithm description

The algorithm description of computing n! is shown in Figure 2.31. In the second refinement, the loop keeps running when $T \leq 5$. Note that expressions of loop condition in both refinements are in fact equivalent.

Top level pseudo code description	First refinement	Second refinement
Compute 5!	Start from 1 * 2	Let product S = 1, multiplier T = 2

	Repeat following operations
	  Store product into S
	  Increase multiplier by 1
		do
		  S*T→S
		  T+1→T
			Pay attention to the description and expression of loop conditions here

	End until multiplier is larger than 5	while (T<=5)	
Output result	Output result	Output S	

Figure 2.31: Pseudo code of n! algorithm.

In accordance with the pseudo code in the second refinement, we may draw the flowchart as illustrated in Figure 2.32.

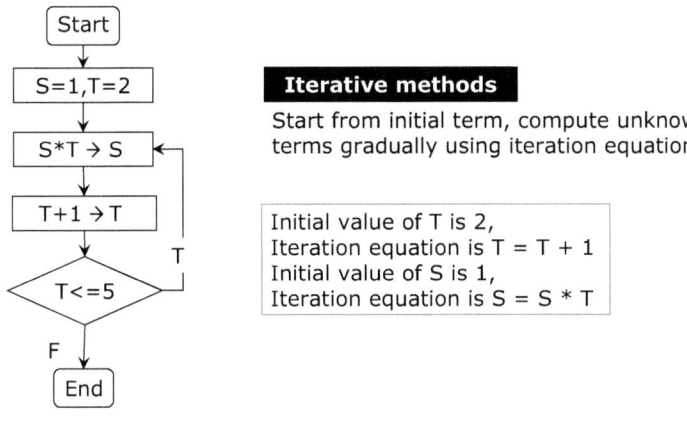

Figure 2.32: Flowchart of n! algorithm.

The execution process of the algorithm is more clearly shown in the flowchart. We first initialize product S and multiplier T. Further, we repeat multiplication and increment until T is larger than 5. At the end of each iteration, we check whether T satisfies the loop condition. The loop continues if the checking yields true and terminates otherwise.

Initial value of T is 2 and iteration equation is $T = T + 1$. The initial value of S is 1 and the iteration equation is $S = S * T$. We call methods "that start from known terms and gradually work out unknown terms using iteration equation" as iterative methods. "Step by step" is one of the features of computers. The result from previous computation is often needed in the next computation.

> **Knowledge ABC** Iterative methods
> An iterative method is a procedure that uses an initial term to generate the required unknown term through a finite series of iterations. Any problem that contains an iteration equation can be solved using iterative methods.
>   Iteration steps:
> (1) List known terms in the problem.
> (2) Write out iteration equation based on relations in the problem.
> (3) Iterate finite times using the iteration equation until we find the solution.

2.5.1.6 Analysis of execution process

Let us check how S and T change during the execution of the algorithm. We may use a table as shown in Figure 2.33, to list values of S and T in each step of the flow so that the execution process can be clearly seen.

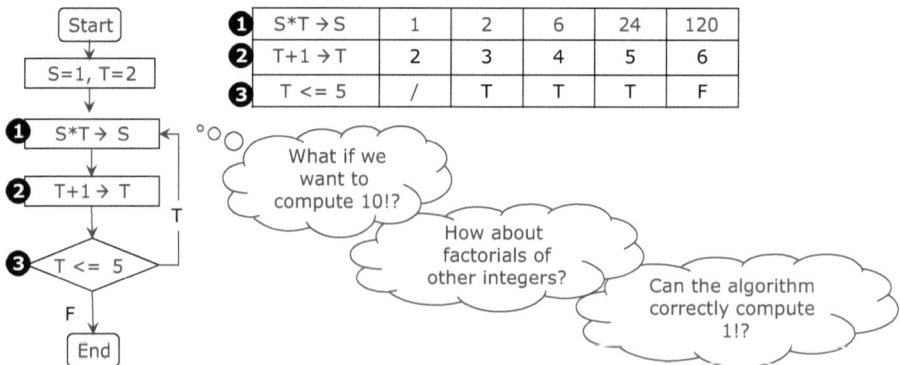

Figure 2.33: Analysis of the n! flow.

In the first column, S and T have initial values 1 and 2. Step 3 has not been executed, so the cell is left blank. Step 1 gets executed, the value of S becomes S times T, namely, 2. Step 2 is executed next and T becomes 3 after increasing by 1. T is then checked in step 3. 3 is less than 5, so the loop condition holds true and the flow goes to branch "True." Consequently, step 1 is executed again and S is 2 *3 = 6. T increases to 4 in step 2. The loop condition is met again, so steps 1, 2, and 3 are executed once again until T is larger than 5 and the flow ends.

This is the algorithm of computing 5!, but computing 5! is not our ultimate goal. To make the algorithm universal, we need to tackle some other challenges.

2.5.1.7 Testing
(1) What if we want to compute 10!?
    We can change the condition in step 3 to T ≤ 10.
(2) How about factorials of other integers?
    We can modify the condition in step 3 to T ≤ n, where the value of n is obtained from keyboard input.
(3) Can the algorithm correctly compute 1!?
    When step 3 is executed, T = 3 and n = 1. The program ends because T is larger than n. The value of S is thus 2, which is incorrect.
    To correct our result, we can change the initial value of T from 2 to 1.
    The refined flow, shown in Figure 2.34, has an input of n and changes initial value of T.
(4) What if the user inputs invalid data, n = −1, for instance?
    In this case, the loop will be executed once and the result will be S = 1, which is once again wrong.

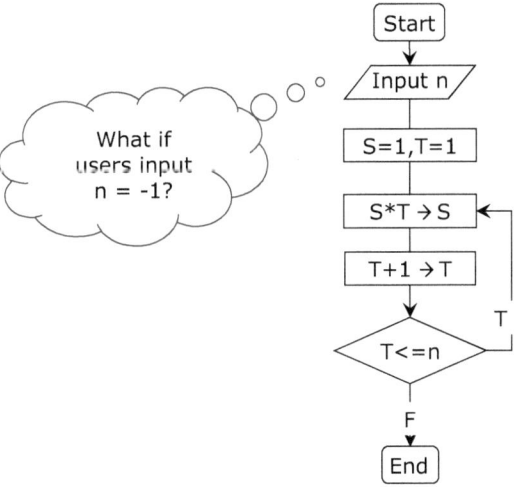

Figure 2.34: The first refined version of n! flow.

To prevent users from providing invalid data, we need to add validation of input. If the input is invalid, a warning will be prompted to users. Further refined flow is shown in Figure 2.35. In general, programs should have a mechanism for errors so that they can correctly handle "illegal" input data.

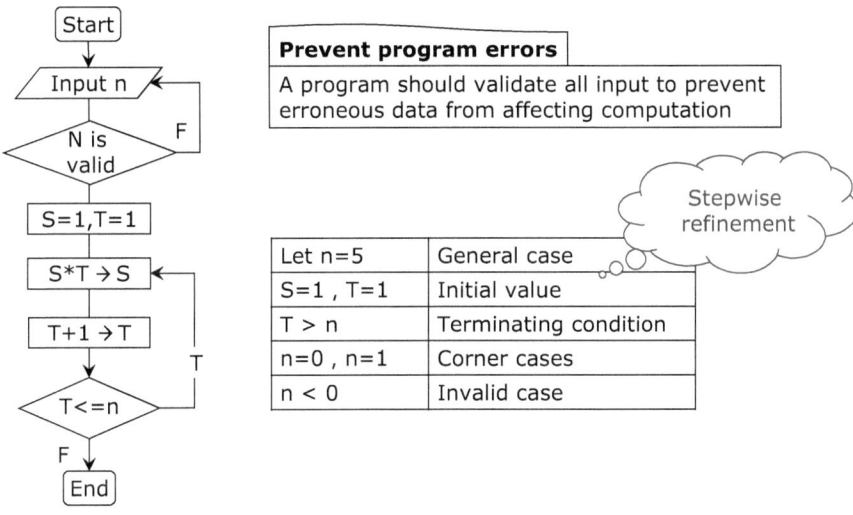

Figure 2.35: The second refined version of n! flow.

2.5.1.8 Summary of algorithm design procedures

Having considered the above cases, we have a comprehensive algorithm. Recall our design process and there are the following key steps:
(1) Let $n = 5$: consider the process starting from a normal case.
(2) S = 1, T = 1: determine initial values in algorithm.
(3) T > n: determine the terminating condition of the algorithm.
(4) n = 0, n = 1: consider corner cases.
(5) n < 0: consider error handling.

In this example, we considered corner cases and error handling after we had set up the basic flow. It is clearer to solve problems step by step like we just did. Algorithm design cannot be done all at once. It is a process of improvement. Good algorithms cannot be created without effort and refinement. This is also true for problem-solving.

2.5.2 Algorithm analysis: Starting from corner cases

We designed our n! algorithm starting from normal cases in the last section. In fact, we may also look at corner cases first as shown in the example given further.

2.5.2.1 Problem description

Neighboring zone problem
An n*m matrix is divided into t rectangle zones, each represented by a number between 1 and t. Cells in the same zone are all represented by the number of that zone. As can be seen from Figure 2.36, a 6*8 matrix is divided into eight zones labeled as 1 to 8. We say that two zones are neighbors if they share an edge. For example, zone 5 has six neighbors, namely, zones 1, 2, 3, 6, 7, and 8, whereas zone 4 is not its neighbor. Please design an algorithm to find all neighbors of zone k.

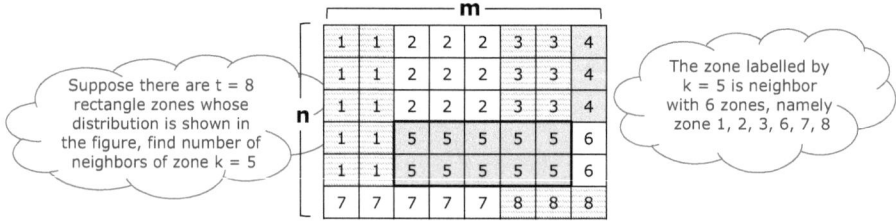

Figure 2.36: Neighboring zone problem.

2.5.2.2 Ideas of algorithm

There are several corner cases in this problem: there may be only one zone, zone k may have only one element, and so on. On the other hand, it would be tricky to start from general cases, where zone k has multiple cells as there could be many neighbor configurations, which makes it challenging to write decision conditions. However, zone k must have four neighbors, each sharing one edge with it, if it has only one cell as shown in Figure 2.37. This is a simple base case. In this case, we can check each cell in the matrix. If the value of a cell is k, we count the number of its neighbors that do not have value k. This is a method that is suitable for computers.

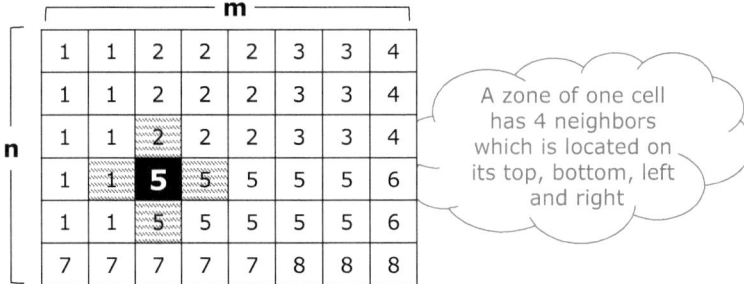

Figure 2.37: Analysis of neighboring zone problem.

Concerning comprehensiveness, we shall consider corner cases in this problem. Readers may have noticed that cells on the edge of the matrix do not have four neighbors. For these cells, we can add cells labeled by 0 outside the matrix as "supplement" and restrict our search in the original matrix as shown in Figure 2.38.

0	0	0	0	0	0	0	0	0	0
0	■	1	2	2	2	3	3	4	0
0	1	1	2	2	2	3	3	4	0
0	1	1	2	2	2	3	3	4	0
0	1	1	5	5	5	5	5	6	0
0	1	1	5	5	5	5	5	6	0
0	7	7	7	7	7	8	8	8	0
0	0	0	0	0	0	0	0	0	0

Figure 2.38: Search region design of neighboring zone problem.

2.5.3 Keys of algorithm design

Having seen these examples, we realize that solutions to problems can be derived from general cases as well as corner cases. Whichever cases we choose, they need to be simple but universal, thus suitable for computers. Unlike solving physics or mathematics problems where we are usually able to apply existing formulas, solving problems with computers requires us to analyze traits of problems. However, there are still some patterns in algorithm design, which are summarized in Figure 2.39.

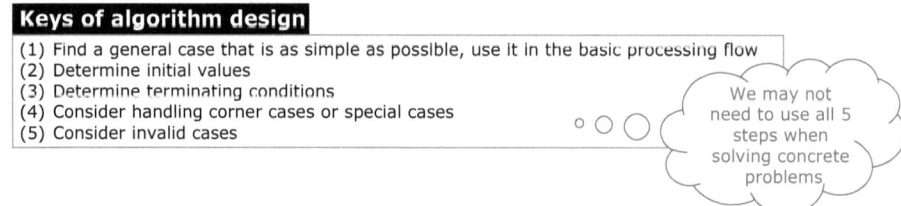

Keys of algorithm design
(1) Find a general case that is as simple as possible, use it in the basic processing flow
(2) Determine initial values
(3) Determine terminating conditions
(4) Consider handling corner cases or special cases
(5) Consider invalid cases

We may not need to use all 5 steps when solving concrete problems

Figure 2.39: Keys of algorithm design.

From these steps, we can see that verification is needed to determine whether an algorithm is well-designed and can function as expected. To verify, we should design test data before designing the algorithm. We can use general methods of software testing to test algorithms. They are covered in detail in the chapter "Execution of Programs."

Knowledge ABC Software testing and test cases
- Software testing: it is the assessment process of running the software under specified conditions, looking for bugs, determining quality, and evaluating whether design requirements are fulfilled. This is a classic definition. We may also consider software testing to be a comparison of actual output and expected output.
- Test cases: these include test input, execution requirements, and expected output designed for a specific target. They are used to test an execution path or to verify if certain requirements are fulfilled.

2.6 Procedures of algorithm design and characteristics of algorithms

2.6.1 Position of algorithms in the process of solving problems with computers

In the chapter "Introduction to Programs," we have seen that the process of solving problems with computers, starting from raising a problem and ending with a solution, has four major steps, namely, modeling, data structure and algorithm design, as well as coding and testing.

Modeling extracts functionalities and data from problems and seeks relations between the information by analysis. Data structure design tries to find a way to organize and store data. Algorithm design attempts to seek a solution. Coding converts the algorithm into code. Last but not least, testing is done to test our code.

2.6.2 General process of algorithm design

We have discussed key points in algorithm design. They are general rules for implementing algorithms with restriction of the three key elements of algorithms, namely input, functionality, and result. From the perspective of algorithm design, there are other challenges in addition to implementing algorithm: whether there is a classic algorithm strategy, which solution of all possible ones are better than the rest, whether there is a universal method to assess algorithms, and others. In accordance with procedures of program development, algorithm design can be done through the following steps:

(1) Determine key elements of the algorithm

We need to correctly understand input and find out what functionalities and results are desired.

(2) Design and describe the algorithm

We use the top-down stepwise refinement approach to design an algorithm following general rules of algorithm design. During the design process, we can learn from existing algorithms. Classic algorithms include brute force, divide and conquer, decrease and conquer, dynamic programming, greedy algorithm, backtracking, branch and bound, approximation algorithm, randomized algorithm, and so on. To design an algorithm for a new problem, we can use these strategies flexibly to create a new one. Finally, we need to choose an algorithm description method to record the procedures in our solution clearly.

(3) Manual check

Logic errors cannot be detected by computers because computers execute programs without understanding the motivation behind. Experience and research suggest that running algorithms manually with test cases is one way to detect logic errors in algorithms. Test cases should be designed in a way that exposes as many errors as possible.

(4) Analyze the efficiency of the algorithm

Efficiency can be assessed by the amount of computing resources used, which can be time or space. Time efficiency indicates the speed of execution, whereas space efficiency shows the amount of extra memory space needed. Details on efficiency are usually covered in the data structure class.

(5) Implement the algorithm

Finally, we convert our algorithm into programs using certain programming languages. Note that a good algorithm is the result of effort and refinement.

2.6.3 Characteristics of algorithms

In previous discussions of example algorithms, careful readers may have noticed that there exist restrictions on the three key elements of algorithms. They are called "characteristics of algorithms" as shown in Figure 2.40. Finiteness, definiteness, and effectiveness are restrictions or requirements on how algorithm implements functionalities required.

Characteristics of algorithms
- Input: have zero or more input
- Output: produce result that fulfills functionality requirement
- Finiteness: an algorithm should contain finite steps
- Definiteness: each step of an algorithm should be precisely defined without ambiguity
- Effectiveness: each step of an algorithm can be done effectively and generate certain result

Figure 2.40: Characteristics of algorithms.

(1) Input: An algorithm has zero or more inputs, which are taken from the information of the problem to solve. An algorithm can have zero input in special cases. For example, no input data are needed when solving an equation following specified procedures and conditions.

(2) Output: An algorithm has one or more output (an algorithm must have at least one input). The outputs are determined by functionalities required.

(3) Finiteness: An algorithm must always terminate after a finite number of steps (for any valid input) and such step must be completed in finite time. In contrast to the same concept in mathematics, the finiteness of algorithms means they are reasonable and acceptable in applications.

(4) Definiteness: Each step of an algorithm must be precisely defined without ambiguity. Identical input should generate identical output under all circumstances.

(5) Effectiveness: The operations in an algorithm can be done by finite basic steps that are already implemented.

2.6.4 Characteristics of good algorithms

There can be multiple solutions to the same problem. To assess these solutions, we need a standard. A generally acknowledged standard is that a good algorithm needs to have the following features in addition to five characteristics of algorithms.

(1) Correctness

The fundamental goal of algorithm design is that algorithms should have required functionalities. "Correctness" may refer to many things, but they can generally be sorted into the following four categories:

    1) Programs do not contain syntax errors.

    2) Programs generate results that meet functionality requirements for multiple input data.

    3) Programs generate results that meet functionality requirements for input data in chosen cases that are typical but strict and difficult at the same time.

    4) Programs generate results that meet functionality requirements for all valid input data.

    We usually use the third definition to evaluate whether a program is correct.

(2) Readability

Suppose an algorithm is correct, then readability should be the most crucial factor. In other words, it is our first priority to make sure that programmers can work efficiently. This is particularly significant as large software systems are usually created by multiple programmers nowadays. Moreover, errors may be hidden in unreadable programs and it can be painful to debug them.

(3) Robustness

It refers to the ability of algorithms to cope with invalid input data. It is also called fault tolerance. A good algorithm should be able to identify erroneous input data and handle them properly.

    Reasonable and effective test cases can help us find as many errors as possible in testing and guarantee the robustness of algorithms.

(4) High efficiency

An algorithm is highly efficient if it executes efficiently. High efficiency may refer to two things.

    1) High time efficiency. Time efficiency is a measure of the execution time of algorithms. An algorithm is highly time-efficient if it can be executed in a short period of time.

    2) High space efficiency. The memory space of algorithms is the maximum memory space that is needed during the execution of algorithms. We mainly focus on the supplementary memory space needed during execution. An algorithm that needs less memory space is called an algorithm with low memory requirements.

Research shows that there is much more room for improvement of time efficiency than of space efficiency in most cases. In addition, we can sacrifice one of them for the other.

2.7 Summary

An algorithm can be considered as a complete solution consisting of basic operations and order of these operations. We may also deem algorithms as finite, definite computation sequences designed under certain requirements to solve a certain type of problem.

Methods and steps in computer algorithms should be consistent with the characteristics of computers. Each step should be simple so that it can be executed by computers tirelessly. There are limitations in computers as well. Feasible solutions in daily life may not work on computers.

Although we may be able to directly write out code for easy problems during the early phase of programming learning, we need to form good programming habits and follow general rules of algorithm design. It prevents us from giving incomplete solutions due to missing steps in the design process when solving complex problems.

Major concepts of this chapter and their relations are shown in Figure 2.41.

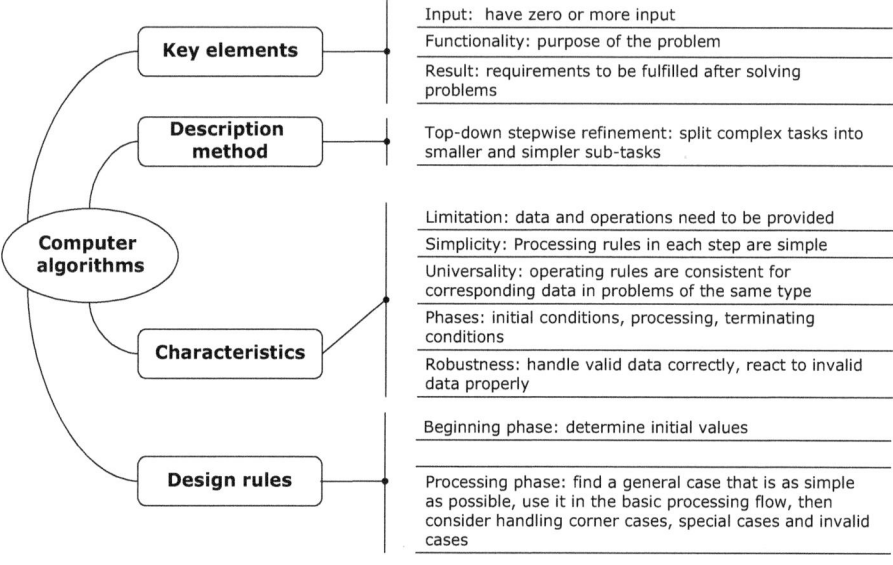

Figure 2.41: Concepts of algorithms and their relations.

Steps of solutions are called algorithms.

We need to base our solutions on the characteristics of computers when using them to solve problems.

Input data, functionality, and result output are the three key elements of algorithms.

We need to start from top level when solving complex problems with the help of classic technique "divide and conquer,"

And use top-down stepwise refinement method to decompose the task,
Into subtasks that can be easily implemented in programming languages.
Finally, we combine these modules, simply and flexibly.

Computers think in a different way from us, which seems stubborn.
Data and operations need to be provided without mistakes.
Each operation is essentially binary code.
We should handle data of the same type consistently, so that computers can
solve different problems using the same method.
The flow of algorithms consists of three phases: initial data, terminating condition,
and processing process where we implement required functionalities.
We may consider general cases first,
and carefully investigate all corner cases.
Test cases should cover edge cases, special cases, and errors,
So that the program becomes more robust.
We should learn and master classic algorithms,
In order to apply them when solving new problems.
There is a standard to assess algorithms,
Where the most important thing being correctness,
Readable, robust,
Time-efficient, and space-efficient,
These are virtues of good algorithms.

2.8 Exercises

2.8.1 Multiple-choice questions

(1) [Concept of algorithms]
Which of the following statements is correct about algorithms? ( )
A) An algorithm is equivalent to the solution to a problem.
B) One algorithm can only solve one problem. It cannot be reused.
C) An algorithm is executed step by step. Operations in each step must be precisely described.
D) There is only one algorithm that solves problems of a certain type.

(2) [Characteristics of algorithms]
Which of the following statements is wrong about the characteristics of algorithms? ( )
A) Finiteness: an algorithm must terminate in finite steps.
B) Input: an algorithm must have at least one input.
C) Definiteness: the steps of an algorithm must be clearly described.
D) Output: an algorithm must have at least one output.

(3) [Principle of programming]

Which of the following is not a fundamental principle of structured programming? ( )

A) Polymorphism   B) Top-down

C) Modularization   D) Stepwise refinement

(4) [Structured design]

Which of the following statements is wrong about modularization of programs? ( )

A) We can use the bottom-up stepwise refinement design method to construct programs using independent modules.

B) Dividing programs into independent and single-function modules makes code reuse easier.

C) Dividing programs into independent modules makes coding and debugging easier.

D) We can use the top-down stepwise refinement design method to construct programs using independent modules.

(5) [Characteristics of algorithms]

Finiteness of algorithms means that ( )

A) The run time of an algorithm is finite.

B) The amount of data an algorithm can process is finite.

C) The length of an algorithm is finite.

D) An algorithm can only be used by finite users.

(6) [Description of algorithms]

Which of the following cannot be used to describe algorithms? ( )

A) Text description   B) Programming statements

C) Pseudo code and flowcharts   D) E-R diagrams

(7) [Algorithm design]

Which of the following statements is correct? ( )

A) The so-called algorithm is a method of computation.

B) Programs are also one way to describe algorithms.

C) We only need to consider how to obtain computation results in algorithm design.

D) We can ignore computation time in algorithm design.

(8) [Software testing]

Which of the following statements is wrong about software testing? ( )

A) We must follow the testing plan to eradicate randomness.

B) We should select test data randomly.

C) We should select test data meticulously.

D) Software testing is an important way to guarantee software quality.

2.8.2 Fill in the tables

(1) [Finding the maximum]

Based on the flow of the maximum finding algorithm, fill in the table in Figure 2.42 with the current value of Max in each iteration.

The input data: 12, −3, 25, 120, 0, 20

Iteration	1	2	3	4	5	6
Max						

Figure 2.42: Algorithms: fill in the tables question 1.

(2) [Finding the minimum]

Based on the flow of the minimum finding algorithm, determine whether we need to update the value of Min in each iteration and fill in the table in Figure 2.43.

The input data: 12, −3, 25, 120, 0, 20

Iteration	1	2	3	4	5	6
Update						

Figure 2.43: Algorithms: fill in the tables question 2.

(3) [Sequential search]

Based on the flow of sequential search, fill in the table in Figure 2.44 with the number of comparisons needed to find 33.

Note: we look for the number in the sequence sequentially.

Data to be searched	32	13	65	77	33	71	93

Figure 2.44: Algorithms: fill in the tables question 3.

(4) [Binary search]
Based on the flow of binary searching, determine whether we can find 33 in the numbers given and fill in the table in Figure 2.45.

Data to be searched	71	32	13	33	93	65	77

Figure 2.45: Algorithms: fill in the tables question 4.

(5) [Recursion]
A monkey picked several peaches on day 1. He ate half of them and ate another later. On day 2, he ate half of the remaining and ate another later. He did this in the following days as well. On day 5, he had only one peach left before eating. Please fill in the table in Figure 2.46.

Day	5	4	3	2	1
Number of remaining peaches					

Figure 2.46: Algorithms: fill in the tables question 5.

2.8.3 Algorithm design

Describe an algorithm for each problem below in the form of pseudo code or flowchart.
(1) Design an algorithm for each of the following problems:
   a) Read two numbers from keyboard input, compute and display the sum of them.
   b) Read two numbers from keyboard input, figure out and display the larger one of them.
   c) Read n positive numbers from keyboard input, compute their sum.
(2) Read several nonzero real numbers, count the number of positive ones and the number of negative ones. The algorithm terminates upon receiving 0.
(3) Read a five-digit integer, and determine whether it is a palindromic number.
(4) A natural number (1, 2, 3, 4, 5, 6, etc.) is called a prime number (or a prime) if it is greater than 1 and cannot be written as the product of two smaller natural numbers. For example, 2, 3, 5, and 7 are prime numbers, whereas 4, 6, 8, and 9 are not. Write an algorithm that determines whether a natural number is a prime.
(5) In the Fibonacci sequence, each number is the sum of two preceding ones. The first few numbers in the sequence are 0, 1, 1, 2, 3, 5, 8, 13, 21... Write an algorithm to compute the nth Fibonacci number.

(6) Charges of a telecommunication company are as follows: local calls cost ¥0.22 per minute if the call lasts less than 3 min; if a call lasts more than 3 min, the part which exceeds 3 min costs ¥0.1 per minute (or part thereof). Design an algorithm to compute charges.

(7) Five people are sitting together. When asked about their age, the fifth person says he is two years older than the fourth person. The fourth person says he is two years older than the third. The third person says he is 2 years older than the second, who is 2 years older than the first. The first person answers that he is 10. How old is the fifth person? Figure out a universal computation formula and algorithm (using recursion).

3 Basic data types

Main contents
- Basic data types, the essence of types, storage mechanism of integers, and floating-point numbers
- Definition of variables, referencing method, and their way of storage in memory
- Operators and their usage, the concept of expressions, categorization of results of operations
- Summary of data elements

Learning objectives
- Understand and master concept of data types, data storage, data referencing, and data operation
- Know how to use common operators and expressions
- Understand and master the usage of constants and variables

3.1 Constants and variables

At the checkout in supermarkets, we are given a receipt by a cashier, on which information of our purchases is written as shown in Figure 3.1.

The column total is computed by multiplying per-unit price with quantity, where the per-unit price is a constant and quantity is a variable. Some values are fixed, whereas others keep changing in many problems. For example, we have speed, time, and distance in moving object problems. In circles, we have a radius, perimeter, and Pi. In the shopping example above, we have the number of purchased goods, per-unit price, and total.

Data in programs can be categorized into two types based on how they are used: constants and variables. The value of a constant cannot be modified during the execution of programs, whereas the value of a variable can be changed during execution.

3.1.1 Constants

There are two kinds of constants: literals and symbolic constants. One can use literal constants directly in programs as needed without having to define in advance. However, if a constant is used multiple times in a program, we can use a symbolic constant instead to allow easy modification. In this case, we only need to modify once if we need to change the value of the constant. Symbolic constants should be defined before being used. To define one, we need to use the define macro. For example, `#define LEN 128` means that every occurrence of LEN represents 128. More on macros will be covered in the chapter "Preprocessing."

https://doi.org/10.1515/9783110692327-003

Case Study

Shopping receipt

```
==============================
Item Name  Price Per-Unit  Quantity  Total
------------------------------------------------
Notebook   15.60           1         15.60
Battery    8.00            2         16.00
Bread      3.60            2         7.20
Milk       26.80           1         26.80
------------------------------------------------
Subtotal                   6         65.60
Discount   3.60            Total     62.00
Received   100.00          Change    38.00
------------------------------------------------
```

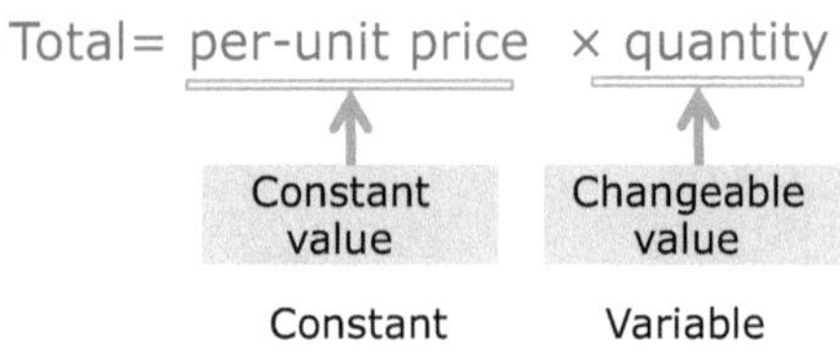

$$\text{Total} = \text{per-unit price} \times \text{quantity}$$

Constant value	Changeable value
Constant	Variable

Figure 3.1: Shopping receipt.

Example 3.1 Example of constants in programs
The per-unit price of a notebook is ¥15.6. Write a program that outputs the per-unit price and total price of two notebooks.

```
01  #include <stdio.h>
02  #define PRICE 15.6 //Define symbolic constant PRICE which represents 15.6
03  int main(void)
04  {
05      printf("Per-unit price: %f\n", PRICE); //PRICE – symbolic constant
06      printf("Total: %f\n", PRICE*2);  //2 – literal
07      return 0;
08  }
```

On line 2, we use macro define to define the symbolic constant PRICE. Thus, every occurrence of PRICE in the program, in lines 5 and 6 for instance, represents value 15.6. To update the price, we can simply modify line 2. This way of representation is more precise than using numeral 15.6 and improves the readability of our program.

Good habit in programming
Instead of numerals, we should use meaningful symbols for constants related to the physical world or with physics meaning. In C language, this is done by using meaningful enumerations or macros. The concept of enumerations will be introduced in the chapter "Composite Data Types."

C has many types of constants as shown in Figure 3.2.

Integer numerals include decimal, octal, and hexadecimal numerals. Decimal numerals are the ones we are familiar with. The octal numeral system only uses digits 0–7 and an octal number is prefixed with 0 to distinguish from decimal numbers. The hexadecimal numeral system uses symbols "0" to "9" to represent values zero to nine and "A" to "F" (or lowercase counterparts) to represent values 10 to 15. To distinguish from decimal numbers, we prefix hexadecimal numbers with 0x or 0X.

	Form	Representation rule	Example
Integer	Decimal	Using digits 0 to 9	23, 127
	Octal	Using digits 0 to 7, prefixed by 0	023, 0127
	Hexadecimal	Using digits 0 to 9 and A to F/a to F, prefixed with 0x or 0X	0x23, 0xc8
Real number	Decimal	Numbers with decimal point	1.0 +12.0 -2.0
	Exponential	Number e/E number	1.8e-3 -23E+6
Character	Printable character	Single printable character wrapped by single quotation marks	'a' 'A' '+' '3'
	Escape character	'\' and printable character wrapped by single quotation marks	' \n '
	String	Character sequence wrapped by double quotation marks	"ABC"、 "123"、 "a"

Figure 3.2: Types of constants.

Real numerals can be written in decimal form or exponential form. Decimal form is a number with a decimal point. Exponential form, on the other hand, is also called "scientific notation," in which $1.8 * 10-3$ is represented by 1.8e-3 and $-23 * 106$ is represented by $-23E + 6$, as shown in Figure 3.2. Herein e and E can be used interchangeably.

Character literals include printable character, escape character, and string literals. Printable character literals are single printable characters wrapped by single quotation marks, where they are characters that can be displayed on a screen. For example, the character "a," symbol "+," and character "3" are all printable characters. Note that character 3, which has ASCII value 51, is different from decimal numeral 3.

There are also special characters that cannot be displayed on screens. A newline is one example of these characters. We use escape characters to represent them in C. The escape character table of C is given in Appendix D. At this stage of learning, we only need to remember newline is represented by "\n."

String literals are sequences of characters wrapped by double quotation marks, for example, "ABC" and "123."

Knowledge ABC Story behind "return and newline"
Before computers were even created, there was a kind of machine called teleprinters. Such machines could print 10 characters in a second. However, it took 0.2 seconds for them to move to the next line, during which two characters could be printed. If new characters were typed during this 0.2 second, they would be discarded.

To solve this problem, the creators of these machines added two unique characters to the end of each line. One of them was "Carriage Return," namely "Return," which instructed the printer to reset the position of carriage to the beginning of a line. The other was "Line Feed," namely "Newline," which fed the paper to advance to the next line. This was how "Newline" and "Return" were created. They were later introduced to computers when computers were invented.

> **i** **Knowledge ABC** ASCII and Chinese character encoding
> All kinds of information, including numbers, characters, sounds, and images, are stored in computers as binary codes.
>   – American Standard Code for Information Interchange (ASCII)
> ASCII is a character encoding standard used in computers to display modern English and other western European languages, which are based on Latin characters. It is the most frequently used single-byte character encoding system nowadays.
>   ASCII uses combinations of seven or eight binary digits to represent 128 or 256 possible characters. Standard ASCII, or basic ASCII, uses seven binary digits to represent all English characters (of both upper and lower cases), number 0 to 9, punctuation marks, and special control characters used in American English.
>   The remaining 128 characters in the 256-character version are called extended ASCII codes. They are supported in many x86-based systems. Extended ASCII codes use the additional eighth position to represent 128 other special characters, characters in other languages, and graphical symbols.
>   – Chinese character encoding
> Chinese character encodings are used in computers to represent Chinese characters. Chinese characters are represented as 16-digit binary codes in computers. In 1981, the Standardization Administration of the People's Republic of China published GB2312, which included 6,763 Chinese characters, as a unified standard for designing input/output devices so that information could be exchanged smoothly.

3.1.2 Variables

3.1.2.1 Key elements of variables

Let us take a look at a storage problem in real life. When Mrs. Brown goes to the supermarket, she needs to store her personal belongings into an electronic locker before shopping. To do so, she needs to press the "Store" button on the locker first. A receipt labeled with a number is then printed and the compartment with the corresponding number is opened automatically. Finally, she puts her belongings in and closes the compartment.

However, the locker at this supermarket is specially designed as shown in Figure 3.3. There is an animal sticker on the door of each compartment. They are useful for the help desk staff to find the correct compartment for customers who lost their receipt and couldn't recall the number or position of their compartment. Customers find animal stickers convenient because they are intuitive and easy to remember. Supermarket staff finds the numbering scheme of compartments convenient because it indicates the location of compartments.

A locker compartment can store belongings of customers. The actual location of a customer's compartment is allocated by the locker system based on the current locker space. Hence, the key elements of locker compartments are a name that can be used to address it, objects that can be stored in it as well as withdrawn from it, and a location that can be allocated.

Locker compartment

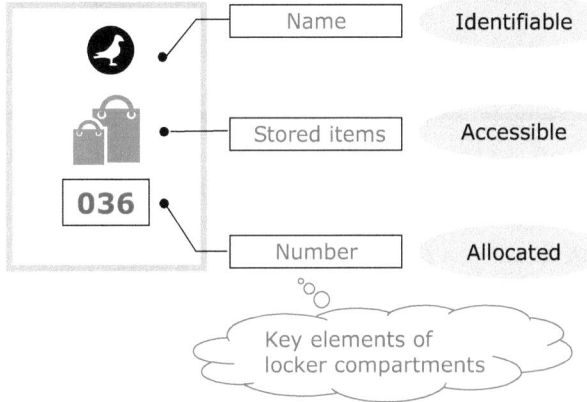

Figure 3.3: Key elements of locker compartments.

The process of programs storing and accessing data is similar to how customers store their belongings into and retrieve them from a locker. The space data are stored in, which is called a storage unit, is similar to a locker compartment as shown in Figure 3.4. The name of a storage unit is called variable name in programming languages, which is "a" in this example. The name can be other words or letters as well. The value of data stored in a storage unit is called the value of the variable. In this example, we say the variable "a" has value 6. The value of a variable can be updated as needed. The location of a storage unit is called an address in computers. In general, variables are objects that associate with memory space where their contents are stored and accessed.

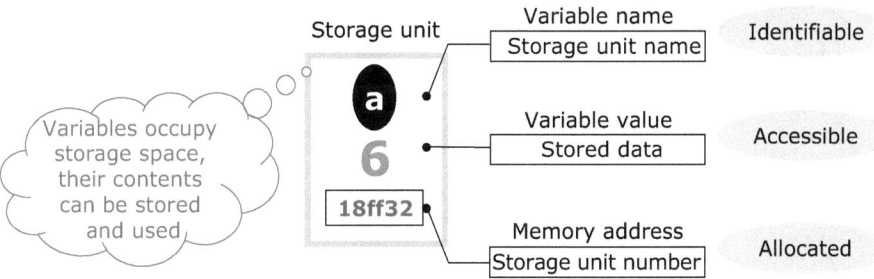

Figure 3.4: Key elements of storage units.

Hence, we may summarize the three key elements of variables: variable name, variable value, and memory location. What rules should we have regarding these elements then?

Basically, we need to determine the rules of naming a variable, requesting memory space, and using allocated memory space.

3.1.2.2 Rules of variable naming

There are rules for naming variables. Names of variables and constants in C are marked by identifiers. As shown in Figure 3.5, identifiers consist of letters, numbers, and underscores with the exception that an identifier cannot begin with a number. Underscore and both cases of letters are used to increase readability. In C programs, variable names are case-sensitive and we usually use lower case letters in identifiers. Variable names should be meaningful to be remembered and read easily. For example, we should try to use English words and their combinations as often as possible. Some variables are named following some conventions. One notable example is using i, j, and k for loop variables. Some keywords have been used as identifiers by the language itself. Thus, we cannot use them as variable names. ANSI C has 32 keywords (or reserved words) that cannot be used otherwise. There are 12 other identifiers used for preprocessing, which should be prefixed by a # sign when using.

Identifier

A symbol used to identify an object. In programs, an identifier is a word with special meaning defined by programmers.

Cannot begin with number

Composition	Naming convention	Beware	Prohibited
Letter, number, underscore	Meaningful	Case-sensitive	Keyword

Keyword

Keywords (reserved words) defined by ANSI C
auto, break, case, char, const, continue, default, do, double, else, enum, extern, float, for, goto, if, int, long, register, return, short, signed, sizeof, static, struct, switch, typedef, union, unsigned, void, volatile, while

Special words in preprocessing
define, elif, else, endif, error, if, ifdef, ifndef, include, line, progma, undef

Figure 3.5: Identifiers and keywords.

Good habit in programming
(1) Variable names consisting of multiple words make a program more readable.
(2) Meaningful identifiers make programs self-explained (have fewer comments).

For example, compare the following variable names:
- variablename
- variable_name
- VariableName

The second one follows the UNIX naming convention, whereas the third one follows the Windows naming convention. Apparently, the first one is less readable and the rest are more obvious at a glance.

3.1.2.3 Method of requesting memory space

A locker compartment is opened by pressing the "Store" button. Requesting memory space is done by defining variables. A variable definition is made up of data-type identifier and variable name as shown in Figure 3.6.

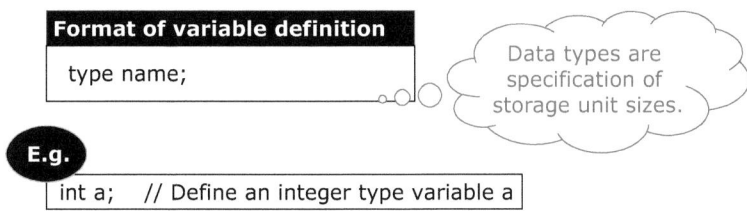

Figure 3.6: Format of variable definition.

Data-type identifiers are names of data types in C. For instance, "int a" defines an integer variable a, where "int" represents integer type. More on data types will be covered later. After we define a variable, computers allocate memory space of a certain size at a suitable location in memory based on this definition.

If we store the value into a storage unit when defining a variable, this process is then called variable initialization in programming languages as shown in Figure 3.7, where the operation on the storage unit is called a variable assignment.

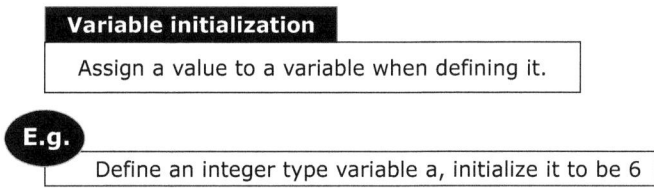

Figure 3.7: Variable initialization.

In essence, a variable definition is the process of programmers requesting a storage unit of a certain size, which is determined by the system based on the type of the variable. When requested, computers allocate memory of that size at a suitable location in memory.

3.1.2.4 Usage of memory space

We put data into memory space for storage and further use. Programmers use variable names to access these data. This is called "variable referencing" as shown in Figure 3.8.

Variable referencing
Programmers access data in storage units by using variable names.

Variables can only be used after definition

Figure 3.8: Variable referencing.

Essentially, a variable is a named block of continuous memory space. We request and name such space by defining variables and use it through variable names. This space is used to store data and the variable type determines its size. Let us see some examples of variables.

Example 3.2 Checking the three key elements of variables
Check the three key elements of variables in a debugger.

[Analysis]
We first write a simple program with only one variable. Then we trace how space, address, and value change when the variable is defined and initialized.

1. Source code

```
01  #include <stdio.h>
02  int main(void)
03  {
04      int a=6;            //Define variable a, initialize it as 6
05      printf("%d\n", a);  //Display value of a onto screen
06      return 0;
07  }
```

2. Tracing and debugging
We define variable a and initialize it to be 6 on line 4. Then we output its value onto the screen on line 5. Here, we reference variable a by using its value.

General methods of tracing and debugging can be found in the chapter "Execution of Programs."

In the Watch window of the debugger, we notice that the value of a is 6, as shown in Figure 3.9. More details of the debugger are covered in the corresponding chapter. "&" sign is used to obtain the address of a, namely, the memory address of the storage unit. sizeof(a) calculates the size of the space variable a takes up in memory, where the size is measured in bytes. In our case, a takes up 4 bytes. The size of the variables is determined by their types. Variable a is an integer, which takes up 4 bytes in this system.

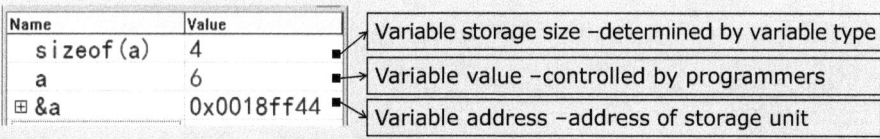

Figure 3.9: Key elements of variables.

Example 3.3 Definition and initialization of variables
Several cases of variable definition and initialization are listed in Figure 3.10. Analyze the attributes of these variables according to their definition and initial value.

Row	Variable definition	Variable name	Content of storage unit	Length of storage unit
1	int sum ;	sum		sizeof(int)
2	int sum=16 ;	sum	16	sizeof(sum)
3	long m, n=12 ;	m		sizeof(m)
		n	12	sizeof(n)
4	double x=23.568, y;	x	23.568	sizeof(double)
		y		sizeof(y)
5	char ch1='a',ch2=66;	ch1	97	sizeof(char)
		ch2	66	sizeof(ch2)

What value is in the storage unit if a variable is not initialized?

Why is the value in the storage unit 97 when we initialize variable ch1 with 'a'?

Figure 3.10: Example of variable definition and initialization.

[Analysis]
In Figure 3.10, the column "content of the storage unit" records the value stored in the storage unit when the variable is defined. This value can be changed as needed during the execution of programs. Note that these variables are all defined inside a function.

On the third row, we define two long integer m and n, where m is not initialized and n is initialized to 12.

On the fourth row, we define two real numbers x and y. To figure out the size of the storage unit for a variable, we can simply put the variable name or variable type inside parentheses of size of operator.

On the fifth row, we define two character variables ch1 and ch2, where ch1 is initialized to character literal a. However, the value stored in its storage unit is 97 instead of a, why is this the case? This is because characters are stored in computers after being encoded in C environment. ASCII value for character a is exactly decimal number 97.

What is stored in the storage unit if we don't initialize the variable then?

In contrast to locker compartments that are empty when not being used, a storage unit not in use still has data in it. However, it is an arbitrary number, which is meaningless to programmers.

Example 3.4 Assignment of variables and memory space viewing

```
1   // Variable assignment
2   #include<stdio.h>
3   int main(void)
4   {
5      char c1,c2;
6
7      c1=97;                       // Assign 97 to c1
8      c2='b';    // Assign 98 to c2
9      printf( "%c %c\n ", c1, c2); //%c: output c1 and c2 as characters
10     printf( "%d %d ", c1, c2); //%d: output c1 and c2 as integers
11     return 0;
12  }
```

Output:
```
    a   b
   97   8
```

[Analysis]

1. Program analysis

The %c and %d on lines 9 and 10 are format specifiers of output function printf. They are used to output data onto screen in certain formats.

Take variable c1 as an example, which has value 97 in its storage unit. When output is as a character, the character for ASCII value 97 is displayed on the screen. When output is as a number, 97 is displayed instead.

The three key elements of variable c1 and c2 are shown in Figure 3.11.

Variable name	Content of storage unit	Length of storage unit	ASCII value
c1	97	1 byte	a
c2	98	1 byte	b

Figure 3.11: Three key elements of variables.

2. Program tracing

In Figure 3.12, we see that variables c1 and c2 have initial value −52, which is an arbitrary number. Because −52 has no corresponding character in ASCII, "?" is displayed instead.

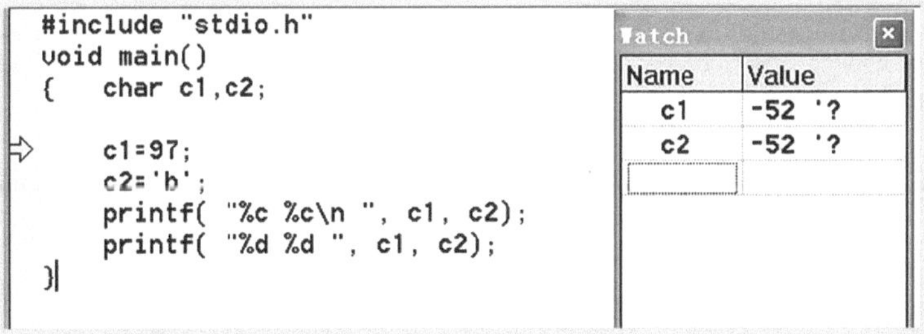

```
#include "stdio.h"
void main()
{    char c1,c2;

⇨   c1=97;
    c2='b';
    printf( "%c %c\n ", c1, c2);
    printf( "%d %d ", c1, c2);
}|
```

▼atch	☒
Name	Value
c1	-52 '?
c2	-52 '?

Figure 3.12: Debugging step 1 of variable assignment.

After we assign values to c1 and c2, c1 has value 97, which is represented by character a in ASCII, whereas c2 has value 98, which is represented by character b as shown in Figure 3.13.

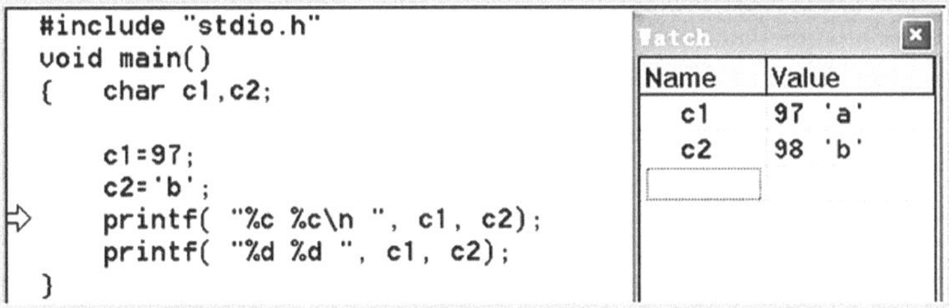

```
#include "stdio.h"
void main()
{    char c1,c2;

    c1=97;
    c2='b';
⇨   printf( "%c %c\n ", c1, c2);
    printf( "%d %d ", c1, c2);
}
```

▼atch	☒
Name	Value
c1	97 'a'
c2	98 'b'

Figure 3.13: Debugging step 2 of variable assignment.

3.2 Data types

Computers can handle all kinds of data, each with different attributes. We may categorize data based on their property, form of representation, storage size, and form of construction.

- Property: integers, decimals, characters, etc.
- Form of representation: data can be represented by constants or variables in programs.
- Storage size: different types of data take up different sizes of memory space.
- Form of construction: data can be of basic types or composite types.

Basic data types in practical problems include the numeral type and character type as shown in Figure 3.14. To solve problems with computers, we need to store data into computers before executing any operations. Hence, we should consider how these basic data should be categorized and stored in computers first.

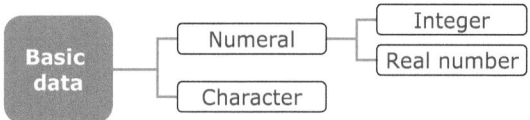

Figure 3.14: Basic data.

3.2.1 Representation of information in computers

3.2.1.1 Binary system
A lightbulb, or a switch, being on or off is two different states, so they can be used to represent 0 and 1 in logic. We call such an information system a binary system as shown in Figure 3.15.

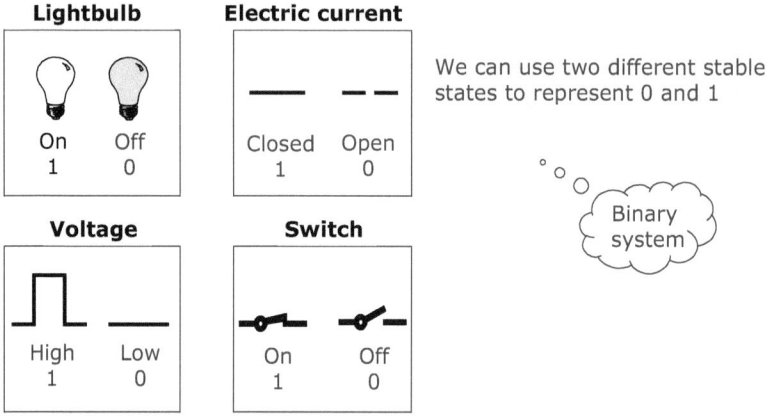

Figure 3.15: Binary system.

3.2.1.2 Binary representation
The combination of states of multiple lightbulbs can be used to represent a sequence of 0 and 1. Computers consist of many electronic components internally, which are controlled by circuits. A switch in these circuits can be set to one of two stable states. Thus, we can use a combination of switch states to represent multiple 0s and 1s as shown in Figure 3.16. We use one 0 or 1 to represent a bit in computers, but the question is: what can be represented by these 0s and 1s?

We are all familiar with decimal numbers. In fact, they represent a positional numeral system as shown in Figure 3.17. Take decimal number 256 as an example, the second digit (counting from right-hand side) 5 represents 50, which is the product of 5 and position value of the second digit, namely 10 raised to the first power. The third digit 2 represents 200, which is the product of 2 and position value of this

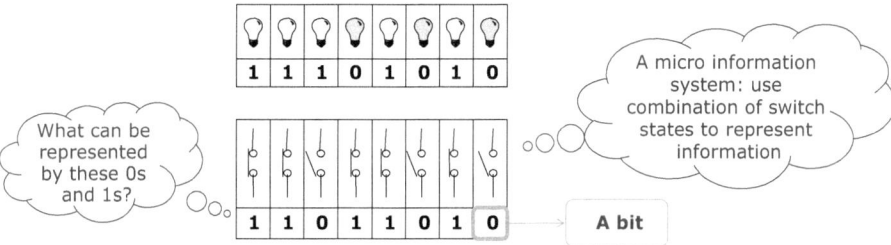

Figure 3.16: Binary sequence represented by switches.

Position	3rd	2nd	1st
Digit 0~9	2	5	6
Position value	10^2	10^1	10^0

Position	7	6	5	4	3	2	1	0
Digit 0~1	1	1	1	0	1	0	1	0
Position value	2^7	2^6	2^5	2^4	2^3	2^2	2^1	2^0

Figure 3.17: Positional numeral system.

digit, namely 10 raised to the second power. The base of position value in the decimal system is 10.

Similarly, digits in the binary system are 0 and 1 while the base of position value is 2. When adding binary numbers, we carry 1 to the next digit when two digits add up to 2. When subtracting binary numbers, we borrow 1 from the previous digit and use it as 2 during subtraction of current digits.

If we list all possible four-digit binary numbers, it is not hard to notice that there are 16 corresponding decimal numbers, namely 0 to 15 as shown in Figure 3.18. We may conclude that n-digit binary numbers can represent 2^n numbers.

Binary number	0000	0001	0010	0011	0100	0101	0110	0111
Decimal number	0	1	2	3	4	5	6	7
Binary number	1000	1001	1010	1011	1100	1101	1110	1111
Decimal number	8	9	10	11	12	13	14	15

n-digit binary numbers can represent 2^n numbers

Figure 3.18: Number of digits in binary numbers and numbers they represent.

3.2.2 Processing of information in computers

Having learned how information is represented in computers, we now focus on how information is handled in computers.

3.2.2.1 Modular system

After a full day of meetings, Mr. Brown returned to his office at 5 pm. He noticed that his clock stopped at 9 o'clock. When winding the clock, he found that the number of hours he needed if winding clockwise and the number of hours he needed if winding counterclockwise added up to 12. For example, turning the short hand 8 h forward has the same effect of turning it 4 h backward. We say that 8 and 4 are the complement of each other in the full cycle of a clock, namely 12 h. A clock can be seen as a counter of time. We call the counting interval of a recurrent counting system "modulus." We can replace subtraction with addition in any counting system with modulus as shown in Figure 3.19.

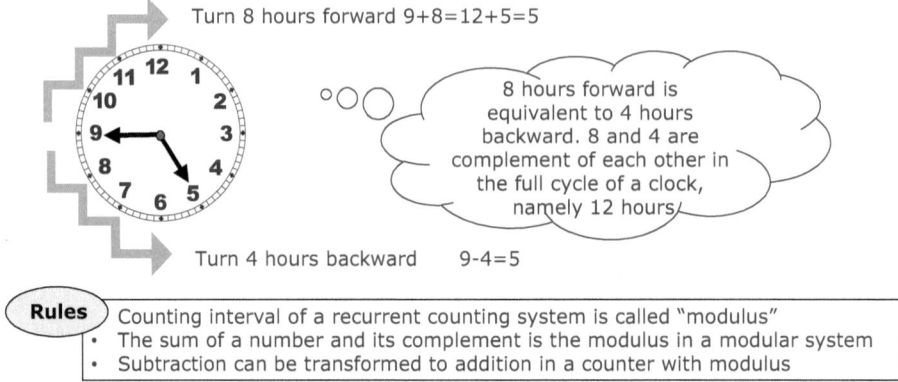

Turn 8 hours forward 9+8=12+5=5

8 hours forward is equivalent to 4 hours backward. 8 and 4 are complement of each other in the full cycle of a clock, namely 12 hours.

Turn 4 hours backward   9-4=5

Rules   Counting interval of a recurrent counting system is called "modulus"
- The sum of a number and its complement is the modulus in a modular system
- Subtraction can be transformed to addition in a counter with modulus

Figure 3.19: Modular system.

3.2.2.2 Binary modular system

There is a domain for data in modular systems, where the data can change and recur. In fact, binary memory space is also a modular system. For instance, in a memory space of four-digit binary numbers as shown in Figure 3.20, the value can change into 1111 from 0000 by repeatedly adding 1, which matches the characteristics of a modular system.

Position	3	2	1	0
Minimum	0	0	0	0
Maximum	1	1	1	1

Modulus=$[1111-0+1]_2=[1,0000]_2=[2^4]_{10}=[16]_{10}$

Figure 3.20: Binary modular system.

We can calculate the modulus of four-digit binary numbers by subtracting the minimum representable number from the maximum and adding one. The result is 16, which is exactly 2 raised to the fourth power.

Mr. Brown wanted to verify whether subtraction can be replaced with addition in this binary modular system. He planned to calculate 0 minus 6 and 0 plus 10 and compare the results as shown in Figure 3.21.

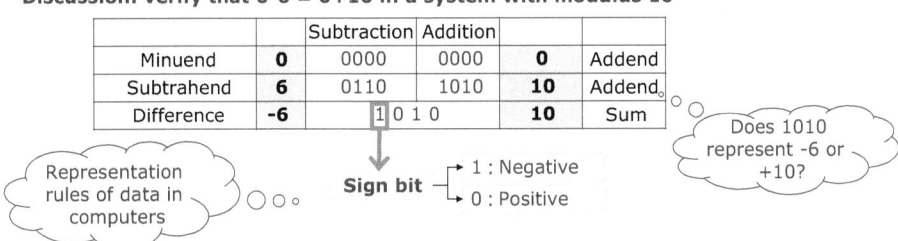

Discussion: verify that 0-6 = 0+10 in a system with modulus 16

		Subtraction	Addition		
Minuend	**0**	0000	0000	**0**	Addend
Subtrahend	**6**	0110	1010	**10**	Addend₀
Difference	**-6**	1 0 1 0		**10**	Sum

Sign bit → 1 : Negative / 0 : Positive

Does 1010 represent -6 or +10?

Representation rules of data in computers

Figure 3.21: Complement in the binary system.

Based on the conclusion above, the complement of −6 is 10 in a system with modulus 16, so the results should be identical. 0 plus 10 is $[1010]_2$, so $[1010]_2$ ought to be the complement of −6.

However, this leads to a question: should $[1010]_2$ represent −6 or 10?

A rule of representing data in computers is thus needed, Mr. Brown thought. We could use the most significant bit to distinguish between positive and negative numbers, where 1 indicates negative and 0 indicates positive. As this bit is used to indicate the sign of a number, it is also called "sign bit."

Two problems are yet to be solved after the verification:
(1) We need to review the domain of four-digit binary numbers (0000 to 1111) after introducing sign bit.
(2) We need to find a pattern of relations between positive and negative binary numbers in a signed system.

3.2.2.3 Representation of numbers in binary modular system

Finding a pattern of relations between positive and negative binary numbers in a signed binary system Mr. Brown decided to tackle the second problem first. To figure out relations between positive and negative numbers, it might be easier for us to consider two numbers with the same absolute value. We could use an actual number, for example, 6, and try to find a relation between binary representations of 6 and −6. As shown in Figure 3.22, Mr. Brown tried flipping and addition to see whether there is any relation between them.

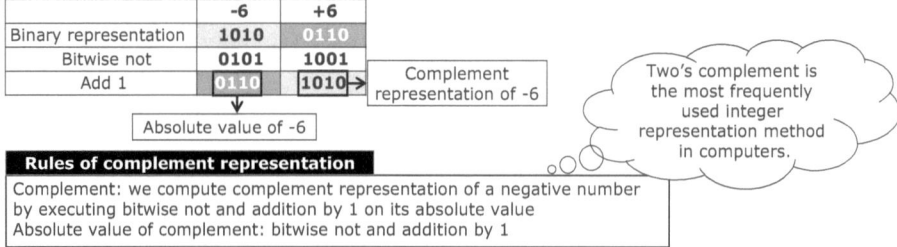

Figure 3.22: Two's complement.

"Bitwise not" on the third row of the table means executing not operation (flip 0 to 1 and flip 1 to 0) on each bit.

After a bitwise not operation and adding by 1, Mr. Brown noticed that −6 became its own absolute value, and 6 became the complement representation of −6. This is the rule of complement representation as shown in Figure 3.22.

As complement representation is used in subtraction, there is no need to use it on positive numbers. However, we define the complement representation of positive numbers to be the same to make our theory comprehensive.

One of the merits of complement representation is that subtraction, multiplication, and division can all be transformed into addition, which largely simplifies circuit design of arithmetic units in computers. Although signed integers can be represented in multiple ways in computers, we usually use 2's complement to represent them.

3.2.2.4 Range of binary system

With complement representation, Mr. Brown tried to solve the first problem, which is to determine the range of signed four-digit binary numbers. As shown in Figure 3.23, 1 or 0 on the most significant bit now indicates sign. Using the fact that the absolute value of the minimal negative number (obtained by applying bitwise not and adding 1) should be the largest, he found the minimum negative number and the maximum

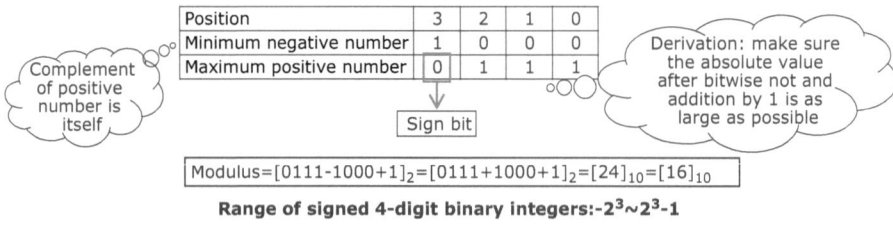

Figure 3.23: Range of binary numbers.

positive number and noticed that modulus was still 16. Note that the sign bit was also flipped and added by 1. Hence, the range of signed four-digit binary integers is -2^3 to 2^3-1. Similarly, we can figure out the range of signed n-digit binary integers.

Think and discuss Does $[1010]_2$ represent -6 or $+10$ in a signed four-digit binary integer system? **Discussion:** The range of such a system is -23 to $23-1$, namely -8 to $+7$, which doesn't include $+10$. Hence, [1010]2 represents -6.

3.2.3 Basic data types in C

Data types indicate the size of the space that data need. We can figure out the domain and operations allowed of data by examining data type. There are three basic types: integer, real numbers, and characters as shown in Figure 3.24.

Unsigned integers have no sign bit, so they only represent positive integers. Real numbers are represented in a different way, which will be covered in remaining later.

Each data type is of a certain size and has its own domain. It is worth noting that the size of a type may vary on different computers. The size of long type in C is always defined as word length of the machine, where word length is the maximum number of bits of binary data a computer can process in an integer operation. PC nowadays usually uses 32 bits for integers.

int, float, and char are the most frequently used basic types.

Although the size of types varies in different computers, there are still some patterns and rules, which are given below.

Rules of data types
(1) The minimum length of a storage unit is 8 bits, which can be used for one character. One byte is equal to 8 bits. Lengths of other storage units are multiples of 8 bits.
(2) The storage unit of pointer type records the "number of a storage unit," which is an integer, thus it has the same length as integers.
(3) Floating-point numbers are usually 2 N times (N is an integer) the length of integers as shown in Figure 3.25.
(4) There are two types of storage rules: integers, characters, and pointers follow integer rules, whereas floating-point numbers have their own rules.

Think and discuss How do we test the size of types?
We mentioned that the size of types might vary in different systems. How do we know the size of a type in the system we are using?
Discussion: C provides the sizeof operator for testing type sizes. sizeof is an operator in C/C ++, which returns number of bytes an object or a type takes up in memory.

Category	Has sign	Keyword	Meaning	Length	Range	
Integer	Yes	int	Integer	16	$-2^{15} \sim 2^{15}-1$	$-32768 \sim 32767$
		short	Short integer	16	$-2^{15} \sim 2^{15}-1$	$-32768 \sim 32767$
		long	Long integer	32	$-2^{31} \sim 2^{31}-1$	
	No	unsigned int	Unsigned integer	16	$0 \sim 2^{16}-1$	$0 \sim 65535$
		unsigned short	Unsigned short integer	16	$0 \sim 2^{16}-1$	$0 \sim 65535$
		unsigned long	Unsigned long integer	32	$0 \sim 2^{32}-1$	$0 \sim 4294967295$
Real number	Yes	float	Single-precision real number	32	$-2^{128} \sim 2^{128}$	
		double	Double-precision real number	64	$-2^{1024} \sim 2^{1024}$	
Character	Yes	char	Character	8	$-2^{7} \sim 2^{7}-1$	$-128 \sim 127$
	No	unsigned char	Unsigned character	8	$0 \sim 2^{8}-1$	$0 \sim 255$

Figure 3.24: Basic data types.

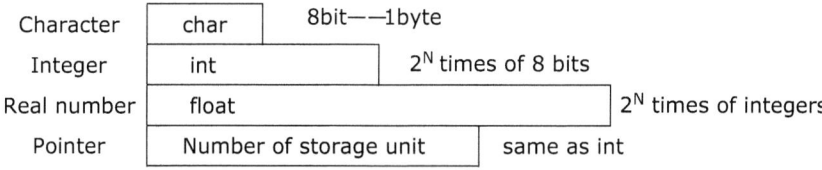

Figure 3.25: Size pattern of data types.

Example 3.5 Using sizeof operator to test type sizes
We can design a program to test sizes of common types.

```
1   //Use sizeof to test type sizes
2   #include<stdio.h>
3   int main(void)
4   {
5   printf("int size = %d\n", sizeof(int));
6   printf("short int size = %d\n", sizeof(short int));
7   printf("long int size = %d\n", sizeof(long int));
8   return 0;
9   }
```

Output:
```
    int size = 4
    short int size = 2
    long int size = 4
```

Explanation: int size = 4 indicates that the size of int in the IDE in which this program is executed is 4 bytes.

3.3 Storage rules of integers

There are four integer types in C:
- Basic type: Keyword is int, which are the first three letters of integer.
- Short type: Keyword is short [int] (note that content inside square brackets can be omitted).
- Long type: Keyword is long [int].
- Unsigned type: There are three unsigned types, namely unsigned [int], unsigned short, and unsigned long. They can only be used to store unsigned integers.

3.3.1 Signed integers

We shall use −12 as an example to learn the characteristics and rules of storage of signed integers. Suppose int type takes 16 bits in the following discussion.

The signed binary form of integer −12 is obtained by applying bitwise not and addition by 1 to integer 12 as shown in Figure 3.26. Careful readers may have noticed how the sign bit turns to 1 during this process. Comparing the binary representation of + 12 and −12, we find that sign bit is not the only difference.

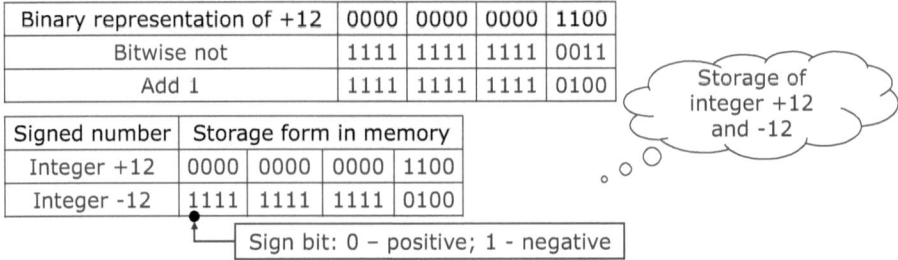

Binary representation of +12	0000	0000	0000	1100
Bitwise not	1111	1111	1111	0011
Add 1	1111	1111	1111	0100

Signed number	Storage form in memory			
Integer +12	0000	0000	0000	1100
Integer -12	1111	1111	1111	0100

Sign bit: 0 – positive; 1 - negative

Figure 3.26: Storage of signed integers.

Storage rules of signed integers can be summarized as follows: positive integers are stored as its binary representation, whereas negative integers are stored as its complement representation as shown in Figure 3.27.

Signed integer	Storage rule
Positive integer	Binary representation
Negative integer	Bitwise not and add 1 to its corresponding positive value

Figure 3.27: Storage rule of signed integers.

3.3.2 Unsigned integers

Again, we suppose the size of unsigned int is 16 bits.

Unsigned integers can only represent positive integers and zero. The sign bit used in signed integers is merely a normal bit in unsigned cases. The complement of 12 in signed representation is now decimal number 65524 in unsigned representation as shown in Figure 3.28.

Hence, we need to pay extra attention when storing and displaying data, as a binary number in the same storage unit can represent different things when used differently.

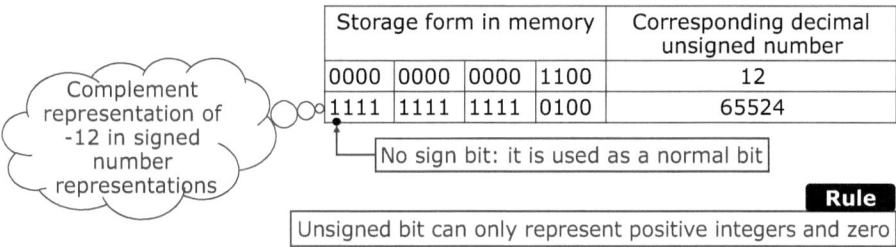

Figure 3.28: Storage of unsigned integers.

3.3.3 Characters

Character type has a size of 8 bits. When storing character A into computers, we are in fact storing its ASCII value 65 (here it is a decimal number) into a storage unit. As a result, character and integer data can be used interchangeably, where "interchangeably" means they share the same storage rules and operation rules as shown in Figure 3.29.

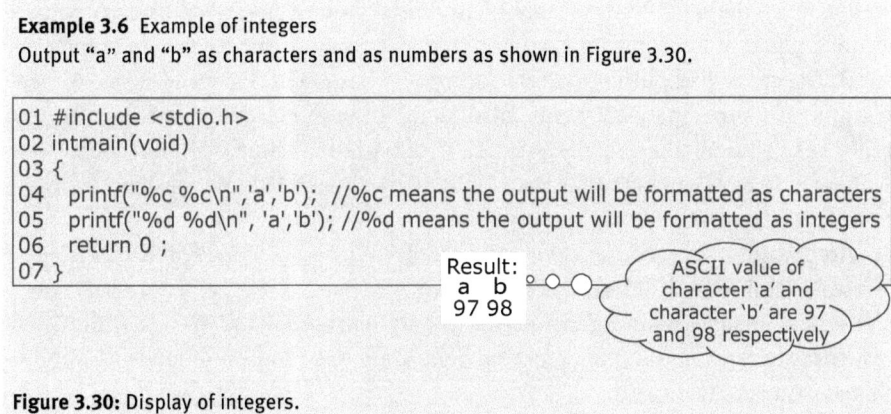

Fig. 3.29: Storage of characters.

Example 3.6 Example of integers
Output "a" and "b" as characters and as numbers as shown in Figure 3.30.

```
01 #include <stdio.h>
02 intmain(void)
03 {
04   printf("%c %c\n",'a','b');  //%c means the output will be formatted as characters
05   printf("%d %d\n", 'a','b'); //%d means the output will be formatted as integers
06   return 0 ;
07 }
```

Result:
a b
97 98

ASCII value of character 'a' and character 'b' are 97 and 98 respectively

Figure 3.30: Display of integers.

[Analysis]

On line 4, %c means the output will be formatted as characters, thus the result is character a and b.

On line 5, %d means the output will be formatted as integers, thus the result is ASCII value of a and b, namely 97 and 98.

Through this example, we learned that the same data can be displayed as different things by changing the output format.

3.4 Storage rules of real numbers

Figure 3.31 shows an example program of displaying real numbers. Floating-point numbers are how real numbers are stored in computers.

Case Study

Trap of floating -point numbers

```
01 #include <stdio.h>
02 int main(void)
03 {
04    float f=123.456;
05    if (f == 123.456) printf("Yes");  //If f = 123.456, output Yes
06    else printf ("No");              //Otherwise output No
07    printf( "f=%f \n",f);            //Output value of f
08    return0 ;
09 }
```

Why is this the case?

Result
No
123.456001

Figure 3.31: Trap of floating-point numbers.

On line 4, we define a float variable f with initial value 123.456. On line 5, two consecutive equal signs form an operator that checks whether its two operands are equal. The entire line outputs "Yes" to screen if f is equal to 123.456. On line 6, "No" is output if the comparison yields false. Combining these statements, we see that either yes or no shall be displayed. Line 7 outputs the value of f onto the screen. Readers may have guessed that yes would be displayed. However, as shown in the figure, the actual output may be surprising.

Aren't computers accurate computing tools? Why is there a deviation in results? Can we trust the results given by computers?

In fact, this is the error generated by binary representation. We use finite 32-bit sequences to represent infinite real numbers. Thus, the representation is an approximated value in most cases.

3.4.1 Representation of real numbers

To figure out the reason behind the error, let us take a look at the representation of real numbers.

When a number is extremely small or large, such as the mass of an electron (9×10^{-28} g) or mass of the sun (2×10^{33} g), we can write it as a real number multiplied by the nth power of 10, where the integer part of the real number has only one digit. This method is simple, convenient, yet accurate. Such representations are called "scientific notation."

3.4.2 Representation of floating-point numbers

Modern computers adopted the floating-point number representation as shown in Figure 3.32. In essence, it uses scientific notation to describe real numbers. Floating-point representations describe significant figures and range of representable numbers separately. More specifically, they use a fraction, a base, an exponent, and a sign bit to represent real numbers.

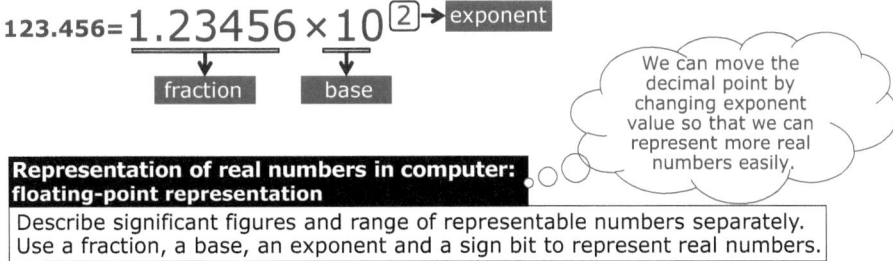

Figure 3.32 illustration: $123.456 = 1.23456 \times 10^{②} \rightarrow$ exponent, with "fraction" labeling 1.23456 and "base" labeling 10.

Speech bubble: We can move the decimal point by changing exponent value so that we can represent more real numbers easily.

Representation of real numbers in computer: floating-point representation
Describe significant figures and range of representable numbers separately. Use a fraction, a base, an exponent and a sign bit to represent real numbers.

Figure 3.32: Representation of floating-point numbers.

We shall use 32-bit float type, whose bit layout is shown in Figure 3.33, as an example. The fraction M occupies 23 bits. The exponent e, together with a bias, forms the biased exponent. Exponent represents the exponential part and occupies 8 bits. Since floating-point numbers are signed, one bit is needed for the sign bit.

The exponent occupies 8 bits, which can be used to represent numbers from −128 to 127. IEEE-754 uses value −128 for special purposes, so the actual range e can represent is −127 to +127 and the exponent bias of float type is 127. Using exponent bias makes the exponent an unsigned number so that operations can be done more quickly.

The IEEE-754 standard regulates the representation method we just described. In addition to 32-bit float type, there is also a 64-bit double type. Formula to compute the real value assumed by these representation methods is shown in Figure 3.34.

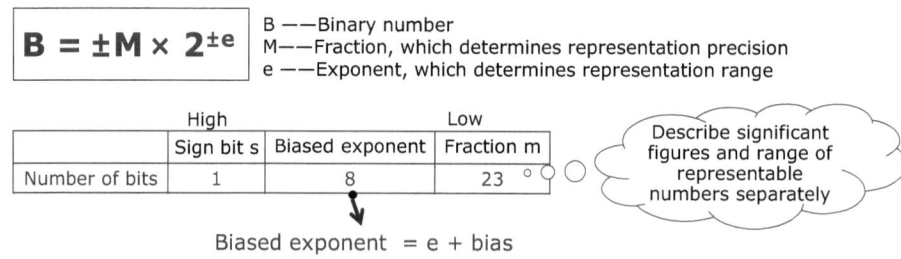

$$B = \pm M \times 2^{\pm e}$$

B ——Binary number
M——Fraction, which determines representation precision
e ——Exponent, which determines representation range

	High		Low
	Sign bit s	Biased exponent	Fraction m
Number of bits	1	8	23

Describe significant figures and range of representable numbers separately

Biased exponent = e + bias

Figure 3.33: Storage of float type.

Type	Storage format			Total number of bits	Bias
	Sign bit	Biased Exponent	Fraction		
Short real numbers (float)	1	8	23	32	127
Long real numbers (double)	1	11	52	64	1023

The bias is subtracted

$$\text{Real value} = [(-1)^{sign}] \times [\ 1.fraction] \times (2^{Biased\ exponent\ -bias})$$

Figure 3.34: Formula to compute the real value.

Knowledge ABC IEEE-754

In the 1960s and 1970s, computer manufacturers use different floating-point representations in a wide variety of computers. It was extremely inconvenient to exchange data and cooperate without a universal standard. To solve this problem, a floating-point number working group in the Institute of Electrical and Electronics Engineers (IEEE) started to work on a standard for floating-point numbers in the late 1970s. In 1980, Intel announced Intel 8087, a floating-point coprocessor with advanced and reasonable floating-point representations and operations. Its floating-point arithmetic was later adopted by IEEE as the standard and published in 1985. In fact, it had already been adopted by various computer manufacturers in the early 1980s and had become a de-facto industry standard.

Floating-point numbers in IEEE-754 consist of three fields: a sign bit on the left, a biased exponent, and a fraction on the right.

Example 3.7 Storage of real numbers

Convert −12 and 0.25 to 32-bit floating-point numbers.

[Analysis]

The conversion process is shown in Figure 3.35.

12 is 1100 in binary, which can be normalized as 1.1 * 2^3, so the exponent is 3. As it is negative, the sign bit is 1. The biased exponent, whose binary form is shown in the "biased exponent" column in Figure 3.35, is calculated by adding exponent 3 and exponent bias 127.

Decimal	Normalization	Exponent	Sign	Biased exponent (exponent + bias)	Fraction
-12.0	-1.1x23	3	1	10000010	1000000 00000000 00000000
0.25	1.0x2-2	-2	0	01111101	0000000 00000000 00000000

$$(12)_{10} \rightarrow (1100)_2 \rightarrow 1.1*2^3$$

Figure 3.35: Floating-point representation of real numbers.

To obtain the fraction, we omit the 1 left to the decimal point in the normalized number and then pad the rest with 0.

Similarly, we can write out the 32-bit floating-point representation for 0.25.

Comparing the floating-point representation of −12 and complement representation of −12 as shown in Figure 3.36, we may conclude that even integers and real numbers of the same value are stored as completely different values inside computers.

Binary representation of +12	0000,0000,0000,0000,0000,0000,0000,1100
Bitwise not	1111,1111,1111,1111,1111,1111,1111,0011
Add 1	1111,1111,1111,1111,1111,1111,1111,0100

	Representation in memory
Integer -12	1111,1111,1111,1111,1111,1111,1111,0100
Real number -12.0	1100,0001,0100,0000,0000,0000,0000,0000

Figure 3.36: Comparison of integer and real number storage.

Conclusion Storage rules of data

Integers and real numbers have different rules of storage. Even the same number can have different values when saved as integer and as real number. When some data are stored as a certain type, we should never use them as another type, unless we know the essence of these data.

Example 3.8 Binary form of 123.456

Analyse the floating-point representation of 123.456.

[Analysis]

The 32-bit floating-point representation of 123.456 can be obtained after normalization and computing biased exponent as shown in Figure 3.37. If we use the real value formula to convert it back to a decimal number, we will find an extra 1 at the end. This is due to the display format of floating-point numbers in programs.

Decimal	Normalization	Exponent	Sign	Biased exponent (exponent + bias)	Fraction
				Representation of 123.456 in memory (32 bits)	
123.456	1.111011 01110100101111001x2⁶	6	0	1000,0101	1110110,11101001,01111001

Convert binary representation of 123.456 into decimal

$[(-1)^\wedge sign] \times [1.Fraction] \times (2^\wedge[Biased\ exponent\ -127\ ])$
$=[(-1)^\wedge 0]*[1.\ 1110110,11101001,01111001]*2^\wedge[1000,0101-0111,1111]$
$=1.\ 1110110,11101001,01111001*2^\wedge 6$
$=1.92900002002716*64$
$=123.456001\textbf{281738}$

Figure 3.37: Floating-point representation of 123.456.

3.4.3 Display precision and range of floating-point numbers

The biased exponent indicates the location of the decimal point in the data and determines the range of floating-point numbers as shown in Figure 3.38. The range of the biased exponent of float type is −127 to +128; therefore, float can represent numbers from -2^{128} to $+2^{128}$.

	Sign bit s	Biased exponent	Fraction m
Number of bits	1	8	23

Use 32-bit float type as an example

	Number of bits	Range	Equivalent range in decimal	Notes
Exponent	8	$-2^{8-1} \sim 2^{8-1}-1$	$-128 \sim 127$	Signed number
Range of float		$-2^{128} \sim +2^{128}$	$-3.40*10^{38} \sim +3.40*10^{38}$	
Fraction	23			Unsigned number
Precision of float		2^{23}	8388608 (7 digits)	At most 7 significant figures

Figure 3.38: Display precision and range of floating-point numbers.

The number of bits in fraction determines the precision of float; 2 to the 23rd power has seven digits when converted to decimal. This means that there are at most seven significant figures, but only the first six are guaranteed to be correct. In other words, the precision of float type is six or seven significant figures.

Similarly, the precision of double type is at most 16 digits.

In conclusion, as shown in Figure 3.39, both display and storage of decimal real numbers have their own set of rules, which we need to understand. It is particularly worth noting that we should avoid checking whether two real numbers are equal as the result may be unexpected.

Storage of decimal real numbers
The system converts them to binary form according to international standard and stores them.

Display of decimal real numbers
The system converts binary form stored in machines to decimal form according to international standard, and then displays in a precision defined by users.

Comparison rule of real numbers
Avoid checking whether two real numbers are equal.

Figure 3.39: Various rules of real numbers.

3.5 Operators and expressions

In the section of algorithms, we have seen problems like scoring, price guessing, and things whose number is unknown. To solve them, we used operations such as addition, subtraction, multiplication, division, comparison, and combination of multiple comparisons as shown in Figure 3.40.

Problem	Data Processing	Operations involved
Scoring by judges	Discard highest and lowest score	Comparison
	Compute average	Addition, division
Guessing price	Guess is higher, lower or equal	Comparison
Things whose number is unknown	Remains 2 when divided by 3, remains 3 when divided by 5, remains 2 when divided by 7	Division (compute remainder), check multiple conditions simultaneously

Figure 3.40: Operations used in data processing.

These can be categorized as the three most important types of operations in C as shown in Figure 3.41.

Type of operation	Major cases	Class	Problems involved
Addition, division, etc.	Addition, subtraction, multiplication, division, compute remainder	Arithmetic operation	• Operators and precedence of operators
Data comparison	Larger than, smaller than, equal to, not equal to	Relational operation	• Associativity of operators and data
Check multiple conditions simultaneously	All conditions hold, not all conditions hold, none of the conditions holds	Logical operation	• Retrieving rule of operation result

3 most common
operations in C

Figure 3.41: Categorization of operations used in data processing.

3.5.1 Operators

Operators of C are shown in Figure 3.42, in which the first four are used more frequently. Their usage will be introduced later.

Type	Operators	Use case
Arithmetic	+ - * / % ++ -- + -	numerical computation
Assignment	= and its extensions	retrieve computation result
Relational	> < >= <= == !=	compare data
Logical	&& || !	check multiple conditions simultaneously
Bitwise	& | ^ ~ << >>	binary number computation
Conditional	? :	easier comparison of data
Comma	,	list multiple expressions
Other	& sizeof	obtain address, size of storage unit

Figure 3.42: Operators in C.

3.5.2 Expressions

Connecting operators and objects to be operated (or operands) following syntax rules, we get statements that are called expressions in C as shown in Figure 3.43. Depending on the operators used, there are various kinds of expressions such as arithmetic expressions and assignment expressions.

Expression

An expression is a statement that connects operands using operators following C grammar rules

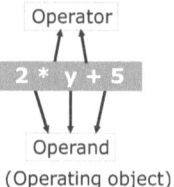

Operator

2 * y + 5

Operand
(Operating object)

Figure 3.43: Expressions.

3.5.3 Precedence of operators

As an expression may contain multiple operators, the order of execution can affect the result. This is why we need to determine which operation should be executed first when there is more than one of them. We call this order precedence of operators as shown in Figure 3.44.

Precedence of operators

The order of evaluation of different operators in an expression.

Operator	Description	Associativity
( )	Parentheses	From left to right
!, ++, --, **sizeof**	NOT, increment, decrement, compute size of type	From right to left
*, /, %	Multiplication, division, remainder	From left to right
+, -	Addition, subtraction	From left to right
<, <=, >, >=	Less than, less than or equal to, greater than, greater than or equal to	From left to right
= =, !=	Equal to, not equal to	From left to right
&&	AND	From left to right
||	OR	From left to right
=,+=, *=, /=, %= ,- =	Assignment operator and compound assignment operators	From right to left

Figure 3.44: Precedence and associativity of operators.

Operators are listed top to bottom in descending precedence, where operators with the highest precedence are listed on the top and operators with the same level of precedence are on the same row.

The last column indicates the associativity of operators, namely which operation gets executed first when given operations have the same level of precedence.

Normally, we don't have to recite the precedence in C, but we need to keep in mind that parentheses have the highest precedence so we can use them to override the precedence of operators as shown in Figure 3.45.

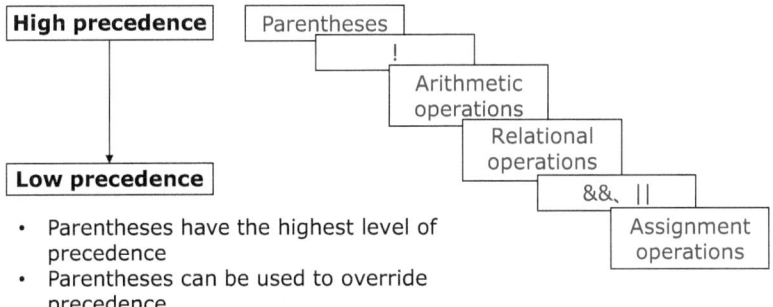

- Parentheses have the highest level of precedence
- Parentheses can be used to override precedence

Figure 3.45: A summary of precedence.

Good habit in programming
When using operators, we need to take good care of precedence. We should use parentheses to determine precedence and avoid using default precedence. This keeps us from misunderstanding a program when reading it and from making mistakes in our own programs by unintentionally using default precedence, which deviates from our design.

3.5.4 Associativity of operators

Associativity defines the order in which operators of the same precedence are evaluated in an expression as shown in Figure 3.46. Take 10/5*2 as an example. The result will be different when we evaluate from left to right and from right to left.

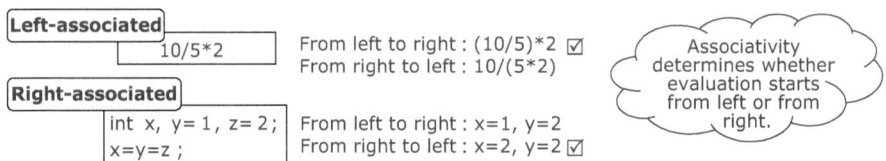

Associativity of operators
- The direction that operators are associated is called associativity.
- When there are multiple operators with the same level of precedence in one expression, the order of evaluation is determined by associativity.

Left-associated
10/5*2
From left to right : (10/5)*2 ☑
From right to left : 10/(5*2)

Right-associated
int x, y= 1, z= 2;
x=y=z ;
From left to right : x=1, y=2
From right to left : x=2, y=2 ☑

Associativity determines whether evaluation starts from left or from right.

Figure 3.46: Associativity of operators.

Arithmetic operators are associated from left to right, that is, operators on the left are evaluated first, which is the order we are familiar with.

In the case of expression x = y = z, if we assign from left to right, then we assign y to x first and then z to y, which yields x = 1 and y = 2. In contrast, if we evaluate from right to left, then we assign z to y first and then y to x, which yields x = 2 and y = 2.

Which one do we choose then, left to right or right to left? People have determined that the associativity of the assignment operator is from right to left, so the second assignment should be executed first. Rules of precedence and associativity are summarized in Figure 3.47.

Rules of precedence and associativity of operators
- Precedence and associativity determine order of execution of operators in an expression.
- Operators are first executed in the order of precedence; operators with the same level of precedence are executed in the order determined by associativity.
- Left-associated means operators on the left are executed first, while right-associated means operators on the right are executed first.

Figure 3.47: Precedence and associativity.

3.6 Numerical operations

When shopping, we need to execute all kinds of operations on prices and quantities to get the result. To do numerical computations in C, we have to define the representation and rules of common mathematical entities, including arithmetic operators, numbers, and character operands as shown in Figure 3.48.

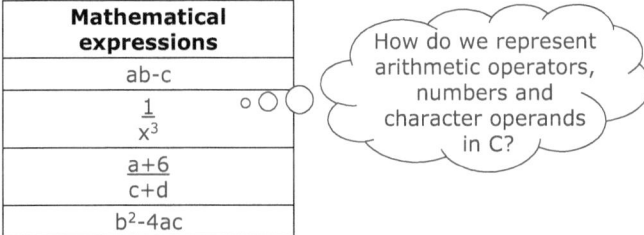

Figure 3.48: Operations and their representations.

3.6.1 Arithmetic operators and expressions

3.6.1.1 Arithmetic operators and expressions

An arithmetic operation involves arithmetic operators and expressions as shown in Figure 3.49. An arithmetic expression is a statement that connects operands using arithmetic operators. Addition, subtraction, multiplication, and division operations are familiar to us, whereas operators of multiplication and division are represented by "*" and "/" in C due to limitation of keyboards. It is worth noting that division of integers yields integer as well. We can use this rule to simplify algorithms in many cases.

Arithmetic expression		
An arithmetic expression is a statement that connects operands using arithmetic operators		
Operator	Meaning	Notes
+	Addition operator or positive sign	
-	Subtraction operator or negative sign	
*	Multiplication operator	
/	Division operator	Division of integers yields integer, the fraction part is discarded
%	Remainder operator	Remainder operation requires operands to be integers

Figure 3.49: Arithmetic operators and expressions.

In addition, C defines a remainder operation, which is used to calculate the remainder in integer division. Essentially, division is merely repeated subtraction. We repeat subtraction until the dividend is smaller than the divisor and the remaining part is called the remainder. Using the remainder operation can simplify algorithms as well.

However, writing arithmetic expressions in C is different from writing them in mathematics as shown in Figure 3.50. In particular, the multiplication operator cannot be omitted and we can use parentheses to override precedence.

Mathematical expressions	Expressions	Notes
ab-c	a*b-c	
$\frac{1}{x^3}$	1/(x*x*x)	• Multiplication operator can't be omitted
$\frac{a+6}{c+d}$	(a+6)/(c+d)	• Use parentheses to override precedence
b^2-4ac=0	b*b-4*a*c	

Figure 3.50: Mathematical expressions and C expressions.

Example 3.9 Example of arithmetic operations
Define two integers a and b, with initial values 7 and 3, respectively. Output sum, difference, product, quotient, remainder, and mean of them onto the screen.

[Analysis]
The program and output are shown in Figure 3.51. We can write expressions inside formatted output function printf() to output results.

```
#include<stdio.h>
int main(void)
{
    int a=7;
    int b=3;
    printf("%d ",a+b);      // Compute and output sum of a and b
    printf("%d ",a-b);      // Compute and output difference of a and b
    printf("%d ",a*b);      // Compute and output product of a and b
    printf("%d ",a/b);      // Compute and output quotient of a divided by b
    printf("%d ",a%b);      // Compute and output remainder of a divided by b
    printf("%d ",(a+b)/2);  // Compute and output average of a and b
    return 0;
}
```

Output : 10 4 21 2 1 5

We can't tell which operation yields which number.

Figure 3.51: Example of arithmetic operations.

The output is a sequence of integers. Without the program, we won't be able to tell which is the sum or the product.

To get clearer results, we can add texts in the output function as shown in Figure 3.52. Note that the integer division on line 10, which is 7 / 3, yields 2. Use of printf() function is covered in detail in the chapter "Input/Output."

```
01 #include<stdio.h>
02 int main(void)
03 {
04     int a=7;
05     int b=3;
06
07     printf("a+b=%d\n",a+b); // Compute and output sum of a and b
08     printf("a-b=%d\n",a-b);  // Compute and output difference of a and b
09     printf("a*b=%d\n",a*b);  // Compute and output product of a and b
10     printf("a/b=%d\n",a/b);  // Compute and output quotient of a divided by b
11     printf("a%%b=%d\n",a%b); // Compute and output remainder of a divided by b
12     printf("average is %d\n",(a+b)/2);  // output average
13     return 0;
14 }
```

Improve output format

```
Result :
a+b=10
a-b=4
a*b=21
a/b=2
a%b=1
average is 5
```

Figure 3.52: Refinement of the example program.

Example 3.10 Time conversion

Convert input time in seconds into minutes and seconds, for example, 500 s is 8 min and 20 s.

[Analysis]

We are asked to write a program that output corresponding minute (variable minute) and second (variable second) given time in seconds (variable time).

If time = 500, then minute = 8 and second = 20. The program is as follows.

```
01   #include<stdio.h>
02   int main(void)
03   {
04     int time; // Define an input variable
05     int minute, second;
06     printf("Please input a time in integer seconds"); // Screen prompt
07     scanf("%d",&time); // Obtain the time input
08     minute = time/60;    // Calculate minute
09     second = time%60;    // Calculate remaining seconds
10     printf("%d minutes %d seconds", minute,second);
11     return 0;
12   }
```

On line 6, we use printf() function to prompt users to input time value.

On line 7, we use scanf() function to obtain users' keyboard input and store it into variable time.

On line 8, we use integer division to calculate the number of minutes in time.

On line 9, we use the remainder operation to calculate the remaining seconds.

Finally, we output the desired result.

3.6.1.2 Increment and decrement operation

When programming, we often need to write "i = i + 1" or "i = i – 1". C provides short-hands for these two operations as shown in Figure 3.53.

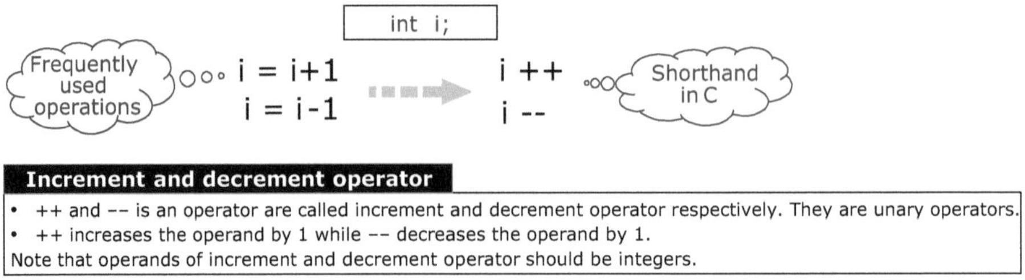

Increment and decrement operator

- ++ and -- is an operator are called increment and decrement operator respectively. They are unary operators.
- ++ increases the operand by 1 while -- decreases the operand by 1.

Note that operands of increment and decrement operator should be integers.

Figure 3.53: Increment and decrement operator.

The ++ operator and –– operator are called increment operator and decrement operator, respectively. They are unary operators. ++ operator adds 1 to its operand, whereas – operator subtracts 1 from its operand. Note that their operands must be integers.

Increment and decrement are not something we have experienced, so we need to be careful and follow the rules when using them. Some examples are shown in Figure 3.54 to help readers become more comfortable with these two operators.

int x, y;

Operator	Example	Meaning	Equivalent statement
++	y = ++x	Increase x by 1, then assign x to y	++x; y=x;
	y = x++	Assign x to y, then increase x by 1	y=x; x++;
--	y = --x	Decrease x by 1, then assign x to y	--x; y=x;
	y = x--	Assign x to y, then decrease x by 1	y=x; x-- ;

Figure 3.54: Increment and decrement example 1.

In the first row, y = ++x will increment x first and then assign x to y. It is equivalent to ++x; y = x.

In the second row, y = x++ means assigning and incrementing, which is equivalent to y = x; x++. Examples of –– operators work in the same way.

Figure 3.55 shows two programs that differ in only one line. The program on the left does increment before assignment while the other does assignment before the increment. This difference leads to different outputs. If the operand of increment or decrement is going to be accessed by other objects in the same statement, we need to determine where to put these operators carefully.

```
int main(void)
{
    int x, y ;
    x=10 ;
    y=++x ;
    printf("%d, %d \n", x, y) ;
    return 0 ;
}
The result is
11 , 11
```

```
int main(void)
{
    int x,  y ;
    x=10 ;
    y=x++ ;
    printf("%d, %d \n", x, y) ;
    return 0 ;
}
The result is
11 , 10
```

Conclusion 1

If the operand of increment or decrement is going to be accessed by other objects in the same statement, we need to carefully determine where to put these operators.

Figure 3.55: Increment and decrement example 2.

We can make outputs of these programs identical by modifying the lines that differ as shown in Figure 3.56.

```
int main(void )
{
    int x, y;
    x=10;
    ++ x ;
    y=x ;
    printf("%d, %d \n", x, y);
    return 0;
}

Result:
11 , 11
```

```
int main(void )
{
    int x, y;
    x=10;
    x++ ;
    y=x ;
    printf("%d, %d \n", x, y);
    return 0;
}

Result:
11 , 11
```

Conclusion 2

If the operand of increment or decrement forms a statement on its own, it doesn't matter where we put the operator.

Figure 3.56: Increment and decrement example 3.

As long as a statement consists of only the ++ operator and its operand (say variable x), the operator simply adds 1 to x, regardless of its position relative to x. In other words, it doesn't matter where we put increment or decrement operators if the object forms a statement on its own.

In C programming, we should use increment and decrement operators with caution. In particular, pay attention to the following issues:

(1) A statement with too many increment or decrement operations is hardly read-able. One reason for writing incomprehensible code is probably that the gener-ated machine code is more efficient. Nevertheless, unreadable programs will only decrease programmers' efficiency.
(2) Different compilers generate different results. This essentially prohibits us from using too many increments or decrements in one statement.
(3) Doing so prevents us from debugging our program.

When a debugger is running, the minimum "execution step" is one line. If multiple statements exist in a line, they are still executed in one "step." Take x = a++*a++*a ++ as an example, the debugger will execute all four statements at once when we trace the program step by step. As a result, it is hard to examine how the value of a changes, which is exactly our purpose of debugging. If we can't do this, then we are denied the chance of debugging.

Good habit in programming
It is recommended to include at most one increment or decrement in a line.

3.6.2 Overflow problems in data operations

Every type has a value range. If operation on a variable yields values outside the range, we won't obtain the desired result.

Example 3.11 A problem of using unsigned number
Suppose we have the following variable:

```
unsigned char size;
```

What is the value of variable size after decrement if it initially has value 0?
Answer: As size is unsigned char, its value can't be negative. In this case, it will become 0xFF.

Example 3.12 A problem of using character
Character type has value range −128 to 127 in C, so the following computation is risky:

```
char chr = 127;
int sum = 200;
chr +=1;
// 127 is the maximum value of char,
// so adding 1 causes overflow and yields -128 instead of 128
sum += chr;  // As a result sum becomes 328 instead of 72
```

Preventing Program Errors
Pay attention to edge values when using variables.

3.7 Logical operations

3.7.1 Relational operations

In the price-guessing game, after a participant makes a guess, the host responds with "too high," "too low," or "exactly." The relation between the guess and actual price can be obtained by comparison, which is one of "larger than," "smaller than," and "equal to." Such a comparison yields either true or false.

Let the participant's guess be value and the actual price be ¥1680. We can write the following relational expressions in C as shown in Figure 3.57. The results of these comparisons can be represented by non-zero and zero.

Case study

Price guessing game

	Possible case	Representation	Possible result	Representation
Comparison of value and 1680	value is greater than 1680	value>1680	Met	Non-zero value
	value is less than 1680	value<1680	Not met	0
	value is equal to 1680	value==1680		

Figure 3.57: Relational operation in the price-guessing game.

"Relational operations" are in fact "comparison operations," which compare two data and determine whether certain relation holds between them. Relational expressions are obtained by connecting operands with relational operators as shown in Figure 3.58.

Relational operation

A relational operation compares two numbers and determines whether they satisfy a given condition. If the condition is met, the operation yields true, which is represented by a non-zero value; otherwise it yields false, which is represented by zero.

Relational expression

A relational expression is a statement that connects two operands (constants, variables or expressions) using relational operators and execute relational operations.

Figure 3.58: Relational operations and relational expressions.

Relational operators in C are shown in Figure 3.59. Some of them deviate from their mathematical representations due to keyboard limitations. Readers should also

Relational operator	Meaning
>	Greater than
>=	Greater than or equal to
<	Less than
<=	Less than or equal to
==	Equal to
!=	Not equal to

Result of relational operation	
0	Non-zero
False condition is not met	True, condition is met

Program error

Relational operator "==" is different from assignment operator "=", they can't be used interchangeably.

Figure 3.59: Relational operators.

keep in mind that == operator, which is a relational operator, should not be confused with = operator, which is an assignment operator.

Relational operations can yield two possible results: true if the relation holds; false otherwise.

There is no Boolean value in C, instead, we use nonzero value to represent "true" and zero for "false." Hence, if a relation expression evaluates to zero, it represents "false"; if it evaluates to a nonzero value, it represents "true" regardless of the sign of that value.

Example 3.13 Determine whether two real numbers are equal
Figure 3.60 shows the code and result of a program that checks whether two real numbers are equal.

```
int main(void)
{
    float x;
    char k;
    x=1.0/10;
    if (x==0.1) k='y';
    else k='n';
    printf( " k=%c,x=%f \n" , k,x);
    return 0;
}
```

Why the result is like this?

Result:k=n,
x=0.100000

Figure 3.60: Determine whether two real numbers are equal.

Is the result unexpected? The reason this happens is that float-type variables have limited precision in floating-point number storage format (IEEE-754).

If we have to compare x with 0.1, we can use expression $(x \geq 0.1 - \varepsilon)$ && $(x \leq 0.1 + \varepsilon)$, where ε is the error bound. We can use $\varepsilon = 10^{-5}$, but we cannot use a value that is too small due to limited precision of float type.

Preventing program errors
Avoid comparing real numbers or floating-point variables using "==" or "!=".

3.7.2 Logical operations

3.7.2.1 Example of Relation Problem

Let us take a look at another relation problem. There is a triangle whose edges have lengths a, b, and c, respectively. We need to classify this triangle.

We need to check relations between edges and classify the triangle based on mathematical definitions. The equilateral triangle requires $a = b$ and $a = c$, so we need to examine two conditions at the same time as shown in Figure 3.61.

> We need to compare multiple data relations, do logical reasoning and determine whether conditions are metTriangle

Case study

Classification of Triangles

triangle	Condition	Analysis
Equilateral triangle	a==b and a==c	Determine whether both conditions are met
Isosceles triangle	a=b or a=c or b=c	Determine whether at least one condition is met
Ordinary triangle	a+b>c and a+c>b and b+c>a	Determine whether all conditions are met
Non-triangle	Doesn't satisfy conditions of ordinary triangle	Determine the opposite of a condition

Figure 3.61: Classification of triangles.

In summary, we need to determine whether multiple conditions are met at the same time or whether one of the conditions is met in practical problems. Sometimes, we need to consider the opposite of the conditions we have. Essentially, we need to compare multiple data relations, do logical reasoning, and determine whether conditions are met.

3.7.2.2 Definition of logical operations

Definition of logical operations and logical expressions are shown in Figure 3.62.

Logical operation

- A logical operation connects one or more conditions using logical operators and determine whether these conditions are met.
- A logical expression evaluates to a Boolean value (true of false).

Logical expression

- A logical expression is a statement that connects one or more expressions using logical operators and executes logical operations. We use logical expressions to express combination of multiple conditions in C.

Figure 3.62: Logical operations and logical expressions.

There are three logical operators, namely AND, OR, and NOT, whose usages are shown in Figure 3.63.

Name	Operator	Operation rule
AND	&&	It yields true if value of both operands are true and yields false otherwise.
OR	||	It yields false if value of both operands are false and yields true otherwise.
NOT	!	It yields false if value of operand is true and yields true if value of operand is false.

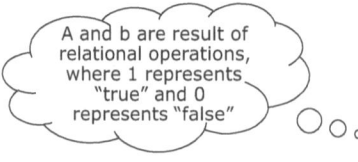

	Operand		Result of logical operation		
	a	b	a && b	a || b	!a
Boolean value	0	0	0	0	1
	0	1	0	1	1
	1	0	0	1	0
	1	1	1	1	0

Memorizing tip: AND yields false if one operand is false, OR yields true if one operand is true, NOT flips its operand.

Figure 3.63: Logical operators and their usage.

(1) AND

AND yields true if two operands are both true, and false otherwise. Suppose the operands are a and b, which are results of relational operations and are either true (1) or false (0). If a and b are both true (or 1), a AND b yields true; if one of them is false, the result is false. AND is similar to multiplication in their way of working.

(2) OR

OR yields false if two operands are both false, and true otherwise. OR is similar to addition in their way of working.

(3) NOT

NOT yields false if the operand is true, and true if the operand is false.

We can remember rules of logical operators easily using this tip: AND yields false if one operand is false, OR yields true if one operand is true, NOT flips its operand.

3.7.2.3 Examples of Logical Operations

Example 3.14 Classify a triangle
There is a triangle whose edges have lengths a, b, and c, respectively. Use logical and conditional expressions in C to describe conditions for different types of triangles.

[Analysis]
The conditions and corresponding C expressions are shown in Figure 3.64.
In the first row, an equilateral triangle requires a == b and a == c, which can be represented by a == b && a == c.

In the last row for nontriangle, the ! sign is used for NOT, which yields the complement of its operand and is often called "logical complement" as well.

triangle	Condition	C expression
Equilateral triangle	a==b AND a==c	a==b && a==c
Isosceles triangle	a=b OR a=c OR b=c	a==b || a==c || b==c
Ordinary triangle	a+b>c AND a+c>b AND b+c>a	a+b>c && a+c>b && b+c>a
Non-triangle	Doesn't satisfy conditions of ordinary triangle	! (a+b>c && a+c>b && b+c>a)

Figure 3.64: Conditions of triangles and their C expressions.

Example 3.15 Classify a character
Given a character from keyboard input, store it into character variable c and determine whether it is a number, an uppercase letter or a lowercase letter.

[Analysis]
According to the ASCII table, if the character is a number, c should be between character 0 and character 9, which can be written as conditional expressions shown in Figure 3.65. Note that $c \geq$ '0' and $c \leq$ '9' are two relational expressions. Only when both of them are met, namely they both yield true, can c be a number. Similarly, we can write out expressions for other cases.

Class of character c	Condition	Expression
Number	c is between character 0 and 9	c>='0'&& c<='9'
Not a number	c is not between character 0 and 9	! (c>='0'&& c<='9')
Upper case letter	c is not between character A and Z	c>='A'&& c<='Z'
Lower case letter	c is between character A and Z	c>='a'&& c<='b'

Figure 3.65: Classify a character.

Example 3.16 Determine result of expression
Let int x = 1, y = 1, z = 1. What are the values of x, y and z after executing operation ++x \|\|++y && ++z?

[Analysis]
There are operators of more than one type with different precedences. We should refer to a manual if we cannot recall their precedence. A better way is to use parentheses to express logical relations described in the problem when programming.

First, we evaluate the expression to the left of OR, namely ++x. The result is 2 and will be used as one operand of OR. In C, nonzero values are treated as "true," represented by "TRUE" in the figure. As OR yields true if one of the operands is true, we no longer need to evaluate the expression to its right.

Similarly, the && operator yields false, if one of the operands is false. Omitting evaluation like this is often called a side effect of logical operators. Thus, we need to use them carefully when programming.

As a result, only increment of x is executed, whereas increment of y and z are skipped. In fact, it is not recommended to use multiple increment or decrement in one statement as different compilers may explain it in different ways and yield different results.

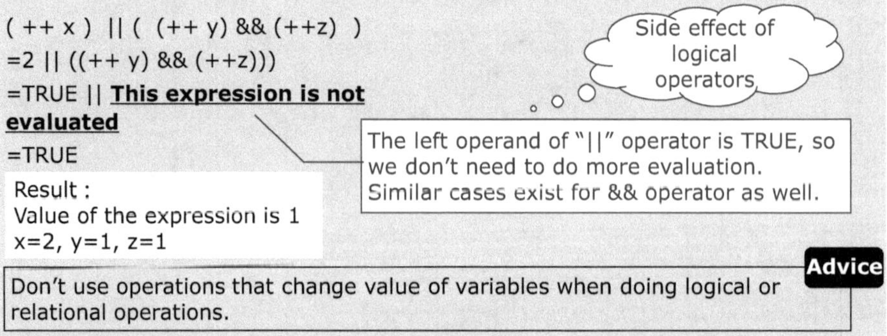

(++ x) || ((++ y) && (++z))

=2 || ((++ y) && (++z)))

=TRUE || **This expression is not evaluated**

=TRUE

Side effect of logical operators

The left operand of "||" operator is TRUE, so we don't need to do more evaluation. Similar cases exist for && operator as well.

Result :
Value of the expression is 1
x=2, y=1, z=1

Advice

Don't use operations that change value of variables when doing logical or relational operations.

Figure 3.66: Determine the result of an expression.

Example 3.17 A detective story

The police are questioning four theft suspects. They already know one of the suspects is the real thief and every suspect either lies or tells the truth. Their answers to the policeman's question are as follows. Please determine who committed the crime.

A says, "B didn't do it, it was D."
B says, "I didn't do it, it was C."
C says, "A didn't do it, it was B."
D says, "I didn't do it."

[Analysis]

Let variables A, B, C, and D stand for these suspects. The value of each variable is either 0 or 1, where 1 means this person is the thief and 0 means otherwise.

We know from the problem description that only one of them is the thief and each of them either lies or tells the truth. We also notice that A, B and C use the same pattern "X didn't do it, it was Y." Hence, regardless of them being honest or not, one of the two people they mentioned must have committed the crime. Consequently, we can write the following expressions without knowing who is being honest and who is not:

"B didn't do it, it was D" can be represented by $B + D = 1$.
"B didn't do it, it was C" can be represented by $B + C = 1$.
"A didn't do it, it was B" can be represented by $A + B = 1$.

We can't determine whether "I didn't do it" is true or not, so we can write it as $A + B + C + D = 1$. One of them is the thief is then equivalent to $(B + D == 1)\&\&(B + C == 1)\&\&(A + B == 1)\&\&(A + B + C + D == 1)$.

3.7.2.4 Rules of logical operations

Having seen these examples, we can summarize rules of logical operations as shown in Figure 3.67.

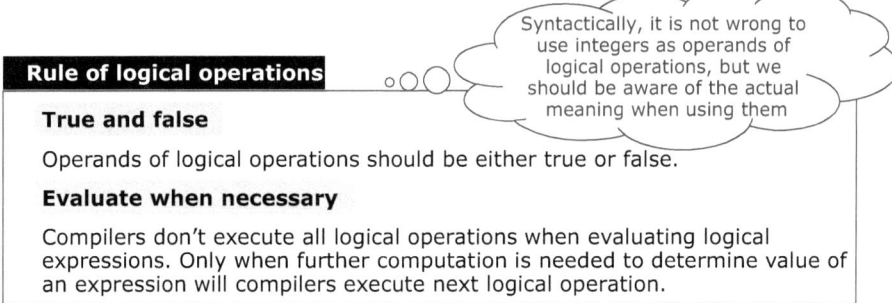

Rule of logical operations

True and false

Operands of logical operations should be either true or false.

Evaluate when necessary

Compilers don't execute all logical operations when evaluating logical expressions. Only when further computation is needed to determine value of an expression will compilers execute next logical operation.

Figure 3.67: Rules of logical operations.

3.8 Type conversion

3.8.1 Computation of data of mixed types in real life

Case Study 1 Computing total on shopping receipt

Looking at the shopping receipt on the table as given in Figure 3.68, Mr. Brown had a question. The total price of each item on the receipt is calculated as price per unit times quantity. The price is a real number, whereas quantity is an integer. In this particular problem, it is only reasonable if the result is a real number as well;

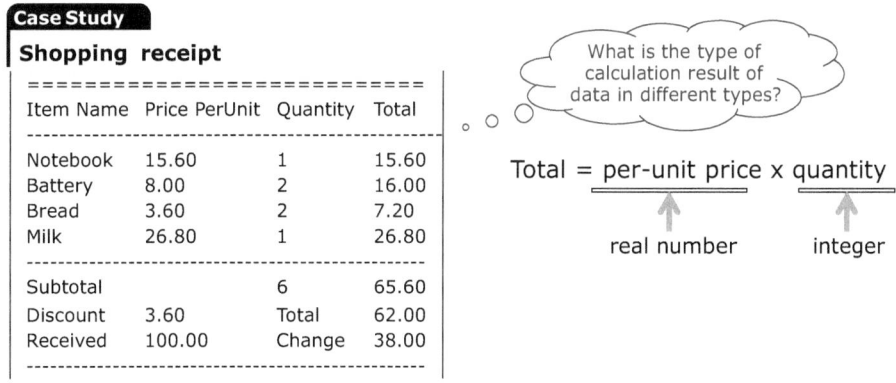

Figure 3.68: Type conversion in shopping receipt.

therefore, a type conversion has happened in our brain naturally. A more general question for computers is: What is the type of the computation result of data of mixed types and how type conversion should be carried out?

Numerical operations have the following rule in C: when assigning result to a variable, the result should be converted to the same type as the variable.

Case Study 2 Computing error of material fee

Mrs. Brown went to a handcraft workshop to learn ceramic art. All students needed to pay for raw materials. There were two types of materials, namely A and B, which were mixed in a 2:1 ratio. Twelve people came to the workshop. They used 18 bags of A and nine bags of B. It is known that A costs ¥32.6 per bag and B costs ¥15.8 per bag. What was the average material fee for each student?

The manual computation result by Mrs. Brown was ¥60.9. To examine rules of mixed-type data computation in computers, Mr. Brown designed a program, which is shown in Figure 3.69.

Case Study

Error in computation of material fee

```
01 #include "stdio.h"
02 #define priceA 32.8
03 #define priceB 15.6
04 int main(void)
05 {
06    intnumA=18, numB=9; //Quantity of material
07    float cost;              //Cost of material
08    cost=(numA/12)*priceA+(numB/12)*priceB;
09    printf("%f", cost);
10    return 0;
11 }
```

Manual computation result :
cost=(18/12)*32.8+(9/12)*15.6
=60.9

Program result : 32.6

The program result is wrong, what is wrong in the program?

Figure 3.69: Computation of material fee.

The calculation formula on line 8 was exactly what Mrs. Brown used. However, the program output 32.6, which was far from the manual result. Why was this the case?

After careful observation, we can see that numA and numB are both integers. We have learned that integer divided by integer, 12 in this case, yields another integer in C. This is why the result was incorrect.

We can change the way of computation to solve this problem without changing existing computation rules. In other words, we only change the type of data as needed. In particular, we can use a real number identifier on integer numA so that it is used as a real number in computation.

3.8.2 Type conversion rules in C

Based on the method we just mentioned, a type conversion grammar was designed in C. It can be used to convert data to another type during computation as shown in Figure 3.70.

Data type conversion

Type conversion converts value of data from one type to another.

Figure 3.70: Data-type conversion.

In programming, type conversion can happen in numerical computation, assignment, output, and function call. C has conversion rules for each case as shown in Figure 3.71. We need to master them through practices. The concept of function is introduced in the corresponding chapter.

Type	Happens when	Processing rule
Operation	Data of different types are computed together	Convert then compute
Assignment	A value is assigned to a variable of different type	Convert to target type
Output	Result needs to be output in certain formats	Output in required formats
Function call	Parameters and arguments are of different types	Use parameter types
	Return value and function are of different types	Use function type

Figure 3.71: Different cases of type conversion.

There are two kinds of type conversion as shown in Figure 3.72. One of them is done by the system automatically, which is called "automatic conversion" or "implicit conversion." The other is done by us programmers manually, which is called "forced conversion" or "explicit conversion."

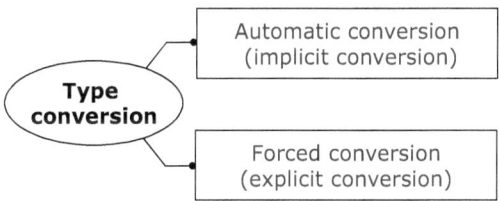

Figure 3.72: Two ways of type conversion.

3.8.3 Forced-type conversion

Forced-type conversion is explicit, which converts the type of an expression to the desired type. The format is shown in Figure 3.73, where we write the type we want in front of the expression to be converted. Explicit means we explicitly write out the type we need. It is worth noting that forced-type conversion changes the type of value instead of the type of the original variable or expression.

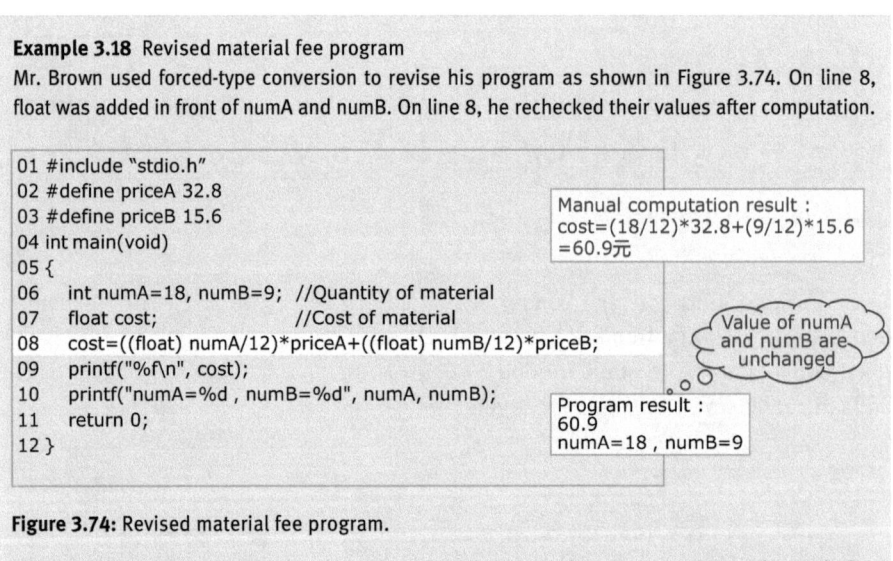

Forced type conversion

Forced type conversions are explicit type conversions that convert type of expressions into a desired one.

Format of force type conversion

(Identifier of desired type) expression

Forced type conversions yield value of desired type. They don't change type of original variables or expressions.

Figure 3.73: Forced-type conversion.

Example 3.18 Revised material fee program
Mr. Brown used forced-type conversion to revise his program as shown in Figure 3.74. On line 8, float was added in front of numA and numB. On line 8, he rechecked their values after computation.

```
01 #include "stdio.h"
02 #define priceA 32.8
03 #define priceB 15.6
04 int main(void)
05 {
06    int numA=18, numB=9;  //Quantity of material
07    float cost;              //Cost of material
08    cost=((float) numA/12)*priceA+((float) numB/12)*priceB;
09    printf("%f\n", cost);
10    printf("numA=%d , numB=%d", numA, numB);
11    return 0;
12 }
```

Manual computation result :
cost=(18/12)*32.8+(9/12)*15.6
=60.9元

Value of numA and numB are unchanged

Program result :
60.9
numA=18 , numB=9

Figure 3.74: Revised material fee program.

The output result was correct this time. The value of numA and numB were also unchanged after forced-type conversion.

Example 3.19 Forced-type conversion
Verify that the value of a variable is not changed after forced-type conversion.

1. Code

```
1  // Example of forced type conversion
2  #include <stdio.h>
3  int main(void)
4  {
5    float x, y;
6    x=2.3;
7    y=4.5;
8
9    printf("(int)(x)+y=%f\n",(int)(x)+y);
10   printf("(int)(x + y)=%d\n",(int)(x + y));
11   printf("x=%f,y=%f\n",x,y);
12   return 0;
13 }
```

Output:

```
     (int)(x)+y=6.500000
     (int)(x + y)=6
     x=2.300000,y=4.500000
```

Notes:
(1)   On line 9: (int)(x) + y = (int)(2.3) + 4.5 = 2 + 4.5 = 6.5.
(2)   On line 10: (int)(x + y) = (int)(2.3 + 4.5) = (int)(6.8) = 6.
(3)   Forced-type conversion changed the type of expression without changing the value of x and y.

2. Tracing and debugging
In the Watch window, as shown in Figure 3.75, we see that the program is going to execute return at this moment. Although forced-type conversion of variable x and y is done, their value is left unchanged.

```
 #include  <stdio.h>             ┌──────────────────────────────┐
 int main()                      │ Watch                     [x] │
 {                               ├──────────────────┬───────────┤
   float  x, y;                  │ Name             │ Value     │
   x=2.3;                        │   x              │ 2.30000   │
   y=4.5;                        │   y              │ 4.50000   │
                                 │   (int)x         │ 2         │
                                 │   (int)y         │ 4         │
   printf("(int)(x)+y=%f\n",(int)(x)+y);  │ (int)(x)+y  │ 6.50000 │
   printf("(int)(x + y)=%d\n",(int)(x + y))  │ (int)(x + y) │ 6  │
   printf("x=%f,y=%f\n",x,y);    │  ┌ ─ ─ ─ ─ ─ ─ ┐           │
⇨  return 0;                     │                           │
 }                               └──────────────────┴───────────┘
```

Figure 3.75: Tracing forced-type conversion program in the debugger.

Preventing program errors
Restrict the use of unnecessary type conversions.

3.8.4 Automatic-type conversion

We are going to introduce the automatic-type conversion in this section. Let us start with an example.

Example 3.20 Discount on material fee
The handcraft workshop offered a holiday discount on the material fee. In addition to a 20% discount, the price after discount was also rounded down.

[Analysis]
To be clearer, Mr. Brown used another variable d_cost for the discounted price as shown in Figure 3.76. Note that it is of int type instead of float.

Manual computation :
cost=(18/12)*32.8+(9/12)*15.6=60.9
d_cost=⌊ cost *0.8⌋=⌊48.72⌋=48

Note that d_cost is truncated instead of rounded.

Program implementation

```
int numA=18, numB=9;        //Quantity of material
float cost;                 //Cost of material
int  d_cost;                //Discounted price
cost=((float) numA/12)*priceA+((float) numB/12)*priceB;
d_cost= cost*0.8;
printf("cost=%f\n", cost);
printf("d_cost=%f\n", d_cost);
```

```
Program result
cost=60.9
d_cost=48
```

Figure 3.76: Discount on material fee.

The manual computation result was cost = 60.9. The value of d_cost was cost times 0.8 and rounded down, which was 48.

The program also output d_cost = 48. Note that d_cost was int type and that the result was truncated instead of rounded.

Automatic-type conversion is done at compile time by the compiler following a set of rules without human interference. It is used in arithmetic operations, assignments, function call, and so on. The most important rule is that the type of value on the right-hand side of = sign is automatically and implicitly converted to the type of the variable on the left-hand side during an assignment (Figure 3.77).

Automatic conversion
Automatic type conversion is done at compile time by the compiler following a set of rules without human interference.

Rules of automatic conversion	
Arithmetic operations	• Convert all data into the longest type in the expression. • Convert then compute
Assignment operations	The type of value on the right-hand side of = sign is automatically and implicitly converted to the type of variable on the left-hand side;
Function call	(a) Arguments are converted into types of parameters. (b) Return value is converted into the type of function.

Figure 3.77: Automatic conversion and its rules.

3.9 Other operations

3.9.1 Conditional expressions

There is a more straightforward yet equivalent representation of if-else statements in C as shown in Figure 3.78.

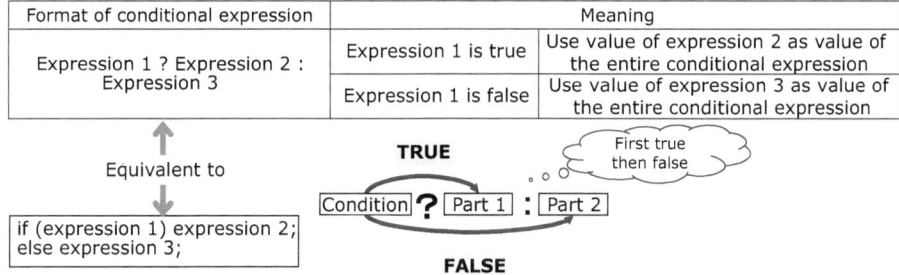

Format of conditional expression	Meaning	
Expression 1 ? Expression 2 : Expression 3	Expression 1 is true	Use value of expression 2 as value of the entire conditional expression
	Expression 1 is false	Use value of expression 3 as value of the entire conditional expression

Equivalent to

if (expression 1) expression 2;
else expression 3;

TRUE

First true then false

Condition ? Part 1 : Part 2

FALSE

Figure 3.78: Conditional operator and expression.

In practice, we often use conditional expressions in assignment statements In C programs. As the conditional operator is also an operator, it can appear multiple times in an expression. C regulates that the associativity of the conditional operator is from right to left as shown in Figure 3.79. Note that the operator consists of a ? sign and a : sign. Using them in a complicated manner makes the logic confusing to programmers, thus affecting the program readability.

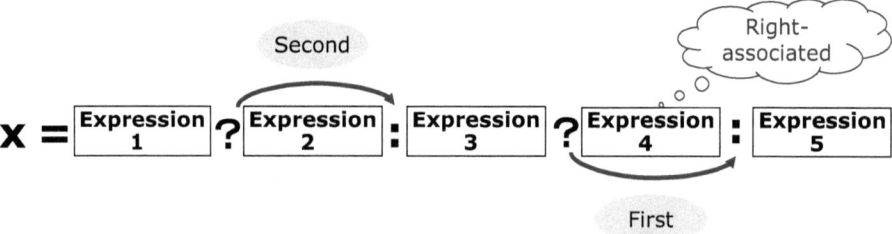

Figure 3.79: Associativity of the conditional operator.

Example 3.21 Determine parity
Given a number input, determine whether it is odd or even.

```
int main(void)
{
    int num;
    printf("Please input a number: ");
    scanf("%d",&num);
    (num%2==0) ? printf("Even") : printf("Odd");
}
```

Example 3.22 Use conditional expression to compute maximum of three numbers
Method 1:

```
int a=90,b=80,c=100,max;
max=a>b?a:b;
max=max>c? max:c;
printf("The maximum of these numbers is: %d",max);
```

Value of max becomes 90 after line 2, and then becomes 100 after line 3.

Method 2:

```
int a=90,b=80,c=60;
printf("The maximum of these numbers is: %d", a>c?a>b? a: b :c );
```

Analysis:
There are two pairs of conditional operators in this method. As conditional operator is right-associated, we should look for the rightmost? Sign and pair it with the closest: sign. Hence, a > c? a > b?a:b:c is equivalent to a > c?(a > b?a:b):c and the result of this expression is 100.

3.9.2 `sizeof` **operator**

`sizeof` operator can be used to compute number of bytes a variable, a constant or a data type takes up in memory as shown in Figure 3.80.

Format	Meaning
sizeof(variable)	Compute number of bytes the variable occupies in memory
sizeof(constant)	Compute number of bytes the constant occupies in memory
sizeof(datatype)	Compute number of bytes the data type occupies in memory

Figure 3.80: `sizeof` operator.

Example 3.23 Example of `sizeof` operator

```c
#include <stdio.h>
int main(void)
{
    int size_constant, size_variable, size_datatype;
    char c;
    size_constant = sizeof(10);
    printf("Number of bytes of constant 10: %d\n", size_constant);
    size_variable=sizeof(c);
    printf("Number of bytes of char variable: %d\n", size_variable);
    size_datatype=sizeof(float);
    printf("Number of bytes of float type: %d\n", size_datatype);
    return 0;
}
```

Output:
    Constant 10 uses 4 bytes; character variable uses 1 byte; float type uses 4 bytes.

3.9.3 Assignment operator and expressions

The basic assignment operator is "=", which assigns value of an expression to a variable as shown in Figure 3.81. Precedence of assignment operator is lower than that of others, so it is often executed last.

There are several things to remember when using assignment operations:
- "=" is an assignment operator instead of the equal sign.
- The object on the left of the assignment operator must be a variable. It cannot be an expression. Assignment should be done from right to left.
- The value of assignment expression is the value of the expression on the right of the assignment operator.

Assignment expression	Meaning	Example	Notes
Variable = expression	1. Evaluate the expression 2. Assign to the variable	a=b+3*c;	Assign value of expression b+3*c to variable b
		x=y=z=100;	"=" is right-associated
		a+(b=3)	Use parentheses to override precedence

Figure 3.81: Assignment operator and expressions.

Format	Meaning of compound operation
Variable binary operator = expression	Variable = variable operator expression

Operator	Name	Expression	Equivalent Expression
=	Assignment operator	a=5	a=5
+=	Addition and assignment operator	a+=5	a=a+5
-=	Subtraction and assignment operator	a-=x+y	a=a-(x+y)
*=	Multiplication and assignment operator	a*=2*x	a=a*2*x
/=	Division and assignment operator	a/=x-y	a=a/(x-y)
%=	Remainder and assignment operator	a%=12	a=a%12

Figure 3.82: Compound assignment operators.

3.9.4 Compound assignment operators

Adding other binary operators in front of the assignment operator " = ", we obtain compound assignment operators as shown in Figure 3.82.

3.9.5 Comma operator and comma expressions

For convenience, programmers want to compute the values of multiple expressions together when only one expression is allowed. For this purpose, C provides a convenient syntax, which is the comma operator that "sticks" multiple expressions together. Grammatically, these expressions become a single expression as shown in Figure 3.83.

In many cases, we use comma expression to obtain the value of each expression instead of obtaining and using the value of the entire comma expression. Comma expressions are most frequently used in for statements. Comma operator has the lowest precedence among all operators.

Comma expression	Evaluation of expression
Expression 1, expression 2, ... expression n	Compute value of each expression starting from expression 1. The value of entire comma expression is the value of expression n

Figure 3.83: Comma expression.

3.10 Summary

The main contents of this chapter and their relations are given in Figure 3.84.

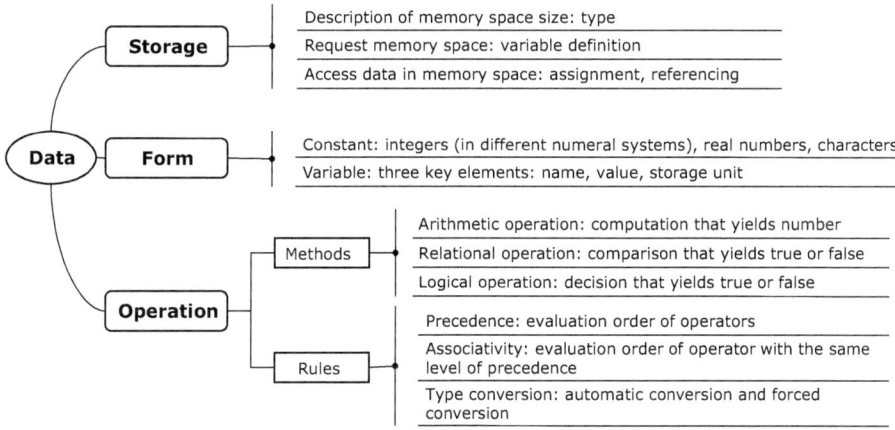

Figure 3.84: Contents of basic data and their relations.

Data are stored, referenced, computed, input, and output in programs.

Many rules exist, so we need to learn more and practice more.

Size of memory is determined by data type and can vary a lot.

Programmers need to choose the suitable type based on characteristics of data.

Constants can be used directly; variables need to be defined and allocated memory.

Numeric literals can be represented as decimal, octal and hexadecimal numbers. Decimal numbers are left unchanged, while octal numbers begin with 0 and hexadecimal numbers begin with 0x.

We should view memory space when debugging and be proficient in number system conversions.

Remainder operations compute the remainder; clever use of them can simplify our algorithms.

Integer divided by integer yields another integer, which is a rule we have to follow.

Different types of data can be computed together; type of result is determined by the type that takes up the most space in memory.

Automatic and forced are the most common types of conversions in computation.

Automatic-type conversions take place in assignments between two storage units.

Data will be kept when converting from a smaller type to a larger one, whereas truncation or rounding happens when converting oppositely.

Forced-type conversions are used if necessary, and data in the original storage unit are unchanged.

3.11 Exercises

3.11.1 Multiple-choice questions

(1) [Concept of variables]

Which of the following statements is wrong about variables in C? ( )

A) The memory unit address of a variable can be changed at any time.

B) The value of a variable can be changed during the program execution.

C) We must define variables before using them in programs.

D) ___ (three underscores) is a valid variable name.

(2) [Identifiers]

Which set of identifiers are all invalid? ( )

A) A P_0 do   B) float la0 _A   C) b-a goto int   D) _123 temp INT

(3) [Data types]

Which of the following statements is wrong about data types in C? ( )

A) To process correlated data of different types (such as "employee information"), we should define our own structure type.

B) We can use double type to store data with multiple digits in the fraction part.

C) We should use logic type to handle "true" and "false."

D) Natural numbers can be accurately represented by int type.

(4) [Symbolic constants]

Which of the following statements is correct about symbolic constants in C? ( )

A) Names of symbolic constants must be identifiers in uppercase.

B) A symbolic constant is a symbol that represents a constant defined by a macro.

C) The value of a symbolic constant cannot be redefined in a program.

D) Names of symbolic constants must be constants.

(5) [Character processing]

```
char c1, c2;
    c1='A'+'8'-'4';
    c2='A'+'8'-'5';
    printf( "%c, %d\n", c1, c2);
```

The ASCII value of character A is 65. What is the output of the program above? ( )

A) E, 68   B) D, 69   C) E, D   D) Nondeterministic

(6) [Forced-type conversion]

Suppose we have the following definition: double x = 5.16894;

The output of statement printf(("%lf\n ", (int)(x*1000+0.5)/1000.) is ( )

A) 5.16900

B) 5.16800

C) 0.00000

D) The type specifier is inconsistent with the output value, so an error message is displayed.

(7) [Logic operations]

Which of the following statements is correct about operands of logic operations? ( )

A) They can be any valid expressions.

B) They must be integers.

C) They can be structure-type data.

D) They must be 0 or 1.

(8) [Range]

Suppose an int type number takes up 2 bytes in the memory. What is the range of unsigned int data? ( )

A) 0–255   B) 1–32767   C) 0–65535   D) 0–2147483647

(9) [Base]

Which of the following bases cannot be used in C source programs? ( )

A) Hexadecimal   B) Octal   C) Decimal   D) Binary

(10) [Conditional operator]

Suppose a = 1,b = 2,c = 3,d = 4. What is the result of expression
a < b?a:c < d?c:d ? ( )

A) 4   B) 3   C) 2   D) 1

3.11.2 Fill in the tables

(1) [Range]

Sort the data types in Figure 3.85 in the order of their sizes on the same platform.

Data type	short	char	int	double	float
Number					

Figure 3.85: Basic data types: Fill in the tables question 1.

(2) [Forced-type conversion]

Fill in the table in Figure 3.86 with variable values after the following statements are executed.

int x,z; float y = 12.4; x = (int)y; z = 3*y;

Variable	x	y	z
Value			

Figure 3.86: Basic data types: fill in the tables question 2.

(3) [Increment and decrement]

Fill in the table in Figure 3.87 with variable values after the following statements are executed.

int x = 10; int y = x−; int z = − x; int a = x++; int b = ++x;

Variable	x	y	z	a	b
Value					

Figure 3.87: Basic data types: fill in the tables question 3.

(4) [Type conversion]

Figure out data types of expressions in Figure 3.88. Variables a, b, c, and d are defined as follows:

char a = 'A'; double b = 12.3; int c = 66; char d;

Expression	a+1	b+2*a	2.0*a+c
Data type			

Figure 3.88: Basic data types: fill in the tables question 4.

(5) [Arithmetic operations]

Figure out values of expressions in Figure 3.89.

mint x = 10; int y = 3; float z = 12.4;

Expression	x/y	x+y-z	z+x/y	z/x*y	x%y
Value (double)					

Figure 3.89: Basic data types: fill in the tables question 5.

(6) [Relational operations]

Figure out the logic values of expressions in Figure 3.90. Variables a, b, c, and d are defined as follows:

char a = 'A'; char b = 'a'; char c = 66; char d = 'A' + 1;

Expression	a==b	a==65	c==d	c=='B'
Logic value				

Figure 3.90: Basic data types: fill in the tables question 6.

(7) [Logic operations]

Figure out the logic values of expressions in Figure 3.91. Variables a, b, c, and d are defined as follows:

int a = 12; int b = 0; int c = −1; int d = 1; int e = 0

Expression	a!=12	b<c	1||b	d&&e	a!=b<c||d&&e
Logic value					

Figure 3.91: Basic data types: fill in the tables question 7.

3.11.3 Programming exercises

(1) Arithmetic operations
Compute the product of three integers.

(2) Arithmetic Operations
Write a program that reads a five-digit integer and outputs the digits delimited by spaces. (Hint: use division and mod operation.)

(3) Data swapping
Write a program that does the following:
1) Read three integers into variables a, b, and c.
2) Assign the initial value of a to b.
3) Assign the initial value of b to c.
4) Assign the initial value of c to a.

(4) Expressions
Figure out the conditional expression that determines whether a character is a digit.

(5) Random function
For each of the following sets of integers, write a statement that displays a random number in the set. (Hint: use the random function in the standard library.)
1) 2,4,6,8,10
2) 3,5,7,9,11
3) 6,10,14,18,22

4 Input/output

4.1 Concept of input/output

Mrs. Brown had a question when logging into a ticket purchasing website as shown in Figure 4.1.

Case Study

How does a program read password?

- When purchasing tickets online, Mrs. Brown was asked to type in password to log into the system.
- Mrs. Brown asked curiously, "The password is entered through keyboard, how is it passed to the program?"
- Mr. Brown commended, "Good question!"

How does a program read information from the real world, and send the processing result back?

Figure 4.1: Login password problem.

A program can read data from keyboard input, so the next question is, naturally, how does program exchange information with the real world? This is related to input/output of program data.

The term input/output is used with respect to computer processors. Sending data from a computer to external output devices is called "output," whereas sending data from input devices into computers is called "input" as shown in Figure 4.2.

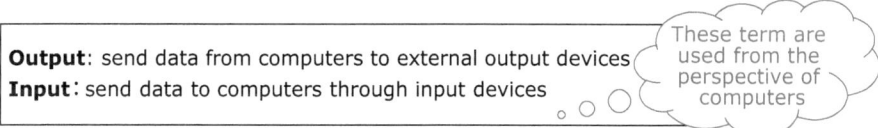

Output: send data from computers to external output devices
Input: send data to computers through input devices

These term are used from the perspective of computers

Figure 4.2: Input/output in computers.

https://doi.org/10.1515/9783110692327-004

4.1.1 Standard input/output

We usually call keyboards and monitors standard input/output devices. Consequently, input/output through these devices are called standard input/output, respectively as shown in Figure 4.3.

Standard Input	Send data into computer memory through standard input devices (keyboard)
Standard Output	Send data from computer memory to standard output devices (monitor)

Input/Output is a complex process that needs special programs to handle

Figure 4.3: Standard input/output.

4.1.2 Standard library functions of C

Recall that functions are child programs that provide certain functionalities. Library functions are functions inside a program library. Frequently used standard library functions of C and related questions are shown in Figure 4.4.

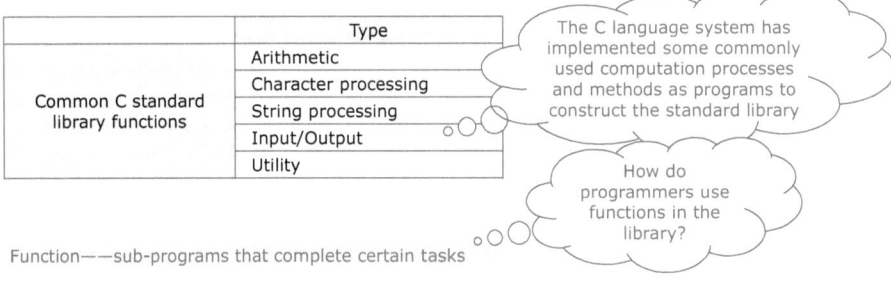

	Type
Common C standard library functions	Arithmetic
	Character processing
	String processing
	Input/Output
	Utility

The C language system has implemented some commonly used computation processes and methods as programs to construct the standard library

How do programmers use functions in the library?

Function——sub-programs that complete certain tasks

Figure 4.4: Standard library functions of C.

The C language system provides its users with function libraries so that programmers can use programs within it directly. Developers should accumulate experience of library functions and use them as frequently as possible rather than starting over on every task. Using library functions shortens the development cycle, thus makes developers' jobs easier. Moreover, it makes programs more portable.

Knowledge ABC Standard library functions

The ANSI (American National Standards Institute) C standard defines standard library functions of C, which includes mathematical functions, input/output functions, string functions, graphical functions, date and time functions, and so on. Each category contains dozens or even hundreds of functions, each of which completes a specific task. They are usually supported, either partially or entirely, by common C compiling environments. Readers may refer to the appendix of this book or manuals of compilers for help on these functions.

The merit of using standard library functions is that users can use them without having to define them again. When we are going to print out something as output, we can simply call an output function with required arguments, as long as we know its functionality, input/output parameters, and return value.

4.1.3 Header files

Each category of standard library functions of C has a corresponding header file that stores declarations of functions in this category as shown in Figure 4.5.

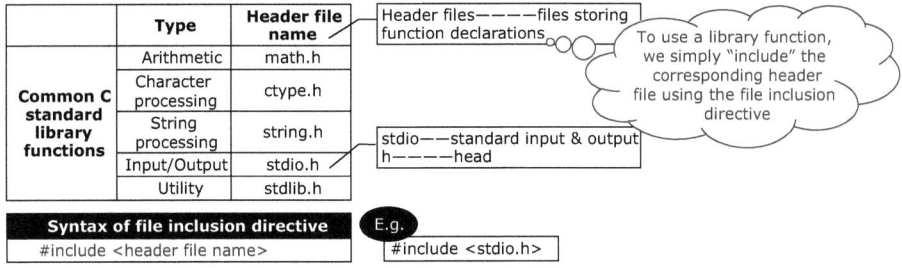

Figure 4.5: File inclusion and header files.

Header files are files storing function declarations. For example, library functions related to input/output are declared in header file stdio.h, where stdio is abbreviated from "standard input/output" and the extension .h is the initial letter of "header."

To use a library function, programmers need to "include" the corresponding header file using the file inclusion directive. The function of the file inclusion directive is to fetch the specified file for use. For instance, to use the sqrt function (a function that computes squared root), we need to add the following line to the beginning of our program:

```
#include "math.h"
```

4.2 Data output

There are two types of data output library functions: character output functions and formatted output functions, which are declared in the standard input/output header file as shown in Figure 4.6.

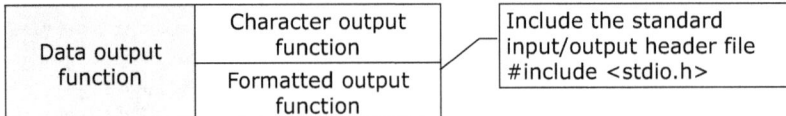

| Data output function | Character output function | Include the standard input/output header file #include <stdio.h> |
| | Formatted output function | |

Figure 4.6: Data output library functions.

4.2.1 Character output functions

putchar is a character output function, whose function signature and functionality are given in Figure 4.7. The "character" inside parentheses, which is either a character variable or a character literal, is the parameter of this function.

Character output function	
Signature	putchar(character)
Functionality	Output a character represented by "character" to standard output devices (monitor)

Output one character at a time

Figure 4.7: Character output function.

> **Example 4.1** Character output function
> The first program we write when learning to program is usually a program that prints on screen the following friendly words: "Hello, world!" We have seen this simplest C program in chapter "Introduction to Programs." Now we are going to add a smiley face to the output to welcome everyone. The revised program is given below, where the comments explain the meaning of each putchar call.
>
> ```
> #include <stdio.h>
> int main(void)
> {
>  printf("Hello, world!\n");
>  putchar (2) ;      //Output a smiley face (ASCII value 2) to screen
>  putchar('\n');    //Output a newline
>  return 0;
> }
> ```

Example 4.2 Character output function
Figure 4.8 shows the program and result, where c1 and c2 are both character variables.

```
01 #include "stdio.h"
02 int main(void )
03 {
04    char c1,c2;        //Define two character variables
05
06    c1='a' ;          //Assign value to c1
07    c2='b' ;          //Assign value to c2
08    putchar(c1);      //Output character a
09    putchar(c2);      //Output character b
10    putchar('\n');    //Output newline
11    putchar(c1-32);   //c1-32='a'-32=97-32=65, which corresponds to 'A'
12    putchar(c2-32);   //c1-32='b'-32=98-32=66, which corresponds to 'B'
13    return 0;
14 }
```

Isn't it cumbersome to output characters like this?

Program result :
ab
AB

Figure 4.8: Character output function example 2.

[Analysis]
On line 8, the character output function is used to output character a stored in the storage unit of variable c1 onto screen.

On line 10, \n stands for a newline. This statement moves the cursor to the beginning of the next line.

On line 11, the argument c1-32 = 'a'-32 = 97-32 = 65 is the American Standard Code for Information Interchange (ASCII) value of character 'A'.

Finally, we obtain the result.

After reading this program, have you noticed the drawback of putchar function?

4.2.2 String output function

puts is another character output function, but it is more convenient. It can print a sequence of characters in one go. Its signature and functionality are given in Figure 4.9.

String output function	
Signature	puts(address)
Functionality	Output a string with newline to standard output devices (monitor)

Figure 4.9: String output function.

Example 4.3 String output function example
We can rewrite the second example of character output function using puts function as shown in Figure 4.10.

```
01 #include"stdio.h"
02 int main(void)
03 {
04    char c[8];     //Define a character array of size 8
05
06    c[0]='a';      //Assign value to c[0]
07    c[1]='b';      //Assign value to c[1]
08    c[2]=\n';      //Assigmewlineto c[2]
09    c[3]=c[0}32;   //c[3}32='a'-32=97-32=65 which corresponds to 'A'
10    c[4]=c[1}32;   // c[4}32='b'-32=98-32=66 which corresponds to 'B'
11    puts(c);
12    return 0;
13 }
```

> Although output efficiency has been increased, why are there Chinese characters?

Program output:
ab
AB烫烫

Figure 4.10: String output function example.

[Analysis]
On line 4, we define a group of eight character variables that are stored sequentially. They are represented by the name of the group c with the corresponding index. More information on such variables are introduced in the chapter "Arrays."

Nonetheless, the result seems to be weird. Some characters we have never used are printed at the end of the result. In fact, this is due to how puts function works. puts will not stop printing characters until a '\0', the terminating character, is met. As we did not assign the value of terminating character in the storage unit after c[4], puts kept searching for it in the memory until one is found.

N.B.: The Chinese characters in Figure 4.10 are garbled output produced by the system, as the system environment uses GBK character encoding by default.

4.2.3 Formatted output function

4.2.3.1 Syntax and signature of formatted output function

Let us look at a formatted output function that can output efficiently: the printf function. There are several parameters inside parentheses. Its function signature and functionality are given in Figure 4.11.

Formatted output function	
Signature	printf (format control sequence, parameter 1,…,parameter n)
Functionality	Output values of parameter 1 to parameter n to standard output devices in format specified by "format control sequence". Parameters are expressions.

Figure 4.11: Formatted output function.

We shall cover the usage of printf through examples and introduce its rules after that.

Example 4.4 Formatted output example: format coordination
Variable definitions and output cases are shown in Figure 4.12.

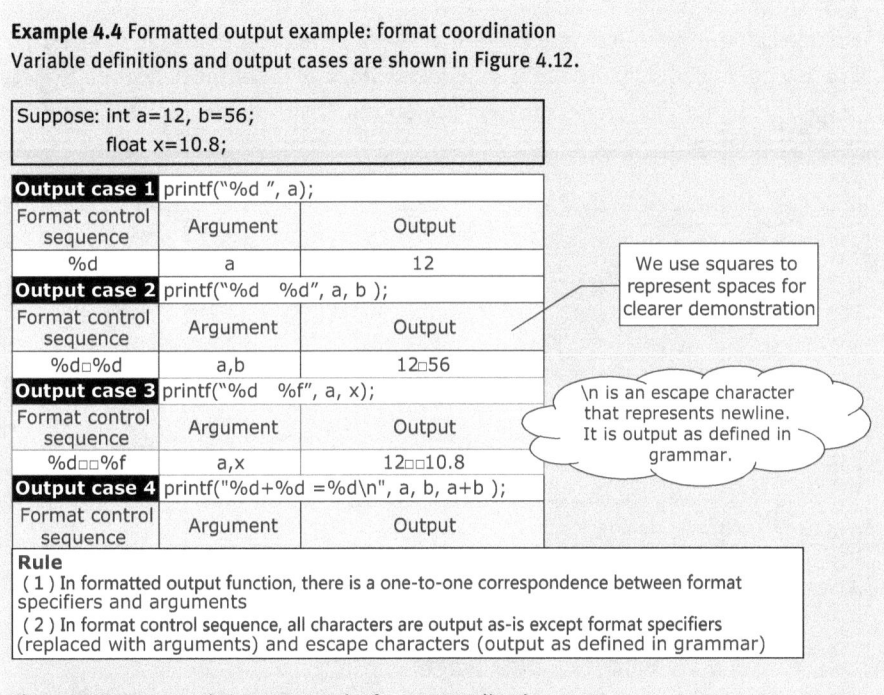

| Suppose: int a=12, b=56; |
| float x=10.8; |

Output case 1 printf("%d ", a);		
Format control sequence	Argument	Output
%d	a	12
Output case 2 printf("%d %d", a, b);		
Format control sequence	Argument	Output
%d□%d	a,b	12□56
Output case 3 printf("%d %f", a, x);		
Format control sequence	Argument	Output
%d□□%f	a,x	12□□10.8
Output case 4 printf("%d+%d =%d\n", a, b, a+b);		
Format control sequence	Argument	Output

We use squares to represent spaces for clearer demonstration

\n is an escape character that represents newline. It is output as defined in grammar.

Rule
(1) In formatted output function, there is a one-to-one correspondence between format specifiers and arguments
(2) In format control sequence, all characters are output as-is except format specifiers (replaced with arguments) and escape characters (output as defined in grammar)

Figure 4.12: Formatted output example: format coordination.

[Analysis]
In output case 1, we print the value of an integer variable a. The format control sequence is the content wrapped by double quotation marks, where %d means that the data will be output as integers. The argument of this function is a, so 12 is output by printf.

In case 2, we print values of a and b, which are both integer variables. There are two %d identifiers in the format control sequence, indicating that two integers will be output. For clearer demonstration, we use a square to indicate the existence of a space between numbers. The arguments are a and b, separated by a comma. The output, in this case, is 12 space 56.

In case 3, we print values of integer variable a and floating-point variable x. %d in the format control sequence corresponds to variable a, whereas %f corresponds to variable x. In the output, two spaces are inserted between 12 and 1.8, namely values of a and x.

In case 4, we print the result of an arithmetic expression. We write the expression in the format control sequence, where actual values are replaced with format specifiers. \n stands for a newline. It is an escape character that cannot be displayed on the screen. The arguments are a, b, and a+b, separated by commas. Hence, the output is 12 + 56 = 68.

The output rules require that all characters are output as-is, except format specifiers, which are replaced with argument values, and escape characters, which are output as defined by the grammar.

4.2.3.2 Output format specifiers

In the format control sequence, there are symbols that indicate types of output data. They are called "type specifiers." Figure 4.13 shows the most frequently used type specifiers, such as %d, %f, %c, %s, and so on. It suffices to know these in an early stage of learning.

	Type specifier	Meaning	Notes
Integer	%d	Output integer in signed decimal form	• The letter can be uppercase in type specifier
	%o	Output integer in unsigned octal form	
	%x	Output integer in unsigned hexadecimal form	
	%u	Output integer in unsigned decimal form	• Sub-specifiers can be inserted between % and type specifiers
Real number	%f	Output real number with fractional part	
	%e	Output real number in exponential form	
	%g	Output real number in the form with smallest width	• '\0' marks the end of a string. It is inserted automatically by the system.
Character	%c	Output a single character	
	%s	Output string (starting from the specified address and ending at '\0')	
Other	%%	Output character %	

Figure 4.13: Output format specifiers.

4.2.3.3 Structure of format control sequence

The format control sequence is a sequence of normal characters and format specifiers wrapped by double quotation marks. It is used to specify types, formats, and number of output data. The structure of the format control sequence of printf function is shown in Figure 4.14. Note that parts inside [] are optional.

%		m	.	n	h/l	Type specifier
Beginning specifier	[Flag specifier]	[Width specifier]	[]	[Precision specifier]	[Length specifier]	Type specifier character

Figure 4.14: Format control sequence of printf function.

4.2.3.4 Subspecifiers

As shown in Figure 4.15, we can insert subspecifiers between % and type specifiers to adjust the number of significant figures (e.g., number of digits in the fractional part) or justification of output data.

Example of subspecifiers:

- %ld: Output as long decimal integers.
- %lf: Output as double type.
- %m.nf: Right-justify output, m indicates the width of the output field, n indicates the number of digits in the fractional part or the number of characters.
- %-m.nf: Left-justify output, m indicates the width of the output field, n indicates the number of digits in the fractional part or the number of characters.

l	Output as long integers (can be used in combination with d, o, x and u)
m	Specify width of output (i.e. total number of digits, where decimal point counts as well)
.n	(1) For floating-point data, output n digits in the fractional part (2) For string, output the first n characters
+	Explicitly output sign of numerical data
-	Left-justify data within the given output field

Figure 4.15: Subspecifiers.

Example 4.5 Output -1 in various forms
Output value of -1 as decimal, octal, hexadecimal, and unsigned numbers.

[Analysis]
Let int m = -1;
    The output statement is printf(("m: %d, %o, %x, %u\n", m, m, m, m);
    The actual output is m: -1, 177777, ffff, 65535
    Are you surprised to see the above output of -1 in different forms?
    The binary representation of -1 is 1111,1111,1111,1111. In other words, it is stored in memory as a number consisting of only "1." If we examine it in different formats, we will obtain different representations. It is similar to how different languages use different words for "apples" while they all describe the same object.

Example 4.6 Output example of character data
Examine the output of character variables and integer variables using type specifier %d and %c.
    Variable definitions and output statements are given in Figure 4.16.

Suppose : int m=97; char ch='A';		
Statement	**Output**	**Notes**
printf("m: %d %c\n",m,m);	m:□97□a	The same variable is output as different values using different type specifiers
printf("ch: %d %c\n",ch,ch);	ch:□65□A	
printf("%s\n","student");	student	%s——output a string

Figure 4.16: Output of character data.

[Analysis]
In the first row of the table in Figure 4.16, the statement prints two values using %d and %c, both of which will be replaced by variable m. However, the first m is output as an integer, whereas the second is output as a character. The value of m is defined to be 97, which is the ASCII value of character a.

Similarly, we see in the second and the third rows that output of the same variable can vary when different type specifiers are used.

In the third row, the type specifier is %s. It is used to print strings. The argument is exactly a string wrapped by double quotation marks.

Example 4.7 Using escape characters
We can learn the roles of escape characters by using formatted output function printf(). The program is as follows:

```
1  //Using escape characters
2  #include <stdio.h>
3  int main(void)
4  {
5    char a,b,c;
6    a='n';
7    b='e';
8    c='\167';      //Octal number 167 stands for character 'w'
9    printf("%c%c%c\n",a,b,c);           //Output as character
10   printf("%c\t%c\t%c\n",a,b,c);       //Jump to next field after a character
11   printf("%c\n%c\n%c\n",a,b,c);       //Jump to new line after a character
12   return 0;
13 }
```

The output is as follows:

```
new
n□□□□□□□e□□□□□□□w
n
e
w
```

Note: \t is an escape character used to advance the cursor to the next field horizontally. Each field takes up eight columns.

4.3 Data input

There are two types of data input functions in C: character input functions and formatted input functions. They are all declared in the standard input/output header file as shown in Figure 4.17.

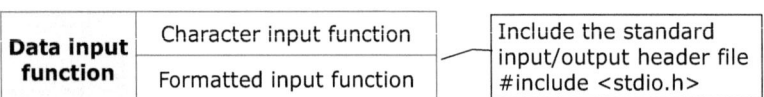

Figure 4.17: Data input function.

4.3.1 Character input function

The function signature and functionality of character input function are shown in Figure 4.18.

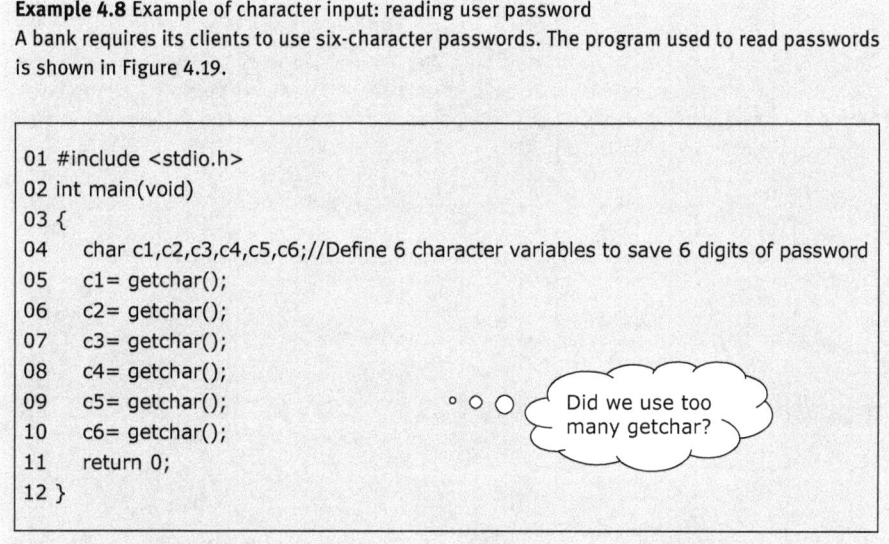

Character input function		
Signature	getchar( )	
Functionality	Read a character interactively from standard input devices (keyboard)	

Interactive input: the program keeps waiting in the console and continues execution until it receives a character input from keyboard

Figure 4.18: Character input function.

Dialogues between two people are often conducted in the form of questions and answers. Likewise, humans and computers need a similar way for information exchange. One such exchanging method is interactive input, where a program keeps waiting in the console and continues execution until it receives a character input from the keyboard. A console window is a window used to display the execution result of programs.

How do programs receive the passwords we type in using keyboards?

Example 4.8 Example of character input: reading user password
A bank requires its clients to use six-character passwords. The program used to read passwords is shown in Figure 4.19.

```
01 #include <stdio.h>
02 int main(void)
03 {
04     char c1,c2,c3,c4,c5,c6;//Define 6 character variables to save 6 digits of password
05     c1= getchar();
06     c2= getchar();
07     c3= getchar();
08     c4= getchar();
09     c5= getchar();
10     c6= getchar();
11     return 0;
12 }
```

Did we use too many getchar?

Figure 4.19: Password reading program.

[Analysis]

On line 4, we define six character variables to store six characters in the password.

On line 5, a character is read from the keyboard and stored into variable c1.

Statements between lines 6 and 10 work in the same way as the one on line 5.

getchar reads one character each time. It is not hard to see that the program is cumbersome due to repeated use of getchar. We will discuss input functions that read data more efficiently later.

Example 4.9 Conversion between different cases of letters

```
1 #include "stdio.h"
2 int main(void)
3 {
4     char ch;
5     ch=getchar( ); //Read a character from keyboard and store into ch
6     printf("%c %d\n",ch,ch); //Display ch and its ASCII value
7     printf("%c %d\n\n",ch-32,ch-32);
8     //Subtract 32 from ASCII value of ch,
      //display corresponding character and the new value
9     return 0;
10 }
```

Output:

```
Input: m
Output: m 109
M 77
```

We can check the values of ch and ch-32 in the Watch window as shown in Figure 4.20. 'm' is the character referencing result of letter m. Its ASCII value is 109. ASCII value of uppercase letter M is 77. After investigating the ASCII table, we can conclude that the difference of ASCII values of the same letter in upper and lower cases is exactly 32.

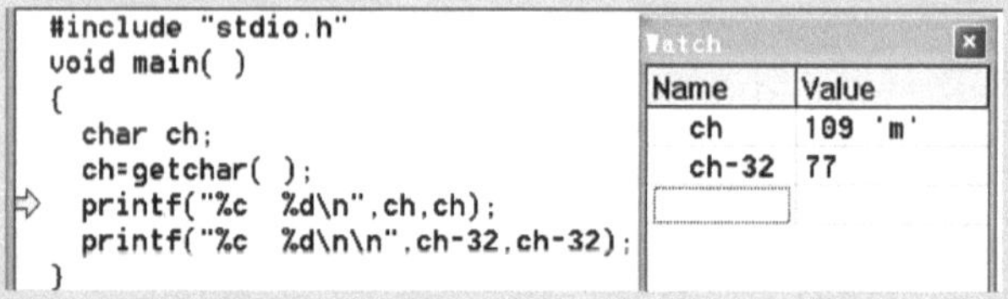

Figure 4.20: Inspection of variable ch in debugger.

Conclusion

We don't have to recite the ASCII values of characters. Instead, we can use "character referencing" to display them.

Knowledge ABC 32 and ASCII value of uppercase and lowercase letters
From the ASCII table, we know that 'A' = 65, 'a' = 97 and consequently 'a'-'A' = 32. Why is the difference of ASCII values of the same letter in upper and lower cases 32? Let us convert them into hexadecimal numbers first:

$$'A' = 65 = 0x41, \ 'a' = 97 = 0x61, \ so \ 'a' - 'A' = 32 = 0x20$$

It is clear that such a design makes it easier to convert letters into the other case in binary and hexadecimal systems.

4.3.2 String input function

gets is a more convenient character input function. It reads a sequence of characters at a time. The function signature and functionality of gets are shown in Figure 4.21.

String input function	
Signature	gets(address)
Functionality	Read a string which ends with a newline from standard input devices

Figure 4.21: String input function.

Note: Function gets can read infinitely many characters, so programmers should make sure the memory space used to store the string is large enough to avoid overflow in read operations.

Example 4.10 Example of function gets
Use function gets() to read a string from keyboard input.

[Analysis]
The program and test results are shown in Figure 4.22.

```
int main(void)
{
        char str1[60];
        gets(str1);
        printf("%s\n",str1);
        return 0;
}
```

Input: hello world!!
Ouput: hello world!!

Function gets reads string from keyboard. It is not affected by spaces in the string.

Figure 4.22: Example of function gets().

Using function gets() to read character sequence is not affected by spaces in the sequence. However, we need to make sure the memory space used to store the sequence is large enough in case the input overflows. Note how it is different from Example 4.14.

Note: The C11 standard proposes a safer function gets_s() to replace gets(). Interested readers may update the above program and test again.

4.3.3 Formatted Input function

The function signature and functionality of the formatted input function are given in Figure 4.23. The function name is scanf. There are several items inside parentheses, of which the format control sequence is similar to the one used in printf. It is worth noting that the address argument is passed to the function by prefixing variable names with "&" sign. The "&" sign is the address-of operator. It is not needed if the variable is already an address.

Formatted input function	
Signature	scanf(format control sequence, address 1..., address n);
Functionality	Read data from keyboard in the format specified by the format control sequence and store them into corresponding variables

Figure 4.23: Formatted input function.

Example 4.11 Example of formatted input function: log into Office of Registrar website
Mr. Brown wanted to log into the Office of Registrar website to upload the grades of students. A username and a password are necessary to log in, where the username (ID) is the payroll number and the password should be a sequence of digits and characters with a length no more than 20, as per university policies. The login program is given in Figure 4.24.

[Analysis]
On line 8, the first scanf stores the username retrieved from keyboard input into the variable id. id is defined as an integer variable, so the type specifier should be %d.

```
01 #include <stdio.h>
02 int main(void )
03 {
04    int id;                  // id is an integer variable
05    char password[20];       // Password is no longer than 20 and stored in an array
06
07    printf("User ID : ");
08    scanf("%d", &id);        //&id is the address of variable id
09    printf("Password : ");
10    scanf("%s", password);   // Array name password is already an address
11    printf("ID=%d\n",id);
12    printf("Password=%s\n", password);
13    return 0;
14 }
```

Program result :
User ID : 2468
Password : abc123
ID=2468
Password=abc123

Both being arguments of scanf, why id is prefixed with an & sign while password is not?

Figure 4.24: Example of formatted input function: log into Office of Registrar website.

On line 10, the second scanf stores the password retrieved from keyboard input into array password. Note that the type specifier here is %s, which is used for reading a sequence of characters. "&" sign is not used here because password is an array name and array names are in fact addresses in C. The concept of arrays is covered in the corresponding chapter.

On lines 11 and 12, the value of variable id and the value of password are printed out for viewing.

Example 4.12 Example of formatted input function: entry of students' grades
1. Enter a student's student number id and grade scores. Sample input: 1602 92.5.
2. Enter grades of course a and course b. Sample input: a = 76, b = 82.

[Analysis]
As per problem description, we can use formatted input function scanf() to read keyboard input as shown in Figure 4.25.

No.	Formatted input	Keyboard input	Variables read	Delimiter
1	scanf("%d%f", &id, &scores);	1601□92.5	id=1601 scores=92.5	Space (default)
2	scanf("a=%d, b=%d", &a, &b);	a=76, b=82	a=76 b=82	Specified character

Figure 4.25: Entry of students' grades.

In the first row of the table, we type in a student number followed by a grade. The student number is an integer and the grade is a real number. We should be careful about the space between the student number 1601 and the grade 92.5. The space is used to separate the inputs. Symbols that are used to separate multiple input data are called "delimiters." Space is the default delimiter.

In the second row, there are other characters in addition to grades in the sample input, so we need to include them in the format control sequence of scanf.

When typing, we need to type in characters like a = and b = as well. In this case, these characters are the specified delimiters.

Forms of format control sequences and examples of delimiters are shown in Figure 4.26.

Two cases of format control sequence	

(1) Type specifier only: use default delimiter
(2) Type specifier + characters:
  · "characters" are read as-is
  · If characters exist between type specifiers, they are used as delimiters	

End of one data item	① When a space, a "newline" or a "tab" is met ② When a certain width is reached ③ When there is an invalid input
End of scanf	Scanf function terminates when every data item is read and Return key is pressed

"Delimiter" is a sign used to separate multiple data items

Note: a data item refers to an address parameter in scanf

Figure 4.26: How to read input and end input in scanf.

i **Knowledge ABC** How input data of scanf are delimited

scanf handles strings, integers, and real numbers the same way, where newline, space, and tab are all considered to be the end of input. However, characters are different from strings as spaces and newlines may also be input as characters, so we need to be careful. When entering strings, integers, or real numbers, these special characters will be treated as delimiters instead of being read into character arrays or variables.

Example 4.13 Example of formatted input function: character input
1. Enter three characters a, b, and c from the keyboard and store them into character variables ch1, ch2, and ch3.
2. Enter two integers and store them into integer variables m and n, and enter a character d and store it into character variable ch.

[Analysis]
Figure 4.27 presents correct and wrong sample inputs.

No.	Statement	Sample input	Notes
1	scanf("%c%c%c",&ch1,&ch2,&ch3);	Wrong: a□b□c✓	A char variable can only store one character
		Correct: abc✓	
2	scanf(" %d%d ", &m, &n); scanf(" %c ", &ch);	Wrong: 32□28✓	Cause of error : newline is read as a character in the second scanf
		Correct: 32□28d✓	

Figure 4.27: Example of formatted input function: Character input.

Let us combine input statements in both questions in the following program and run the program to test why errors occur.

```
01   #include <stdio.h>
02   int main(void)
03   {
04       char ch1,ch2,ch3;
05       int m,n;
06       char ch;
07
08       scanf("%c%c%c",&ch1,&ch2,&ch3);
09       scanf("%d%d", &m, &n);
10       scanf("%c", &ch);
11       printf("%c%c%c\n",ch1,ch2,ch3);
12       printf("%d,%d\n", m, n);
13       printf("%c\n", ch);
14       return 0;
15   }
```

The inputs and values of corresponding variables are shown in Figure 4.28, where <cr> stands for a newline. The analysis of test data is given in Figure 4.29.

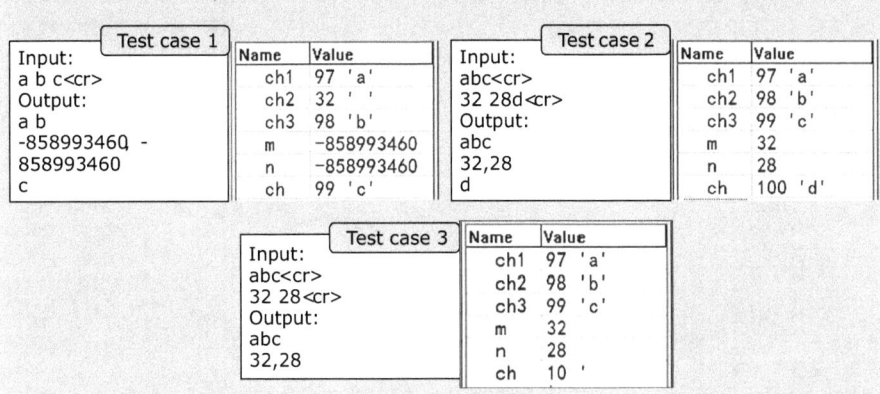

Figure 4.28: Example of formatted input function: Test input of characters.

	ch1	ch2	ch3	m	n	ch
a□b□c✓	a	□	b	Does not accept characters	c
abc✓32□28d✓	a	b	c	32	28	d
abc✓32□28✓	a	b	c	32	28	✓

Figure 4.29: Analysis of test input of characters.

In test case 1, we add spaces in input characters abc. As a result, values of variable ch2 and ch3 on line 8 are not as expected. The root cause is that spaces in input are also characters. The program ended after we typed abc, as scanf on line 9 tried to read contents left by variables on line 8. Type specifiers of variable m and n are both integer type, thus they don't accept character input. Hence, scanf on line 10 read the last character c in the test result.

In test case 3, we add a newline after number 28. In this case, it is no longer possible to enter character d due to the similar fact that newline is also read as a character.

Test case 2 shows the correct input that produces expected result.

When using consecutive scanfs, we should be careful as the current input data may affect the following input statements.

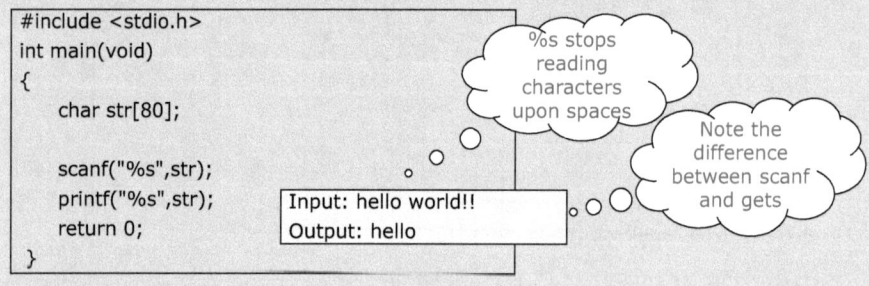

Knowledge ABC scanf and buffer

Because people type characters on the keyboard much slower than a CPU processes them, a storage block (called a buffer) that collects and stores keyboard input temporarily is designed in the system to reduce CPU's time of waiting. ASCII values of characters we type in are stored in the buffer, and the CPU fetches characters that comply with the format specified by input function from the buffer in one go when a newline is input. The remaining characters are still stored in the buffer.

As all input functions share the buffer, errors may occur when input includes both integers and characters. Suppose the current input is an integer, if we don't know whether the next one is a character, it is recommended to add while(getchar() != '\n') continue; Use font for code. which clears the buffer.

Example 4.14 Example of formatted input function: string input
Use scanf to read a sequence of characters. The program is given in Figure 4.30.

```
#include <stdio.h>
int main(void)
{
    char str[80];

    scanf("%s",str);
    printf("%s",str);
    return 0;
}
```

%s stops reading characters upon spaces

Note the difference between scanf and gets

Input: hello world!!
Output: hello

Figure 4.30: String input.

[Analysis]
When using scanf to read character sequences, we should note that a single scanf cannot read the entire sequence if there are spaces in it.

The input in this example is hello□world. Because %s stops reading at spaces, only hello is stored into array str while world is not. Note how it is different from using gets function to read strings.

Think and discuss How to find errors in scanf input quickly?
We have seen in these examples that data are often read incorrectly when reading input with scanf. Is it possible to find errors without tracing in debuggers or printing out all input?

Discussion: In fact, scanf function provides a mechanism to check the correctness of input arguments. Basically, it can tell its caller the number of correct inputs. If all inputs are wrong, scanf will return 0. If we press Ctrl+z to exit, scanf will return -1 (which is character literal EOF).

The function signature of scanf is as follows:

```
int scanf(format control sequence, address 1, address 2, … address n);
```

A test program is given in the following example.

Example 4.15 Return value of scanf function
Examine the correctness of input data using the return value of scanf function.

[Analysis]
1. Test program
We use an integer count to store the return value of scanf. The program is as follows:
```
#include <stdio.h>
int main(void)
{
    int a,b,c;
    int count;
    printf("Enter values of a, b and c, separated by space\n");
    count=scanf("%d%d%d",&a,&b,&c);
    printf("a=%d,b=%d,c=%d,count=%d\n",a,b,c,count);
    return 0;
}
```

2. Test result

The test result is given in Figure 4.31. The value of count indicates the number of data that are correctly read. If we don't want to enter anything, we can press Ctrl + Z to exit.

Sample input	Result
2 3 6	a=2,b=3,c=6,count=3
2 3 a	a=2,b=3,c=-858993460,count=2
2 3,6	a=2,b=3,c=-858993460,count=2
2,3,6	a=2,b=-858993460,c=-858993460,count=1
a b c	a=-858993460,b=-858993460,c= 858993460,count=0
^Z	a=-858993460,b=-858993460,c=-858993460,count=-1

Note: ^Z——Press ctrl+z to exit input mode

Figure 4.31: Test of formatted input function.

> ℹ️ **Knowledge ABC** EOF sign
>
> End of file (EOF) is a literal defined in header file stdio.h, which means "no more data for input." The reason for calling it the end of a "file" is that the program system treats standard input and output as "files." The ANSI standard emphasizes that EOF should be a negative integer, which is usually -1 (though it is not necessary). As a result, the value of EOF may vary in different systems. We test whether the return value is EOF rather than -1 in programs for better portability.
>
> In UNIX and many other systems, EOF sign is input by typing in <Return> combined with <Ctrl+D>. In Microsoft's Window systems, EOF is input by typing in <Ctrl+Z>.

4.4 Typical problems of using formatted input function

scanf() function is a tricky topic in C learning. We shall cover in the following sections some typical problems that often arise.

4.4.1 Typical problems of scanf input

It is not rare that beginners run into the following situation when practicing programming on computers: the machine doesn't continue program execution and waits in the console although required input for scanf has been provided. The root cause is that the machine doesn't consider the input complete while users believe they have entered input as required.

Figure 4.32 lists the most common mistakes beginners may make when using scanf in practice. Although they seem to be minor mistakes, it is hard to find them

No.	Errors encountered during inputting	Sample statement
1	Error window "access violation" pops up	int a; scanf("%d", a);
2	Error window "debug error" pops up	int a; scanf("%f"\n", &a);
3	Can't return to program execution screen after pressing Return	int a; scanf("%d"\n", &a);
4	Data overflow with no error message	char c; scanf("%d", &c);

Figure 4.32: Common problems of using scanf.

during program execution. Beginners often obtain wrong results due to wrong input when practicing, which eventually takes them a long time to figure out why results are wrong and affects their efficiency. We will analyze each of these mistakes now.

4.4.1.1 Common mistake of using scanf 1: wrong address argument

4.4.1.1.1 Sample program

```
int a;
scanf( "%d ", a );
```

4.4.1.1.2 Phenomenon

(a) Compilation: A warning is shown, but compilation succeeds (a program is executable if compilation succeeds).

```
Warning C4700: local variable 'a' used without having been initialized
```

(b) Execution: When executing the program, a dialog box is shown, and the program terminates after we enter an integer as shown in Figure 4.33 (note: "test. exe" is the generated executable file).

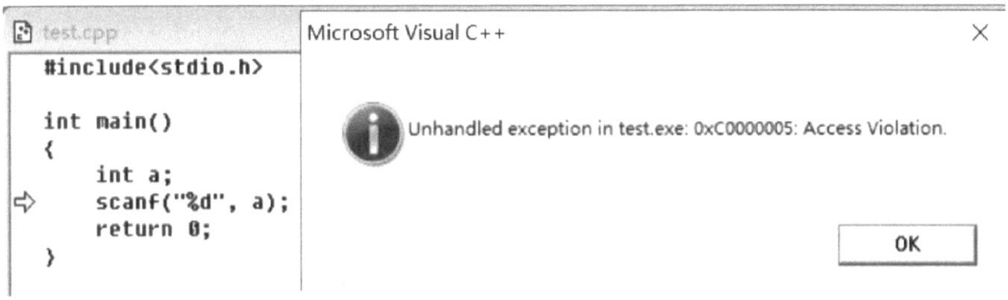

Figure 4.33: Access violation dialog box.

Note:
- Unhandled exception: An exception that is not handled.
- Access violation: Illegal access. An access violation error occurs when the program that is currently executed by the computer tries to access memory that doesn't exist or is not accessible.

4.4.1.1.3 Analysis

The above error occurs because we forgot to enter "&" sign in front of the address argument a. During execution, the value of a is treated as an address. For instance, if the value is 100, the integer we input will be written into memory at address 100. Consequently, the system protection mechanism interferes and terminates the program.

> **Programming error**
> "&" sign in front of an address argument of scanf is missed. (Note: if the argument is a pointer, & is not needed as it is already an address.)

4.4.1.2 Common mistake of using scanf 2: argument type not compatible with type specifier

4.4.1.2.1 Sample program

```
int a;
scanf("%f",&a);
```

4.4.1.2.2 Phenomenon
- No error occurs in compilation and linking.
- Execution: When entering data required by scanf, for example, number 6 (regardless of being integer or real number), an error dialog box pops up and the program terminates as shown in Figure 4.34. After selecting "Ignore," text in the User screen window is shown in Figure 4.35.

4.4.1.2.3 Analysis

The error message of runtime error R6002 is as follows:
- A format string for a printf or scanf function contained a floating-point format specification and the program did not contain any floating-point values or variables.
- The argument type is not compatible with the type specifier, thus the program can't continue execution.

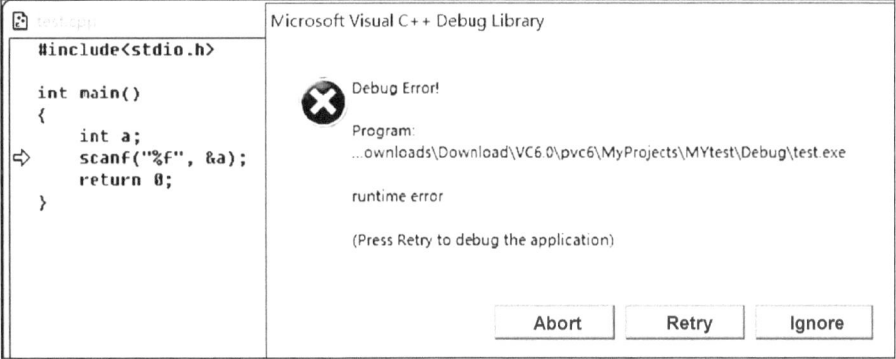

Figure 4.34: Debug error dialog box.

```
6

runtime error R6002
- floating point not loaded
Press any key to continue_
```

Figure 4.35: Abnormal execution result message.

Programming error
An argument type is not compatible with the type specifier in scanf.

4.4.1.3 Common mistake of using scanf 3: argument type compatible with type specifier

4.4.1.3.1 Sample program

```
char c;
scanf("%d", &c);
```

4.4.1.3.2 Phenomenon

Entering an integer overrides the memory space after the character variable.

4.4.1.3.3 Analysis

The memory space after variable c is overridden because the storage unit of it is not large enough for an int variable. In Figure 4.36, the address of character variable c is 0x18ff44, the initial value starting from this address is four CC, where each CC takes up 1 byte. Figure 4.37 shows the memory space of variable c after inputting number 10. We notice that the 4 bytes after address 0x18ff44 are now "0A 00 00 00," but variable c is of char type and only takes up 1 byte in memory.

```
#include <stdio.h>

int main(void)
{
    char c;
⇨   scanf("%d", &c);

    return 0;
}
```

Watch
Name	Value
c	-52 '?
⊞ &c	0x0018ff44 "烫烫?↑"

Memory
Address: 0x18ff44
| 0018FF44 | CC CC CC CC | 烫烫 |
| 0018FF48 | 88 FF 18 00 | .... |

Figure 4.36: Tracing error 3 in debugger 1.

```
#include <stdio.h>

int main(void)
{
    char c;
    scanf("%d", &c);

⇨   return 0;
}
```

Watch
Name	Value
c	10 '

Memory
Address: 0x18ff44
| 0018FF44 | 0A 00 00 00 | .... |
| 0018FF48 | 88 FF 18 00 | .... |

D:\MyWin32App\Win32App\Debug\
10

Figure 4.37: Tracing error 3 in debugger 2.

We can design another test case, where we use another input statement in character type scanf("%c", &c). What we want to inspect is how the first byte and the first four bytes after the address of variable c 0x18ff44 change. In Figure 4.38, we first enter character a, whose ASCII value is 97 or 0 × 61 in hexadecimal. We can see this 61 in the memory window. It is the only byte that has been changed. In Figure 4.39, we enter an integer 6 and it is clear that 4 bytes after address 0x18ff44 are all changed. The above results prove that c is allocated 1 byte of memory by the system.

```
#include <stdio.h>

int main(void)
{
    char c;
    scanf("%c", &c);
⇨│  scanf("%d", &c);

    return 0;
}
```

Watch
Name	Value
c	97 'a'
⊞ &c	0x0018ff44 "a烫蔫 ↑"

▶\ Watch1 ∕ Watch2 ∖ Watch3 ∖ Wa

Memory
Address: 0x18ff44
| 0018FF44 | 61 CC CC CC | a烫. |
| 0018FF48 | 88 FF 18 00 | .... |

Figure 4.38: Tracing error 3 in debugger 3.

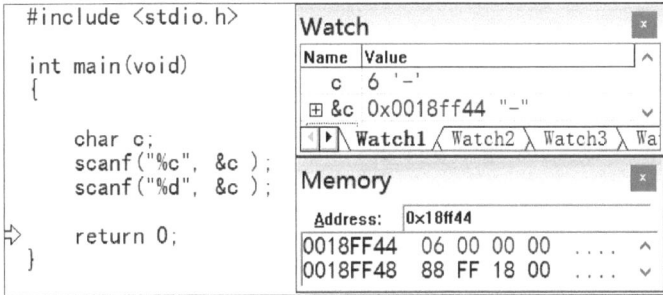

Figure 4.39: Tracing error 3 in debugger 4.

If we enter a nonnumerical character, variable c will not be assigned a value. Readers can try this test case on their own.

> **Programming error**
> Although the execution of the program continues when the argument type and type specifier are compatible but not consistent, data read are incorrect.

4.4.1.4 Common mistake of using scanf 4: '\n' used as newline

4.4.1.4.1 Sample program

```
int a;
scanf("%d\n",&a);
```

4.4.1.4.2 Phenomenon
After entering 5 ↙, the computer pauses execution and waits at console window.

4.4.1.4.3 Analysis
Except from %d, everything in the format control sequence should be input as-is. "\n" is not considered a "Carriage Return" in this case.

> **Programming error**
> "\n" is used in the format control sequence of scanf.

4.5 Summary

Relations between the main concepts of this chapter are shown in Figure 4.40.

People need to communicate with computers using a keyboard and screen,

Keyboard for input, screen for output, they are called standard devices.

The actual communication is carried out by input/output library functions,

Programmers only need to fill in the arguments.

There are specialized functions to handle character and strings for reading and displaying,

putchar and getchar handle character one at a time,

Whereas puts and gets can handle a series of characters efficiently.

Formatted functions printf and scanf are versatile and can handle all kinds of input,

But we need to match argument type with type specifier.

%c is for char, namely a single character,

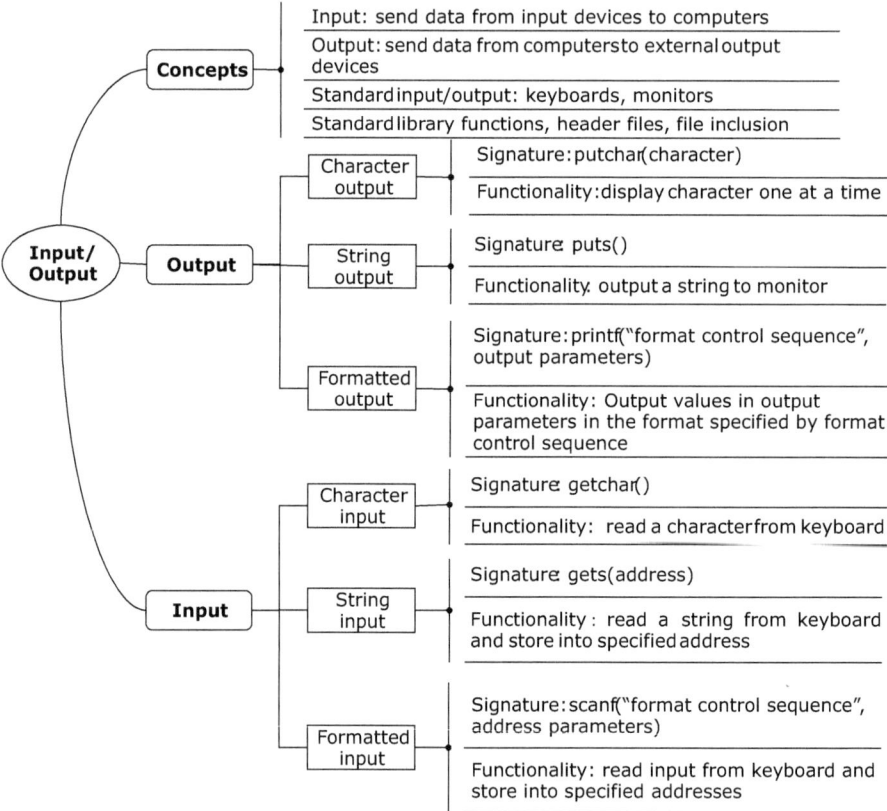

Figure 4.40: Concepts related to input/output and their relations.

%d is for int, %ld is for long,

%f is for float, %lf is for double,

We also have %s, which is used for strings.

Inside double quotation marks in scanf is the format control sequence,

Everything except the % specifier should be input as-is, and & is necessary for address arguments,

Otherwise, data will be misread and results won't be right.

After mastering scanf, printf should be easy-peasy.

4.6 Exercises

4.6.1 Multiple-choice questions

(1) [printf: ASCII values]

The ASCII value of character A is decimal number 65. What is the output of the following program? ( )

char ch1,ch2;

ch1='A'+'5'-'3';

ch2='A'+'6'-'3';

printf("%d,%c\n",ch1,ch2);

A) 67,D   B) B,C   C) C,D   D) Nondeterministic

(2) [printf: % sign]

What is the output of the following program? ( )

int a=2,c=5;

printf("a=%%d,b=%%d\n",a,c);

A) a=%2,b=%5   B) a=2,b=5   C) a=%%d,b=%%d   D) a=%d,b=%d

(3) [printf: width and precision]

Suppose we have the following definitions:

int n = 1234;

double x = 3.1415;

What is the output of the statement printf("%3d,%1.3f\n", n, x) ? ( )

A) 1234,3.142   B) 123,3.142   C) 1234,3.141   D) 123,3.141

(4) [Delimiter in scanf]

Suppose the variables are correctly defined as int variables. We want to assign 1 to a, 2 to b, and 3 to c with the statement scanf("%d%d%d",&a,&b,&c). Which of the following inputs is correct? ( )

(Note: □ represents a space)

A) 123<Return>   B) 1,2,3 <Return>   C) 1□2□3<Return>   D) 1;2;3<Return>

(5) [Delimiter in scanf]

Suppose we have defined int i and float f. We want to assign 100 to i and 765.12 to f with the statement scanf("i=%d,f=%f",&i,&f). Which of the following inputs is correct? ( )

(Note: □ represents a space)

A) 100□765.12<Return>

B) i=100,f=765.12<Return>

C) 100<Return>765.12<Return>

D) x=100<Return>,y=765.12<Return>

(6) [scanf with characters]

char c1,c2,c3;

scanf("%c%c%c",&c1,&c2,&c3);

We want to assign letters A, B, and C to variables c1, c2, and c3, respectively. Which of the following statements is correct about the input format? ( )

A) We should use spaces as delimiters.

B) We shouldn't use any delimiters.

C) We should use newlines as delimiters.

D) We should use tabs as delimiters.

4.6.2 Fill in the tables

(1) [putchar]

Fill in the table in Figure 4.41.

Statement	Output
putchar('D');	
putchar(67);	
putchar('a' + 10);	
putchar('\\');	
char ch_a = 'A'; putchar(ch_a);	

Figure 4.41: Input/output: fill in the tables question 1.

(2) [printf]

Fill in the table in Figure 4.42.

int a = 12; int b = 3;

Output statement	Output
printf("%d , %d\n", a,b);	
printf("%d\n", a, b);	
printf("%d , %d\n", a);	
printf("*%2d*\n", a);	
printf("*%10d*\n", a);	
printf("*%10D*\n", a);	
printf("*%+10.5d*\n", a);	
printf("*%+10d*\n", a);	

Figure 4.42: Input/output: fill in the tables question 2.

(3) [getchar]

Suppose we type in "boy", <Return> and "g i r l" when executing the following program. Fill in the table in Figure 4.43.

```c
#include <stdio.h>
int main(void)
{
  char ch_b, ch_o, ch_y, ch_g, ch_içh_r, ch_l;
  ch_b = getchar();
  ch_o = getchar();
  ch_y = getchar();
  ch_g = getchar();
  ch_i = getchar();
  ch_r = getchar();
  ch_l = getchar();
  return 0;
}
```

Variable	ch_b	ch_o	ch_y	ch_g	ch_i	ch_r	ch_l
Value							

Figure 4.43: Input/output: fill in the tables question 3.

(4) [scanf]
Fill in the table in Figure 4.44 with final values of the variables.

```c
#include <stdio.h>
int main(void)
{
  int age, count, grade;
  float salary;
  count = 100;
  count = scanf("age=%d,grade=%d", &age, &grade);
  scanf("%f", salary);
  return 0;
}
```

Input	12□21□123.21			
Variable	count	age	grade	salary
Value				
Input	age=12,grade=21 □123.21			
Variable	count	age	grade	salary
Value				

Figure 4.44: Input/output: fill in the tables question 4.

4.6.3 Programming exercises

(1) We use the following rule to convert a hundred-mark system grade into a letter grade: 'A' = larger than 90, 'B' = 80~89, 'C' = 70~79, 'D' = 60~69, and 'E' = < 60. Write a program that reads a number grade, calculates the corresponding letter grade, and displays both grades (the number should have two digits in its fraction part, for example, 78.5 should be displayed as 78.50).

(2) Write a program that does the following:
1) Read three integers into variables a, b, and c
2) Assign the initial value of a to b
3) Assign the initial value of b to c
4) Assign the initial value of c to a

(3) Write a program that does the following:
1) Read three double numbers
2) Calculate and output the average of them. The average should be rounded and have one digit in its fraction part.

(4) Write a program that reads three characters and outputs them along with their ASCII values.

5 Program statements

Main contents
- Usage and rules of expressions that construct branches in programs
- The basic concept of loops
- Three constructions of loops and their use cases
- Nested loops
- Characteristics of statements of the same type, their relations and selection criteria
- Exercises of efficiently analyzing programs using flowcharts
- Exercises of top-down algorithm design
- Exercises of program reading
- Debugging methods and techniques of program statements

Learning objectives   !
- Know how to use basic statements to write programs in sequential, branch, and loop structures.
- Know the syntax of expression statements and difference between expressions and expression statements.
- Know how to use fundamental control structure and control statements in C.
- Know basic debugging techniques and how to select test cases.

5.1 Sequential structure

A program of sequential structure consists of statements that are executed in sequential order. Such programs have the simplest structure among all structured programs as shown in Figure 5.1.

Example 5.1 Data swap
Define two integer variables, input two integers, swap values of these variables and output them.

[Analysis]
1. Algorithm description
The process flow is shown in Figure 5.2. We first define two integer variables a and b, then input two integers and store them into a and b. Next, we swap their values and output them. Method of swapping values of a and b has been covered in the section "Effectiveness of Algorithms" in Chapter 2.

https://doi.org/10.1515/9783110692327-005

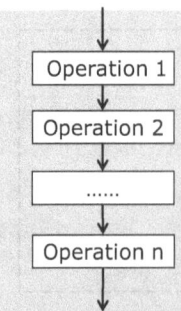

Figure 5.1: Sequential structure.

2. Program implementation

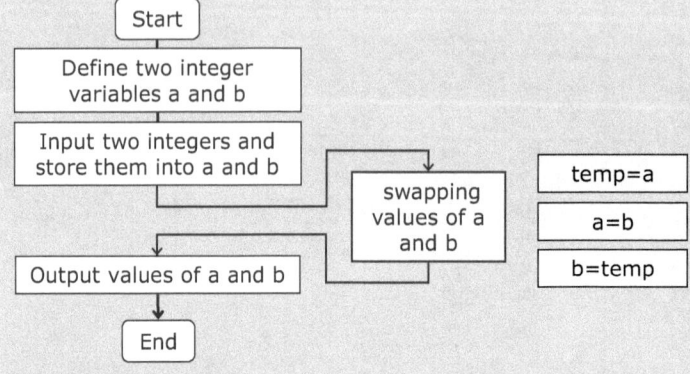

Figure 5.2: Data swap.

```
01 #include<stdio.h>
02 int main(void)
03 {
04     int a,b,temp;
05     printf("Enter a,b: ");
06     scanf("%d,%d",&a,&b);
07     printf("Before swap: a=%d,b=%d\n",a,b);
08     temp=a;
09     a=b;
10     b=temp;
11     printf("After swap: a=%d,b=%d\n",a,b);
12     return 0;
13 }
```

Note that on lines 7 and 11, we print values of both variables before and after swap to conveniently check whether execution result is correct.

Example 5.2 Grade processing

Type in student numbers and English exam grades of four students, print the data and output average grade.

[Analysis]

1. Algorithm description

As shown in Figure 5.3, the program executes the following operations in order: input, computation, and output. The program is given below.

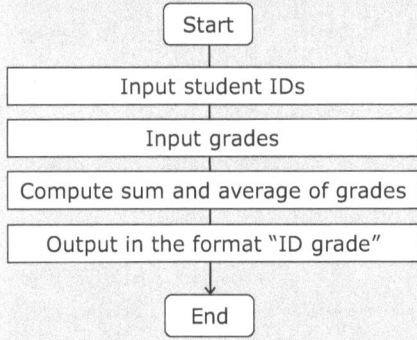

Figure 5.3: Flow of grade processing.

As the program needs to handle grades of four students, four variables are declared on line 4 and four are declared on line 5 to store student numbers and grades. We also noticed that four repeated formatted output functions are used between lines 13 and 16 to output grades.

2. Program implementation

```
01 #include <stdio.h>
02 int main(void )
03 {
04     int number1,number2, number3, number4; //Declare 4 student numbers
05     float grade1, grade2, grade3, grade4;  //Declare 4 grades
06     float ave;                             //Average grade
07
08     printf("input 4 numbers:\n ");         //Student number input prompt
09     scanf("%d%d%d%d",&number1,&number2,&number3, &number4);
10     printf("input 4 grades:\n ");          //Grade input prompt
11     scanf("%f%f%f%f", &grade1, &grade2, &grade3, &grade4);
12     ave=(grade1+ grade2+ grade3+ grade4)/4; //Compute average
13     printf("%d: %f\n ", number1, grade1);
14     printf("%d: %f\n ", number2, grade2);
15     printf("%d: %f\n ", number3, grade3);
16     printf("%d: %f\n ", number4, grade4);
17     printf("average=%f\n ", ave);
18     return 0;
19 }
```

Result:
input numbers:
1 2 3 4
input grades:
86 92 75 64
1: 86.000000
2: 92.000000
3: 75.000000
4: 64.000000
average=79.250000

3. Discussion
Are there any shortcomings in this program?
Discussion: There are repeated statements for similar variables. It would be tedious to write in this way if there were 100 students. We can improve it by using loops.

5.2 Double branch structure

5.2.1 Syntax of double branch structure

In the section of the basic structure of algorithms, we have seen that we need to make judgments when configuring a washing machine. The results of such judgments have two branches, namely "yes" and "no." We extracted a generalized model from this practical problem and used a flowchart or pseudo code to describe the process. In practice, the conditional decision is implemented by expressions in C. Such a branch structure with two exits is represented by an if-else statement as shown in Figure 5.4.

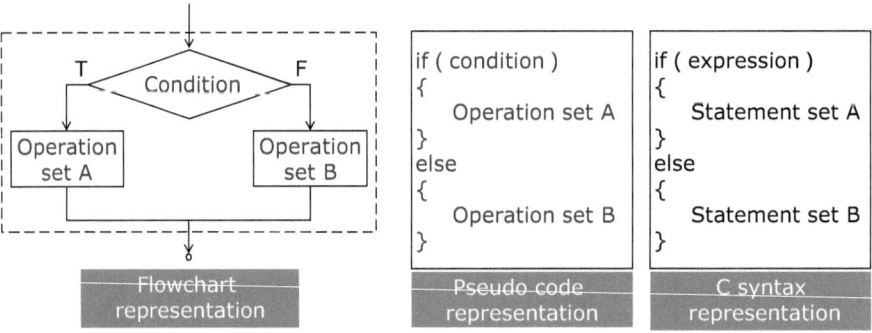

Figure 5.4: Representation of branch structure.

The syntax and flowchart representation of if-else statement are shown in Figure 5.5. Essentially, if the value of the expression is true, statement set A is executed; otherwise, statement set B is executed.

Think and discuss What type of expressions can be used as the "expression" in if-else statements?

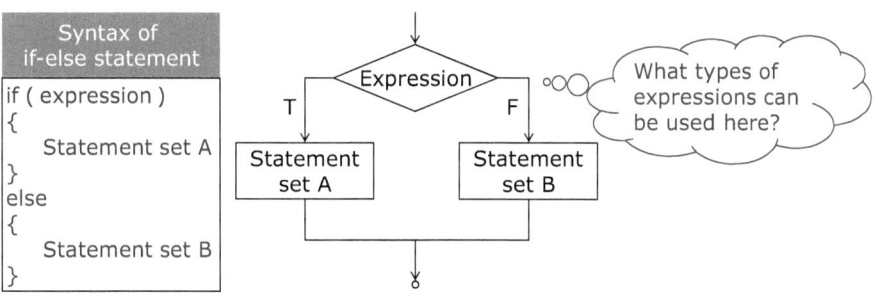

Figure 5.5: Syntax of if-else statement.

From a practical perspective, the answer should be conditional decision, namely relational and logical expressions. Grammatically, however, all correct C expressions can be compiled without errors. Although there are few restrictions put by grammar, we should still design our algorithm based on the logic of practical problems.

There is also a special case of if-else statement as shown in Figure 5.6. The else branch can be omitted to form a single branch structure.

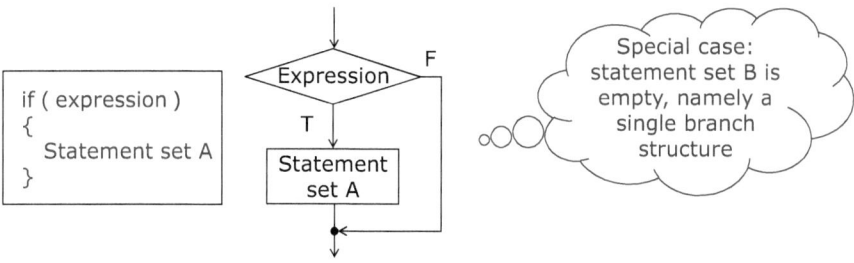

Figure 5.6: Special case of if-else statement.

5.2.2 Role of compound statements

In grammar rules of C, a single statement is sometimes required in a syntax. However, one statement may not be sufficient to complete a task in practice. Hence, a workaround is needed in the grammar, which is called "compound statements."

Formally, a compound statement is a set of multiple statements, but it is a whole entity and works as a single statement grammatically. Whenever a single statement is needed, a compound statement can be used instead as needed. The definition and usage of compound statements are shown in Figure 5.7.

<table>
<tr><td>
Compound statement

  A compound statement is formed by surrounding a group of statements with {}.
  As defined by grammar of C, a compound statement is treated as a single statement instead of multiple.
</td><td>
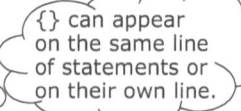

{} can appear on the same line of statements or on their own line.
</td></tr>
</table>

Figure 5.7: Compound statement.

Example 5.3 Role of compound statements
Analyze the role of compound statements in the two programs given in Figure 5.8.

[Analysis]
In the first program, the condition after "if statement" checks whether 1 is larger than 2. According to grammar rules, the first statement after if is executed if the condition evaluates to true, or skipped otherwise. Hence the first printf statement is not executed and the output is "The second statement The third statement."

    The second program uses brackets, thus the "if statement" has two printf statements in its true branch. Hence the output is "The third statement."

```
# include <stdio.h>
int main(void)
{
    if (1 > 2)
    printf("The first statement\n");
    printf("The second statement\n");
    printf("The third statement\n");
    return 0;
}
```
Result:
The second statement
The third statement

```
# include <stdio.h>
int main(void)
{
    if (1 > 2)
    {
        printf("The first statement\n ");
        printf("The second statement\n ");
    }
    printf("The third statement\n ");
    return 0;
}
```

{} can control scope of if statement

Result:
The third statement

Figure 5.8: Role of compound statements.

5.2.3 Example of if statements

Example 5.4 Use if statements to implement result check in the price guessing game
In the price guessing game, the host responds with "too high," "too low," or "exactly" for each guess. Suppose the actual price of an item is ¥168 and let value denote the guesses of a participant. Write statements for the result check.

[Analysis]

The program implementation is shown in Figure 5.9. The three if statements print corresponding results when the value of variable value is greater than, less than, or equal to the actual price 168.

```
if (value>168)  printf("Too high");
if (value<168)  printf("Too low");
if (value==168) printf("Exactly");
```

> {} can be omitted if there is only one statement after if

Figure 5.9: Result check in the price guessing game.

What are the execution paths of these three branching statements?

Let us draw and study the flowchart as shown in Figure 5.10.

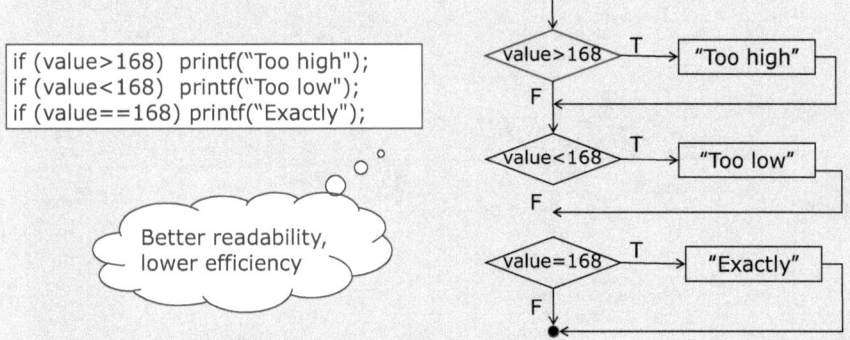

Figure 5.10: Result check solution of the price guessing game.

Logically, if value is greater than 168 and "too high" is printed, it is redundant to check whether value is less than or equal to 168. A program like this has better readability but worse efficiency.

Following the logic of the problem, we first revise the flowchart. As long as a check yields true, we can print the result and terminate the flow. The corresponding program statements are shown in Figure 5.11.

We first check whether value is greater than 168. If true, "too high" is printed. Otherwise, the flow enters the branch for values less than or equal to 168. In this branch, we once again check whether value is less than 168. If true, "too low" is printed. Otherwise, we check whether it is equal to 168. In this case, the result has to be true, so "exactly" is printed.

Note how brackets of if-else statements are aligned in the figure. It is not easy to figure out the execution logic of this refined program without a flowchart to refer to. In other words, we have compromised readability for higher efficiency.

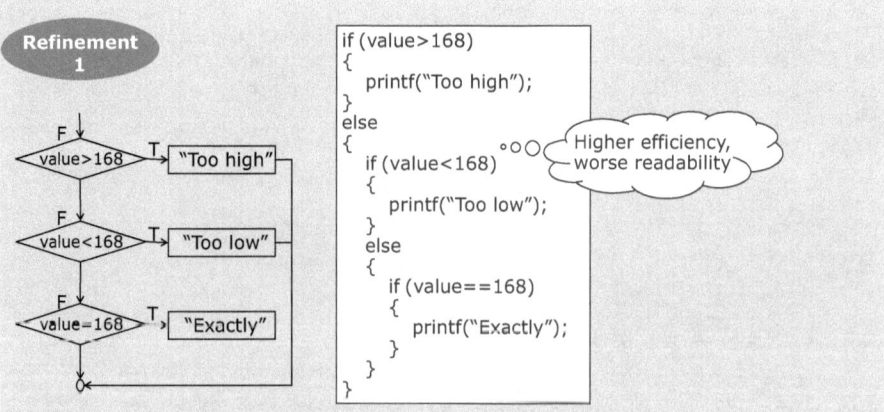

Figure 5.11: Refinement of the price guessing game solution 1.

In fact, the last equality condition in the flow is logically redundant, thus we can further improve the flow. The refined flow and corresponding statements are shown in Figure 5.12.

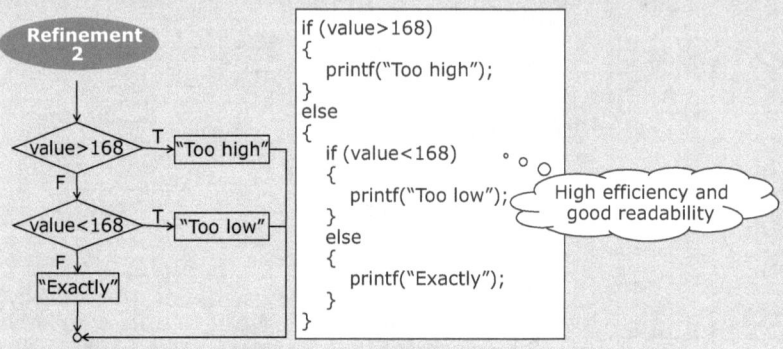

Figure 5.12: Refinement of the price guessing game solution 2.

We first write the case of value greater than 168, where "too high" is printed. If the comparison yields false, we continue to process the cases where value is less than or equal to 168. If it is less than 168, "too low" gets printed; otherwise, "exactly" is printed.

It is clear from this refinement process that flowcharts help us study the execution of programs more intuitively and clearly compared with program statements.

5.2.4 Nested if-else statements

5.2.4.1 Nesting rule of if-else

In examples above, we have seen that an if-else statement can be written inside another if statement, constructing a nested if statement.

However, if two if statements are written in the way illustrated on the left side of Figure 5.13, which if should be matched with else?

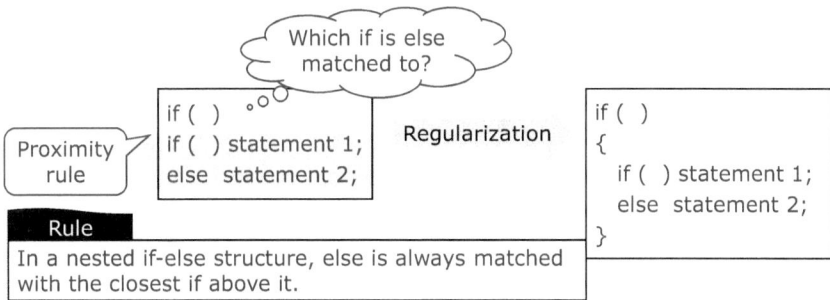

Figure 5.13: Matching of nested if-else case 1.

To avoid ambiguity, we need to design a rule for such situations in grammar. C requires that an else is always matched with the closest if above it in a nested if statement. This is also known as the "proximity rule."

To achieve better readability in this example, a more formal way is to add brackets to the first if to show its scope clearly.

What if the logic of a problem requires that the else in Figure 5.13 should be matched with the first if?

In this case, we can apply the rule of compound statements to write it in the way shown in Figure 5.14. In other words, we need to wrap the true branch with brackets so that it is considered to be a compound statement and in the scope of the first if. Now the first if is the closest one to else, with respect to the nesting rule of if-else statement.

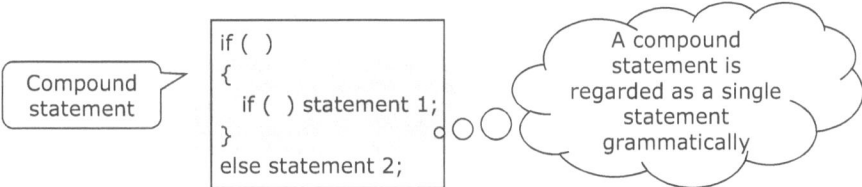

Figure 5.14: Matching of nested if-else case 2.

5.2.4.2 Note on using nested if-else

It is hard to read a program using nested if-else with too many layers. As a fundamental conditional statement in all programming languages, if-else statements are used excessively in programming. However, in the experience of programming, it is not recommended to use a nested if-else with more than three layers. Otherwise, the program becomes extremely unreadable and hard to maintain later.

Good programming habit
Avoid using too many nested if-else statements.

Example 5.5 Compute maximum of three numbers
Compute the maximum of three integers a, b, and c (whose values are obtained from keyboard input).

[Analysis]
1. Data analysis
As per procedures of algorithm design, we should first analyze relations between the data to be processed; in this case, a, b, and c. Possible cases are listed in Figure 5.15.

Possible cases of data	General case	a, b and c have distinct values
	Special or edge case	At least two of them are equal

Figure 5.15: Analysis of data in the maximum problem.

Following the first step of algorithm design, we should design the flow starting from general cases. After designing a draft algorithm, we test it against special cases and edge cases and update it accordingly.

2. Algorithm design solution 1
Based on the requirements in the problem description, we can write out top-level and refined pseudo code as shown in Figure 5.16, with which we can draw the flowchart of execution in Figure 5.17.

Top level pseudo code	First refinement	Second refinement
	Input three numbers a,b and c	Input three numbers a,b and c
	Compare a with b	if a>b
Compare three numbers a, b and c, find the maximum of them	If a is larger, compare a with c, where the larger is the maximum	if a>c max=a
		else max=c
	Otherwise, compare b with c, where the larger is the maximum	else
		if b>c max=b
		else max=c
	Output result	Output max

Figure 5.16: Maximum problem algorithm design solution 1.

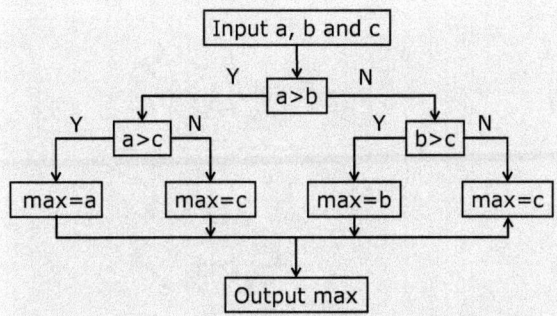

Figure 5.17: Flowchart of maximum problem algorithm design solution 1.

3. Testing
We have designed the algorithm flow based on general cases; therefore, the next step is to test it against special cases of data. Figure 5.18 shows the test result.

Case	Result
a=b=c	max=c
a=b	max is the larger between b and c
a=c	max is the larger between b and c
b=c	max is the larger between a and c

Figure 5.18: Test result of algorithm for maximum problem.

After passing all tests, we can start coding with the second refinement pseudo code or the flowchart.

4. Program implementation

```
1    //Compute maximum of a, b and c, store it into max
2    #include <stdio.h>
3    int main(void)
4    {
5        int a, b, c, max;
6
7        scanf("%d,%d,%d",&a, &b, &c); //Type in a, b and c
8        if (a>b)
9        {
10           if (a>c)
11           {
12               max=a;
13           }
14           else
15           {
16               max=c;   //max is the bigger one between a and c
17           }
18       }
19       else
20       {
```

```
21    if (b>c)
22    {
23       max=b;
24    }
25    else
26    {
27       max=c;  //max is the bigger one between b and c
28    }
29  }
30  printf("max=%d", max); //Output result
31  return 0;
32  }
```

5. Algorithm design solution 2
In the first solution, as max records the maximum value, we can start to use it in the comparison of a and b. The refined pseudo code is given in Figure 5.19.

First refinement
Input three numbers a,b and c
Compare a and b,store the larger one into max
Compare max and c,store the larger one into max
Output result

Figure 5.19: Maximum problem algorithm design solution 2.

6. Algorithm design solution 3
We can also use conditional expression:

```
max=(a>b)? a : b;
max=(max>c)? max : c;
```

5.3 Multiple branch structure

5.3.1 Introduction of multiple branch problems

Case study 1 Multiple branch problem in washing machine settings
In the discussion of fundamental structures of algorithms, we have seen that we may need to choose from multiple options when configuring a washing machine. As shown in Figure 5.20, there are multiple branches for the decision. Using the double branch structure to describe the problem gives us the flow in Figure 5.21, in which it is not hard to see that we can only check one condition at a time. We can imagine that it is necessary to check all possible conditions layer by layer to cover all possible cases if there are many of them in a problem.

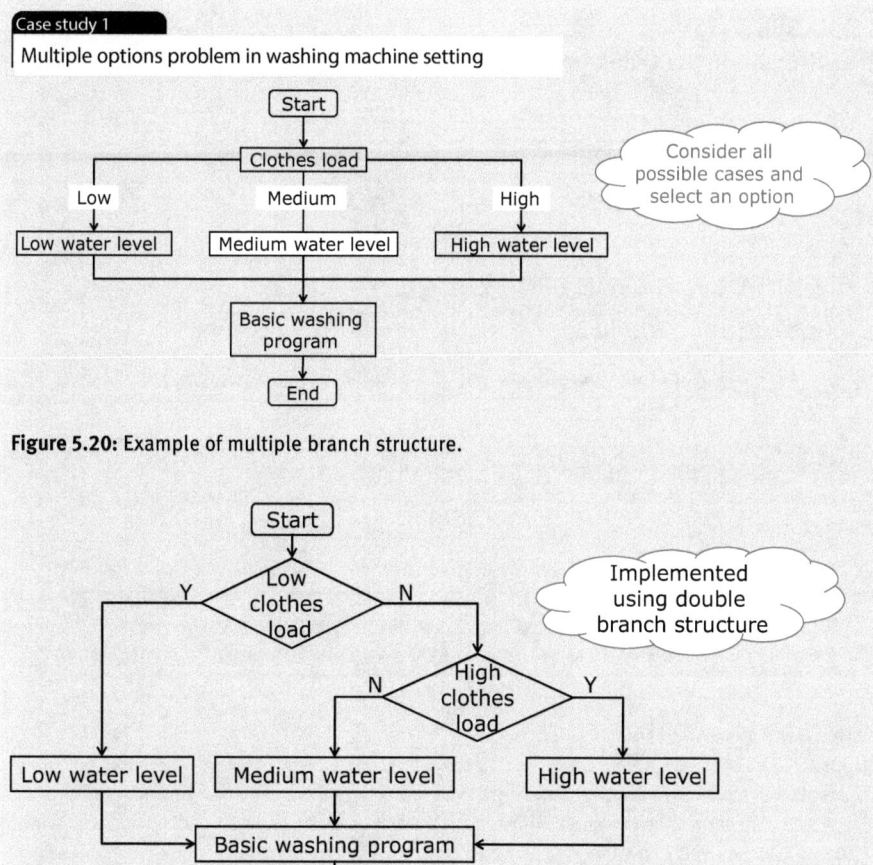

Case study 1
Multiple options problem in washing machine setting

Figure 5.20: Example of multiple branch structure.

Figure 5.21: Double branch structure implementation of multiple branch problem.

Case study 2 Multiposition switches in real life

As shown in Figure 5.22, multiposition switches in real life are all examples of multiple-level selection. Recall how we toggle a switch in practice: we determine a target position and toggle the switch to it at one go.

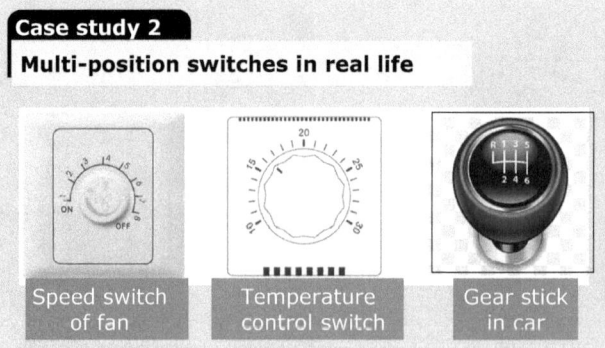

Figure 5.22: Multiposition switches in real life.

Think and discuss Differences between multiposition and multibranch
Essentially, what are the differences between toggling a multiposition switch and a multilayer double branch structure?

Discussion: When toggling a multiposition switch, we simply toggle it to the desired position. Meanwhile, a judgment is needed in each step when using a double branch structure for a multiple branch problem, which is inconvenient. If we have a mechanism similar to the multiposition switch in C language, it would be easier to deal with multiple branch problems in practice.

Case study 3 Mr. Brown's memo book
Mr. Brown is pretty busy, so he usually writes down the schedule for the next week in an electronic memo book in advance. There was a week where Mr. Brown had the schedule shown in Figure 5.23. Figure 5.23 also shows the flowchart he made for the schedule, drawing on multiposition switches. In such a flow, he could know the schedule on a certain day by simply querying the day of the week.

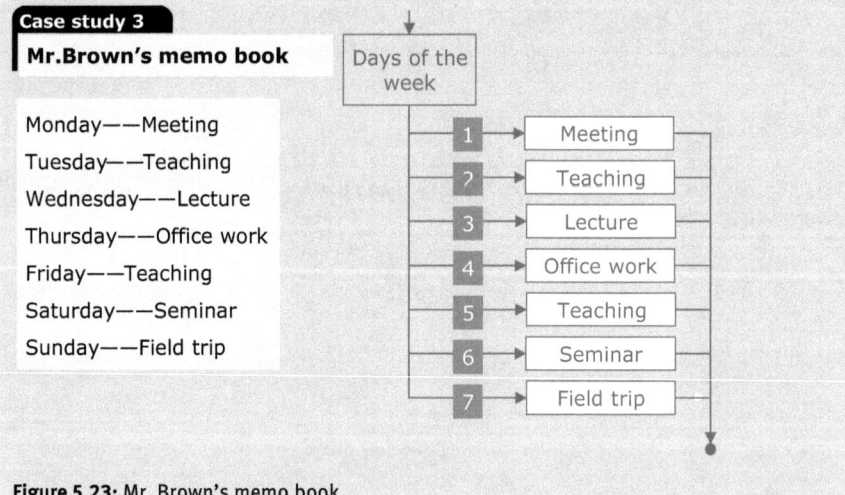

Figure 5.23: Mr. Brown's memo book.

5.3.2 Syntax of multiple branch structure

5.3.2.1 Multiple branch structure model and grammatical representation

Based on the flowchart above, we can summarize the abstract model framework of multiple branch structures. Note that there is a case for "exceptions" after listing all normal cases. It works as a processing path for exceptions after considering all possible cases. It is designed to make the system complete. Switch statements are introduced to C as an implementation of this model. Syntax of switch statements is shown in Figure 5.24.

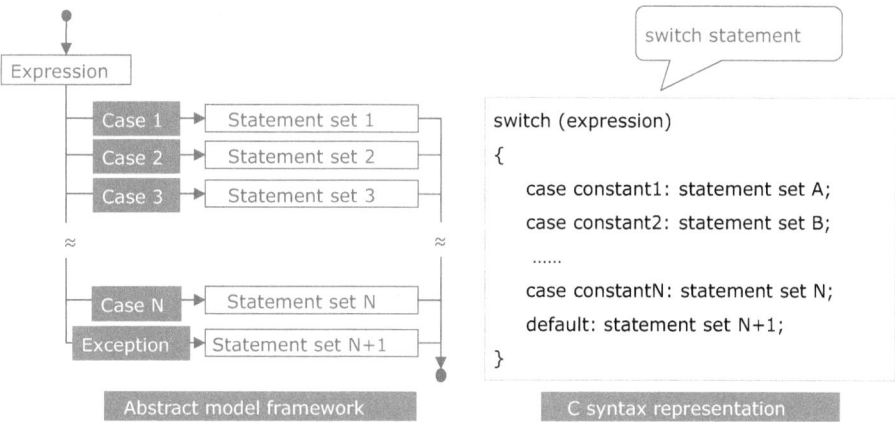

Figure 5.24: Multiple branch model and its representation.

A switch statement works as follows: the value of the expression is evaluated and compared with values of constant expressions below. In the case of a match, statements after that constant expression are executed. The remaining comparisons are then skipped and statements in these cases are executed as well. If none of the constants is equal to the value of the expression, statements in the default case are executed.

5.3.2.2 Grammar test of switch statements

Mr. Brown wrote a schedule querying program using switch statements and tried to obtain his schedule for Wednesday. However, the result he obtained was as shown in Figure 5.25. In addition to the schedule for Wednesday, the program output schedule for the remaining days of the week as well as a warning. What was wrong with the program?

After checking his program, Mr. Brown found that all printf statements in case 3 and cases below were executed given input 3. Comparing the grammar of switch statements in Figure 5.24 and his flowchart, he realized that the execution process

```
#include <stdio.h>
int main(void)
{
   int a;
   printf("Input day of the week: ");
   scanf("%d",&a);
   switch (a)
   {
      case 1: printf("Monday: meeting\n");
      case 2: printf("Tuesday: teaching\n");
      case 3: printf("Wednesday: lecture\n");
      case 4: printf("Thursday: office work\n");
      case 5: printf("Friday: teaching\n");
      case 6: printf("Saturday: seminar\n");
      case 7: printf("Sunday: field trip\n");
      default: printf("Invalid input\n");
   }
   return 0;
}
```

Program result :
Input day of the week: 3
Wednesday: lecture
Thursday: office work
Friday: teaching
Saturday: seminar
Sunday: field trip
Invalid input

Figure 5.25: Mr. Brown's weekly schedule querying program.

of the switch statement was not consistent with the logic of the flowchart. Hence, the grammar model of switch statements needed to be updated.

5.3.2.3 Refined switch statements model and grammar representation

Mr. Brown thought that an interruption mechanism was necessary for switch statements. After statements in a case were executed as required, the program should be able to jump out of the switch statement as shown in Figure 5.26. This interruption

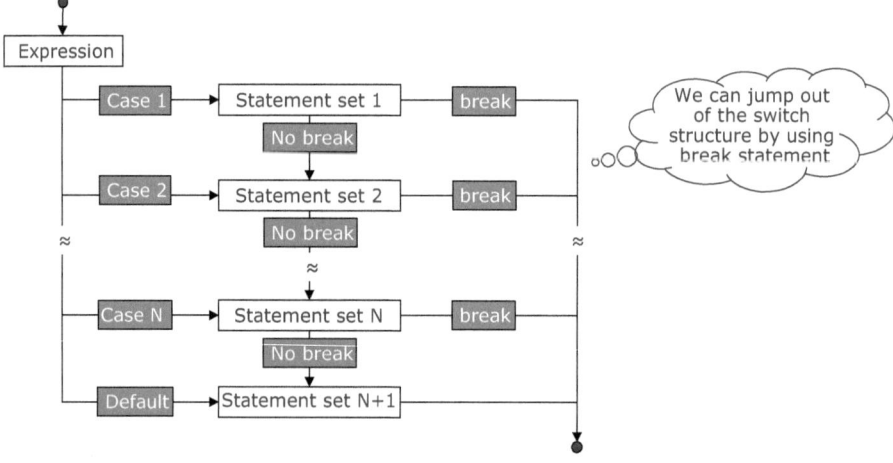

Figure 5.26: Logical flow of switch statements.

mechanism is implemented by break statements in C. Programmers can choose to use break or not based on the logic of the problem. By now, the switch statement model is completed and can handle multiple-branch cases properly.

The syntax of switch statements based on the updated model is shown in Figure 5.27. Break statements can be added if necessary. We put them into square brackets in the grammar description, meaning that they are optional.

Syntax of switch statement

```
switch (expression)
{
  case constant1: statement set A;  [ break; ]
  case constant2: statement set B;  [ break; ]
  ......
  case constantn: statement set N; [ break; ]
  default: statement set N+1;
}
```

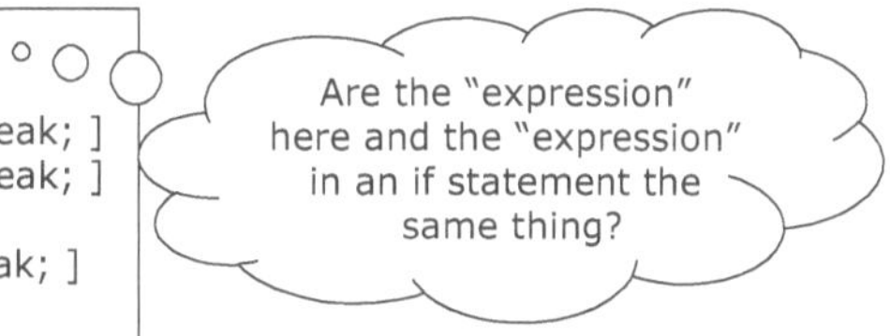

Figure 5.27: Syntax of switch statement.

5.3.2.4 Execution process of switch statements

The execution process of switch statements is as follows:

(1) When executing a switch statement, the expression inside the parentheses that follow it is evaluated first. The program looks for a case value, from constant 1 to constant n, in the body of the switch statement that matches.
(2) If there is a match, statements between that case and the end of the switch statement are executed, including all the remaining cases and default case.
(3) If there is no matching case and a default case exists, statements between default and the end of the switch statement are executed.
(4) If there is no matching case and no default case, the program jumps out of the switch statement and executes statements after it.

Note:
(1) Break statement is used to jump out of the switch statement.
(2) Constants 1 through n are distinct numerical literals or character literals.
(3) Each case branch can comprise multiple statements. {} are optional.
(4) Do not omit the default case. In case an exception occurs, namely, the value of the expression does not match any case, the program may crash.
(5) The value of the expression after switch can be of any type except floating-point types. Why is this the case?

When we introduced data types, there was a rule on real numbers, which stated that "avoid checking whether two real numbers are equal," as such comparisons may yield wrong results.

Think and discuss Are the "expression" in switch and the "expression" in if the same thing?

The expression in if: It is designed to check a condition and the result is either true or false, thus it should be a relational or logical expression.

The expression in switch: It is designed to match one of all possible cases and the result is an integer, thus it should be an arithmetic expression.

5.3.2.5 Testing the refined program

Using the refined grammar of switch statements, Mr. Brown added a break in his querying program, which passed further tests as shown in Figure 5.28.

```
#include <stdio.h>
int main(void)
{
    int a;
    printf("Input day of the week: ");
    scanf("%d",&a);
    switch (a)
    {
        case 1: printf("Monday: meeting\n");    break;
        case 2: printf("Tuesday: teaching\n");    break;
        case 3: printf("Wednesday: lecture\n");    break;
        case 4: printf("Thursday: office work\n");    break;
        case 5: printf("Friday: teaching\n");    break;
        case 6: printf("Saturday: seminar\n");    break;
        case 7: printf("Sunday: field trip\n");    break;
        default: printf("Invalid input\n");
    }
    return 0;
}
```

The refined program uses break statements

Program result:
Input day of the week: 3
Wednesday: lecture

Figure 5.28: Refined querying program.

5.3.3 Example of multiple branch structure

Example 5.6 Grade conversion

Enter a 100-mark system grade, convert it into the corresponding grade in a five-level system. and output. The conversion rule is shown in Figure 5.29.

grade=	A	90≤score≤100
	B	80≤score<90
	C	70≤score<80
	D	60≤score<70
	E	score<60

Figure 5.29: Grade conversion table.

1. Problem analysis

This is a segment problem which can be solved using if statement. It is not hard to write the following program.

```
1   #include <stdio.h>
2   int main(void)
3   {
4       int score;
5       printf("Please input score: ");
6       scanf("%d", &score);    //Input grade
7       if ( score>100 || score <0 )
8       printf("input error! "); //Error handling
9       else if (score >= 90) printf("%d--A\n", score);
10      else if (score >= 80) printf("%d--B\n", score);
11      else if (score >= 70) printf("%d--C\n", score);
12      else if (score >= 70) printf("%d--D\n", score);
13      else if (score >= 0 ) printf("%d--E\n", score);
14      else printf("Input error\n");
15      return 0;
16 }
```

Review: This program uses nested if-else statements. As there are many segments in the problem, the program uses many branches. As a result, it is hardly readable.

We shall discuss a solution using switch statement now.

2. Solution design

We first write out the skeleton of a switch statement:

```
switch ( expression )
{
   case constant1: printf("%d-----A\n", score); break;
   case constant2: printf("%d-----B\n", score); break;
   case constant3: printf("%d-----C\n", score); break;
   case constant4: printf("%d-----D\n", score); break;
   case constant5: printf("%d-----E\n", score); break;
   default:        printf("Input error\n");
}
```

The key is to determine the expression in switch (expression). The range of score is 0–100, it is impossible to list all 100 cases. It is also tricky to find a formula to split the range unevenly into five levels. Based on characteristics of this problem, we can divide score by 10 and split the range into 10 levels. Consequently, the skeleton can be updated as follows:

```
switch ( score/10 )
{
  case 10:
  case 9:  printf("%d-----A\n", score); break;
  case 8:  printf("%d-----B\n", score); break;
  case 7:  printf("%d-----C\n", score); break;
  case 6:  printf("%d-----D\n", score); break;
  case 5:
  case 4:
  case 3:
  case 2:
  case 1:
  case 0:  printf("%d-----E\n", score); break;
  default: printf("Input error\n");
}
```

Note:
(1) The result of score/10 is integer as score is integer.
(2) When score = 100, score/10 = 10, the program jumps to the branch case 10. Because there is no statement in this branch, the program executes the statement below, namely printf ("%d-----A\n", score) as per grammar of switch statement. The program jumps out of the switch when a break is encountered.
(3) The program works similarly for the case score <60.

3. Testing and refinement
Based on data characteristics of this problem, we can design the following test cases as shown in Figure 5.30.

score	score/10
>=110	default
100<score<110	default
100	10
90<= score<100	9
80<= score<90	8
70<= score<80	7
60<= score<70	6
0<− score<60	5/1/3/2/0
score<0	default

Figure 5.30: Test cases of grade conversion problem.

When using these test cases to verify the result, we found an error: when 100 < score < 110, score/10 = 10, and the program outputs "A."

4. Refined program
Part of the refined program is given below:

```
scanf("%d", &score);
if (score>100 && score<110) score=110;//Treat numbers between 100 and 110 as 110
```

```
switch ( score/10 )
{
   case 10:
   case 9:  printf("%d-----A\n", score); break;
   case 8:  printf("%d-----B\n", score); break;
   case 7:  printf("%d-----C\n", score); break;
   case 6:  printf("%d-----D\n", score); break;
   case 5:
   case 4:
   case 3:
   case 2:
   case 1:
   case 0:  printf("%d-----E\n", score); break;
   default: printf("Input error\n");
}
```

Example 5.7 Convert if statements into switch statement
Rewrite the following statements using switch statement (a is an integer).

```
if ( a<5) && (a>=0)
{  if (a>2)
   { if (a<4)   x=1;
     else x=2;
   }
   else x=3;
}
```

[Analysis]

1. Program analysis

We want to use the switch statement to show what the value of x is given various values of a, so we should find the relation between a and x first. The given statements have poor readability; therefore, it is better to use a coordinate system and make a relation table of a and x as shown in Figure 5.31. It is easier to write switch statements in this way.

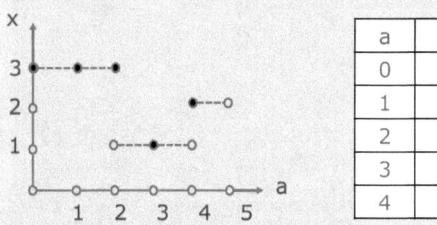

a	x
0	3
1	3
2	3
3	1
4	2

Figure 5.31: Relation between a and x.

2. Program implementation

```
switch (a)
{  case 0:
```

```
   case 1:
   case 2:   x=3;  break;
   case 3:   x=1;  break;
   case 4:   x=2;  break;
   default:  printf("a is error\n");
}
```

3. Discussion

What if a is a real number?

Analysis: The value of the expression of a switch statement must be discrete rather than continuous. If a is a real number, it changes continuously. Based on the coordinate system graph
given above, we can update the relation table of a and x as shown in Figure 5.32.

a	x
0<=a<=2	3
2<a<4	1
4<=a<5	2

Figure 5.32: Updated relation between a and x.

We can use truncation to handle values of a as if they were discrete as shown in Figure 5.33.

a	(int)a	x
	0	
0<=a<=2	1	3
	2	
2<a<4	2	1
	3	1
4<=a<5	4	2

Figure 5.33: Updated relation between a and x with truncation.

It is clear that $x = 3$ when $a = 2$ and $x = 1$ when $2 < a < 3$. The program needs to distinguish these
two situations.

Program implementation:

```
switch ( (int) a )
{ case 0:
  case 1:
  case 2:   if (a>2)  x=1;
            else x=3;  break;
  case 3:   x=1;  break;
  case 4:   x=2;  break;
  default: printf("a is error\n");
}
```

4. Testing

According to principles of testing, we need to test normal values, corner values, special values
as well as error cases. Possible test data are shown in Figure 5.34.

a	<0	0	1	2	2~3	3	4	>4
x	Error	3	3	3	1	1	2	Error

Figure 5.34: Test data.

During testing, we need to rerun the program for each test case, which is inconvenient. Ideally, the program should be executed once and terminated after being tested against all test data. We can achieve this goal after learning loop statements.

Example 5.8 Calculator for arithmetic operations
Design a program that does addition, subtraction, multiplication, and division given expressions entered by users through the keyboard.

[Analysis]
1. Program analysis
Concerning the characteristics of the input data, the operator is the only information that can be used to distinguish between expressions. The operators are of character type and hence integers as well. They can be used to distinguish between different cases, thus we use them as the case values in the switch statement. Figure 5.35 shows all cases of input and output data.

case	Input			Output
	float	char	float	float
'+'	a	+	b	a+b
'-'	a	-	b	a-b
'*'	a	*	b	a*b
'/'	a	/	b	a/b

Figure 5.35: Data analysis.

2. Program implementation
```
1    #include <stdio.h>
2    int main(void)
3    {
4      float a,b;   //Define the operands
5      char c;     //Operator
6
7      printf("input expression: a+(-,*,/)b \n");//Input prompt
8      scanf("%f%c%f",&a,&c,&b); //Enter the expression in order
9      switch(c)      //Compute based on type of operator
10     {
11       case '+':
12       printf("%f\n",a+b);
13       break;
14       case '-':
15       printf("%f\n",a-b);
16       break;
17       case '*':
18       printf("%f\n",a*b);
19       break;
20       case '/':
21       printf("%f\n",a/b); //Can't handle division by 0
23       default:
```

```
24      printf("input error\n");
25  }
26  return 0;
27  }
```

5.3.4 Comparison of various branch structure statements

Branch statements include double branch statements and multiple branch statements, which have different features and use cases as shown in Figure 5.36. Switch statements are used to distinguish multiple cases. Using them makes program structure clearer. However, they can only be used when the case expression evaluates to an integer.

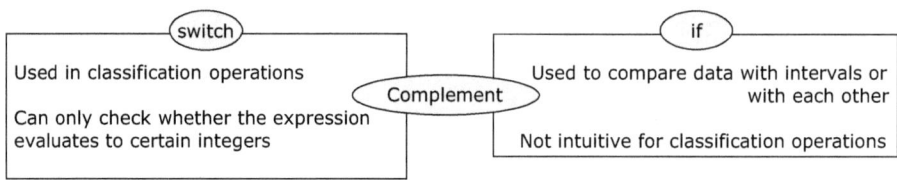

Figure 5.36: Comparison of if and switch.

Conditional statements are used for comparing data with intervals or with each other. Switch statements and if statements complement each other in functionalities.

5.4 Introduction of loop problems

5.4.1 Analysis of key elements in loops

In the discussion of fundamental structures of algorithms, we have seen that repeated operations may be necessary when setting up a washing machine. Repeated operations form a loop as shown in Figure 5.37.

In the grade processing program, discussed in the section of sequential structure, we have seen that statements for similar variables are repeated multiple times. It would be tedious to write the program in this way if we have 100 students. As shown in Figure 5.38, we can write the grade processing program in loops in the same way as how we dealt with the loop flow of washing machines.

Case study 1

Repeated operations in washing machine settings

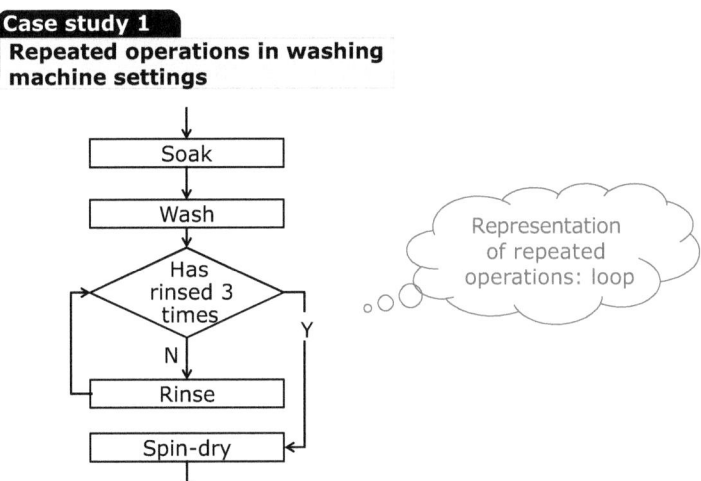

Figure 5.37: Repeated operations in practice.

Case study 2

A problem in grade processing

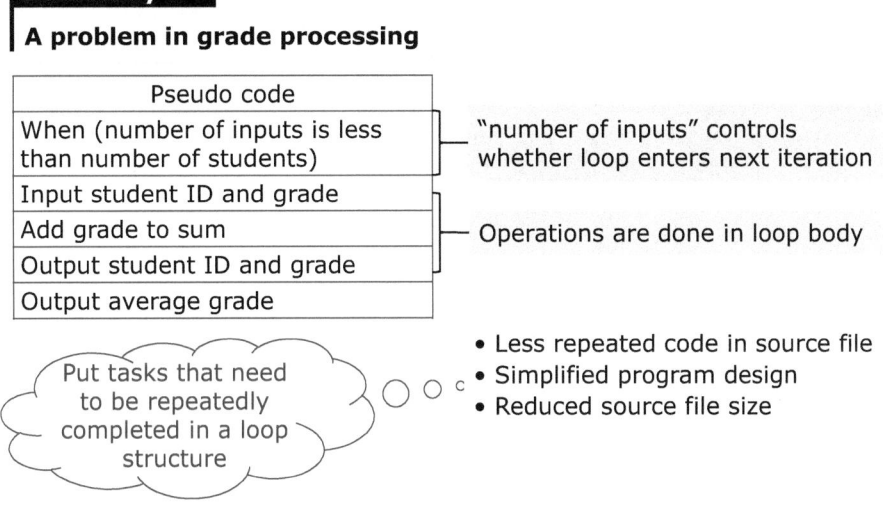

Figure 5.38: Loop in the grade processing problem.

The condition checked in this loop is whether number of inputs is less than number of students. Actual input, addition and output are done inside the loop body. It is clear that it is beneficial to do work that requires multiple executions with loops.

Let us study some concrete examples of loops.

Example 5.9 Highest score in the scoring problem
Enter 10 numbers, find and display the maximum of them.

[Analysis]

The pseudo code of the algorithm and the loop control analysis is shown in Figure 5.39.

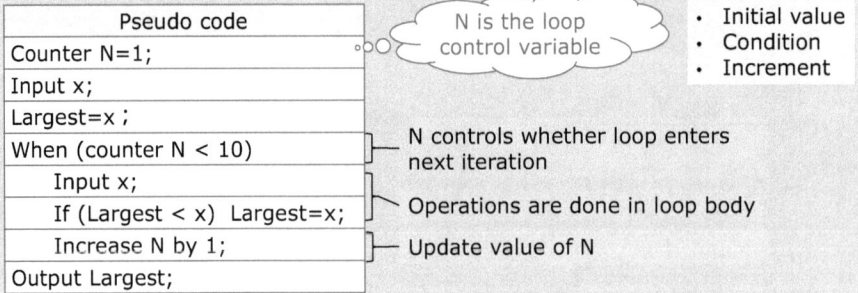

Figure 5.39: Loop in the maximum score problem.

A counter N is used to count the number of score inputs. The loop keeps executing while N < 10. In the loop body, the work to be done includes score inputting, comparing with the largest number Largest and increment of counter N.

The increment of N changes the value used in the loop condition. N is called the loop control variable, which has the following features: it has an initial value, it is used in the condition check, and its value changes in the loop.

Example 5.10 Sum of scores in the scoring problem

The detailed problem description is as follows: enter a series of positive integers through the keyboard, compute and display the sum of them. Suppose that users use "-1" to mark "the end of data input."

[Analysis]

The pseudo code and the loop control analysis are given in Figure 5.40. The condition of loop execution is whether the input is -1. The work to be done in the loop is repeatedly adding the input score. The newly input x changes the value in the loop condition. Herein x is the loop control variable that has an initial value, is used in the condition check, and is updated in the loop.

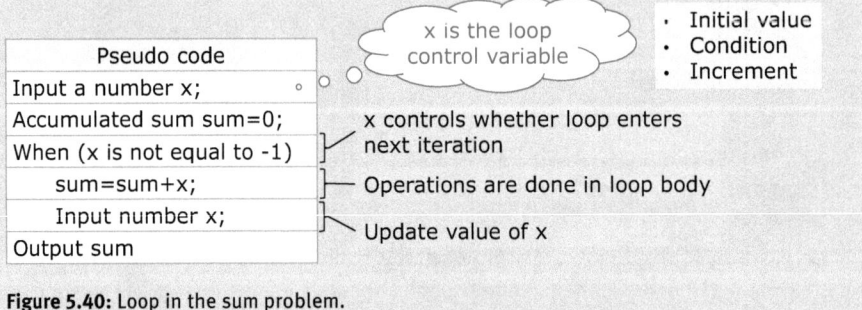

Figure 5.40: Loop in the sum problem.

5.4.2 Three key elements of loops

After studying these examples of loops, we find that whether the loop is executed is determined by the loop control variable, which has three key elements as shown in Figure 5.41. The work to be done in the loop is a set of statements that are executed repeatedly, which is called the loop body in C.

Three key elements of loops
(1) Initial condition: the initial value of loop control variable when the loop is started
(2) Execution condition: the condition that controls whether the loop enters next iteration
(3) Increment: how loop control variable is updated after each iteration

Loop body
A group of statements that are repeatedly executed construct the loop body.

Figure 5.41: Three key elements of loops and the concept of loop body.

We shall now analyze and extract key elements in one of the loop examples above.

Example 5.11 Reanalyze the grade processing problem
Analyze the three key elements of loops in the grade processing problem.

[Analysis]
The pseudo code was given in Figure 5.38.

First, it is clear that the loop control variable is the number of data inputs. The initial condition is the initial value of the number of data inputs, which should be 0 based on the logic of the problem although not explicitly given in the pseudo code.

The execution condition is whether the number of inputs is less than the number of students, whereas the loop increment is the increment of the number of inputs.

Hence, the complete description of the algorithm now includes all three key elements of loops with the addition of initial condition. The refined pseudo code is shown in Figure 5.42.

Initial condition	Number of inputs = 0
Execution condition	Number of inputs is less than number of students
Increment	Increase number of inputs by 1

Original pseudo code	Refined pseudo code
When (number of inputs is less than number of students)	Number of inputs =0
	While (number of inputs is less than number of students)
Input student ID and grade	
Add grade to sum	Input student ID and grade
Output student ID and grade	Add grade to sum
Output average grade	Increase number of inputs by 1
	Output student ID and grade
	Output average grade

The complete design

Figure 5.42: Key elements analysis of the loop in the grade processing problem.

Based on the discussion above, we may summarize the general flow of loops with three key elements as a flowchart as shown in Figure 5.43. An algorithm should contain all three key elements as long as a loop is involved. Otherwise, the algorithm description is incomplete.

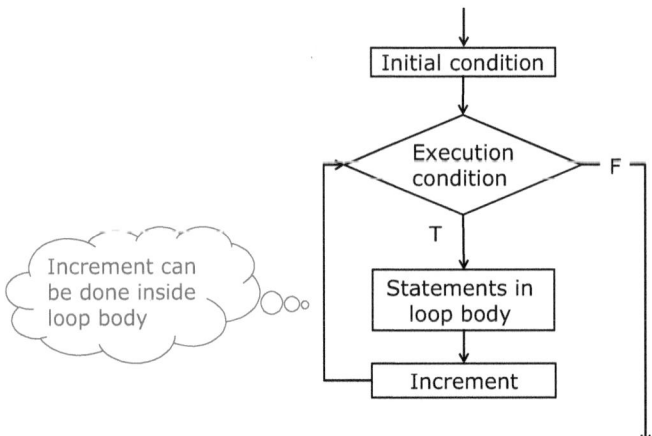

Figure 5.43: General form of flow of loops.

5.4.3 Loop statements

We have learned in the chapter "Introduction to Programs" that there are two categories of loops based on whether the loop body is executed before the condition check. They are while loops and do-while loops.

Syntactically, C has four statements that can be used to implement loops. As shown in Figure 5.44, there are goto loops, while loops, do-while loops, and for loops. We shall focus on the last three in the following sections.

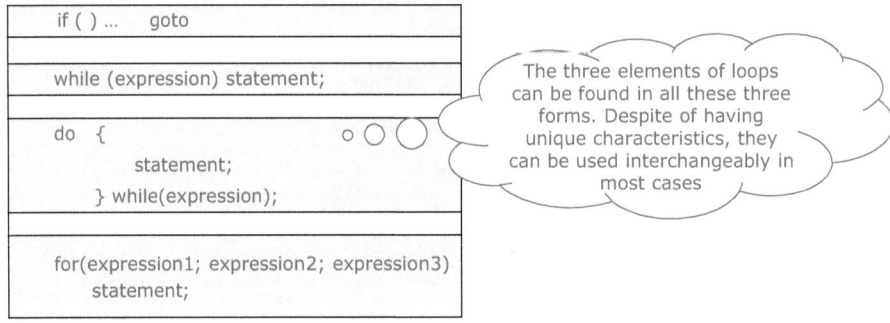

Figure 5.44: Loop statements in C.

5.5 While loops

5.5.1 Syntax of while loops

The syntax and flowchart of while statements in C are shown in Figure 5.45. One can learn the execution process of while loops by observing the flowchart: if the expression evaluates to true, the loop body is executed. Otherwise, the loop is skipped.

The three key elements are not clearly shown in the syntax of while loops, so programmers should look for them based on characteristics of loops used in practical problems. Otherwise, their code would not be complete. We shall learn this again in subsequent examples.

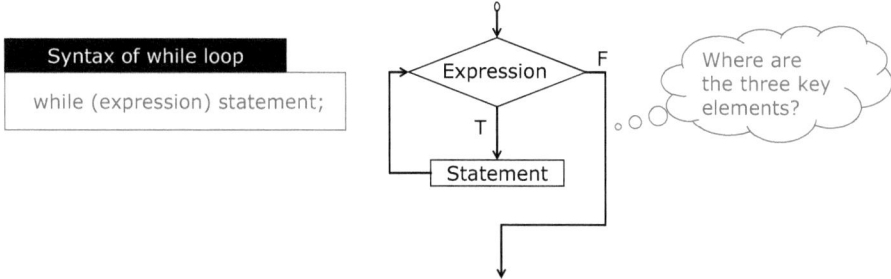

Figure 5.45: Syntax and representation of while loop.

5.5.2 Validation of necessity of the key elements

Example 5.12 Printing numbers with a pattern 2, 4, 6, 8, 10 using while loop

[Analysis]
1. Problem analysis
We first do a tabular analysis on the data we are going to process as shown in Figure 5.46. Clearly, there is a correspondence between the number being printed and the number of prints i. This is a loop process that ends when i becomes 6.

Number of prints i	1	2	3	4	5	6
printf	2	4	6	8	10	End

Figure 5.46: Data analysis of printing numbers with a pattern using while loop.

What are the three key elements in this case? It is easy to see that the elements are as shown in Figure 5.47. The initial value is 1, the execution condition is if i less than 6 and the increment is increasing by 1.

Three key elements		
Initial condition	i=1
Execution condition	i< 6
Increment	i++

Figure 5.47: Three key elements of printing numbers with a pattern using while loop.

2. Algorithm description
- Based on the key elements, we can write the following pseudo code and draw the flowchart shown in Figure 5.48. Comparing these two, we can infer that i less than 6 is the loop execution condition and the loop body is composed of printing i * 2 and i increments by 1. The initial value of the loop control variable i is not shown in the flowchart.

    We can test what happens if we forget to initialize i when running the program.

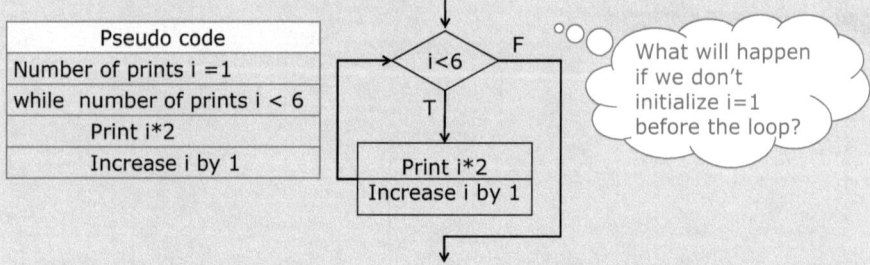

Pseudo code
Number of prints i =1
while number of prints i < 6
Print i*2
Increase i by 1

What will happen if we don't initialize i=1 before the loop?

Figure 5.48: Pseudo code and flowchart of printing numbers with a pattern using while loop.

3. Implementation and debugging
The complete implementation can be found in Figure 5.49. Before debugging, it is vital to determine what to examine. Notably, we want to focus on:
- the entire loop from the beginning to the end,
- what happens if we do not initialize i,
- what happens if we initialize i, and
- what is the value of i when the loop ends.

```
//Use while loop to print 2, 4, 6, 8, 10
#include <stdio.h>

int main(void)
{
    int i;
//  int i=1;
    while (i< 6)
    {
        printf( " %d ",2*i );
        i++;
    }
    return 0;
}
```

Debugging plan

Key points in debugging
- The entire loop from the beginning to the end
- *i* is not initialized
- *i* is initialized
- Value of *i* when the loop ends

Figure 5.49: Implementation and key points in debugging.

(1) Without initialization

In Figure 5.50, i is not initialized. Its value turns out to be a random negative number with a large absolute value. As a result, the output of 2*i is not the expected value 2 and the loop is executed far more than five times. With this value of i, the loop will not terminate until i is greater than 6 after over 800 million iterations.

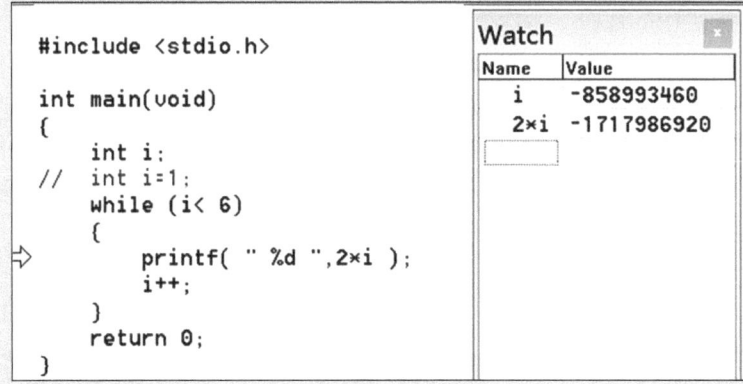

```
#include <stdio.h>

int main(void)
{
    int i;
//  int i=1;
    while (i< 6)
    {
        printf( " %d ",2*i );
        i++;
    }
    return 0;
}
```

Watch	☒
Name	**Value**
i	-858993460
2*i	-1717986920

Figure 5.50: Loop control variable not initialized.

(2) With initialization

In Figure 5.51, i is initialized to 1. In Figure 5.52, its value becomes 2 in the second iteration. In Figure 5.53, its value reaches 6 when the loop is completed. This proves that the loop body has been executed five times.

```
#include <stdio.h>

int main(void)
{
//  int i;
    int i=1;
    while (i< 6)
    {
        printf( " %d ",2*i );
        i++;
    }
    return 0;
}
```

Watch	▾
Name	**Value**
i	1
2*i	2

Figure 5.51: Loop control variable initialized 1.

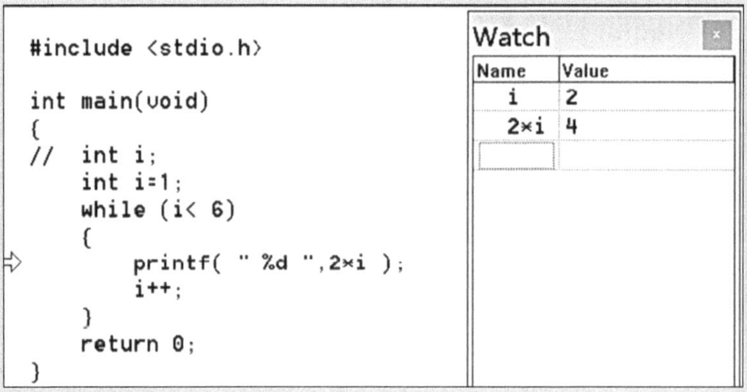

```
#include <stdio.h>

int main(void)
{
//   int i;
     int i=1;
     while (i< 6)
     {
         printf( " %d ",2×i );
         i++;
     }
     return 0;
}
```

Watch	⊠

Name	Value
i	2
2×i	4

Figure 5.52: Loop control variable initialized 2.

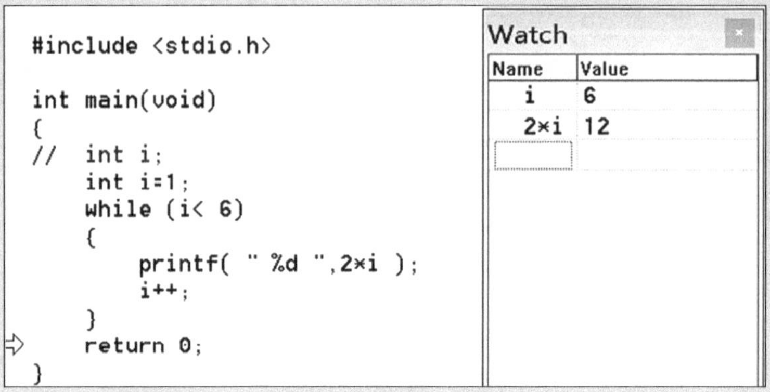

```
#include <stdio.h>

int main(void)
{
//   int i;
     int i=1;
     while (i< 6)
     {
         printf( " %d ",2×i );
         i++;
     }
     return 0;
}
```

Watch	⊠

Name	Value
i	6
2×i	12

Figure 5.53: Loop control variable initialized 3.

4. Conclusion

Using the debugger, we can conclude that the three key elements must work together to ensure the loop body is executed a desired number of times. It is uncertain how many times the loop will be executed if the loop variable is not initialized.

Example 5.13 Scoring problem with known number of judges

Type in scores given by eight judges, output the sum and average.

[Analysis]

1. Algorithm design

The pseudo code is shown in Figure 5.54.

Algorithm description in pseudo code
Initialize the sum to be 0
Initialize the counter to be 0
while counter < 8
Input next score
Add the score to the sum
Counter increases by 1
The average is sum divided by 8
Output the sum and average

Figure 5.54: Algorithm of scoring problem with known number of judges.

2. Program implementation

```
1  int main(void)
2  {
3      int counter;      //Counter
4      int grade;        //Score
5      int total;        //Sum
6      int average;      //Average
7
8      //Initialization phase
9      total = 0;        //Initialize sum
10     counter = 0;      //Initialize counter
11     //Processing phase
12     while ( counter < 8 )   //8 iterations
13     {
14        printf("Enter grade: "); //Input prompt
15        scanf("%d", &grade);    //Read score
16        total = total + grade;  //Add score to sum
17        counter = counter + 1;  //Counter increases by 1
18     }
19     average = total / 8;
20  //Output result
21     printf( " total is %d\n", total );
22     printf( " average is %d\n", average );
23     return 0;
24 }
```

Think and discuss Is variable initialization necessary?
Why is initialization necessary for some variables but not for the others?

Discussion: Beginners often ignore this problem, but ignoring it can lead to errors in program results. In the program above, variables counter and total need initialization, whereas grade and average do not. Variables that need initialization are those that need a value before being used the first time. Their values affect the following computation. In other words, the first operation on them is a "read operation." Initialization is not necessary for variables that are first used in "write operations."

Example 5.14 Scoring problem with unknown number of judges
Type in scores given by several judges, output the average score.

[Analysis]
1. Algorithm design
As the number of judges is unknown in this problem, we need to reconsider the condition of loop execution. We can use a number that is not a normal score as the termination mark for score input, for example, "-1." Then we can write out the pseudo code shown in Figure 5.55.

Algorithm description in pseudo code
Initialize the sum to be 0
Initialize the counter to be 0
Enter a score
while input data is not the termination mark
Enter a score
Add the score to the sum
Increase the counter by 1
Average=sum/counter
Output the average

Figure 5.55: Algorithm of the scoring problem with an unknown number of judges.

2. Testing and refinement of the program
What cases we need to consider when testing the algorithm given in Figure 5.55?
We should consider normal and abnormal cases.
   (1) Normal case: The first input is a score.
   (2) Abnormal case: The first input is the termination mark. In this case, the loop body of the while loop would not be executed, thus the value of the counter is 0. This leads to a division by 0 situation when computing the average, which is a severe logic error and will cause the program to crash.

Programming error
If we have not initialized the counter or the sum, the program result may be incorrect. This is a logic error.
   The refined pseudo code is given in Figure 5.56. Readers should implement the program themselves.

Algorithm description in pseudo code
Initialize the sum to be 0
Initialize the counter to be 0
Enter a score
while input data is not the termination mark
Enter a score
Add the score to the sum
Increase the counter by 1
if counter is not 0
Average=sum/counter
Output the average
Else output "No input"

Figure 5.56: Refined algorithm of the scoring problem with an unknown number of judges.

5.5.3 Example of while loops

Example 5.15 Calculate the sum of integers
Type in a series of positive integers, compute and display the sum of them. Suppose users type in "-1" to indicate "end of input."

[Analysis]
1. Problem analysis
Computing the sum is the process of adding numbers repeatedly, whose algorithm was given in the section "Representation of Algorithms." As a loop exists in the algorithm, there should be the three key elements of the loop as shown in Figure 5.57. However, the key elements, in this case, are not as easy to identify as those in the example "printing numbers with a pattern."

It is clear that the loop execution condition is x not equal to 1.

What is the initial condition of this loop? Because the loop execution condition checks the value of x, it should exist before doing the check. Hence, the initial condition should be "input the value of x."

The increment of loop is done inside the loop body. The value of x we input has been used at the beginning of the loop, thus we need to enter another value for x before rechecking the execution condition. Hence, the increment here is "re-input the value of x."

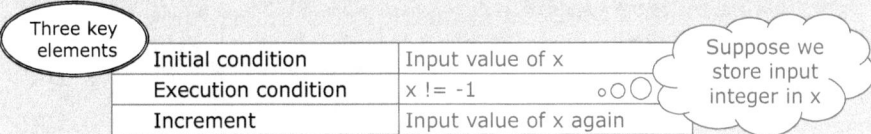

Figure 5.57: Key elements of the loop in calculating the sum of integers.

2. Algorithm description and program implementation

Given the second refinement, we can write the corresponding program statements and eventually convert them to a complete program as shown in Figure 5.58.

Second refinement	Program statements
Accumulated sum sum=0;	sum=0;
Input a number x;	scanf("%d",&x);
When (x is not equal to -1)	while (x != -1)
sum = sum+x;	{ sum=sum+x;
Input number x;	scanf("%d",&x); }
Output sum	printf("sum=%d",sum)

```
#include <stdio.h>
int main(void)
{
    int x, sum=0;

    scanf("%d",&x);
    while (x != -1)
    {
        sum=sum+x;
        scanf("%d",&x);
    }
    printf("sum=%d",sum);
    return 0;
}
```

Figure 5.58: Program of computing sum of integers.

Example 5.16 Read and analyze a program

Analyze the execution process of the following program, describe the intermediate value of the key variable and the final result of the program, and eventually figure out the functionality of the program.

```
1  int main(void)
2  {
3      char ch;
4
5      while (( ch=getchar( ))!='@')
6      {
7          putchar(('A'<=ch && ch<='Z') ? ch-'A'+'a' : ch);
8      }
9      putchar('\n');
10     return 0;
11 }
```

[Analysis]

1. Make a table of the key variable

The variable used inside the loop in this example is ch, where the output is controlled by expression ('A' ≤ ch && ch ≤'Z') and done by character output function. We can list them in Figure 5.59 and use an uppercase letter, a lowercase letter, a nonletter character, and a predetermined termination mark as input data.

Variable ch=getchar()	('A'<=ch && ch<='Z') ?	Output of putchar in the loop	
		ch-'A'+'a'	ch
a	no		a
E	yes	e	
&	no		&
@	End of loop		

Figure 5.59: Analysis table of the character processing program.

Then we can list the value of the expression and the output result based on the program, which helps us examine changes that happen during the execution clearly and find a pattern.

2. Functionality analysis:
We can conclude based on Figure 5.59 that the functionalities of this program include:
  (1) If the input is an uppercase letter, output its lowercase counterpart; otherwise, it is output without being changed.
  (2) The process repeats until the character @ is met.

3. Discussion
What are the three key elements of the loop in this example?
  Discussion:
  (1) Initial condition: ch = getchar()
  (2) Execution condition: ch! = '@'
  (3) Increment: ch = getchar()

The "increment" here is the loop control variable ch reading a new character input from the keyboard. This is also one way of updating the loop control variable.

Knowledge ABC Methods of reading and analyzing programs

To read and analyze a program, we usually list variables, expressions, and operations related to changes in the loop in a table. This helps examine changes that happen during the execution clearly and find a pattern.

Example 5.17 Chickens and rabbits in the same cage
Write a program to find a solution to this problem.

[Analysis]
1. Algorithm analysis
We have introduced this problem in the section of the universality of algorithms, where the stepwise refined algorithm was also given.
  Suppose there are x chickens and y rabbits, which have 35 heads and 94 legs in total.
  In Figure 5.60, the loop is a nested one of two layers. It is worth noting that logically y should be initialized inside the first while loop but outside the second.

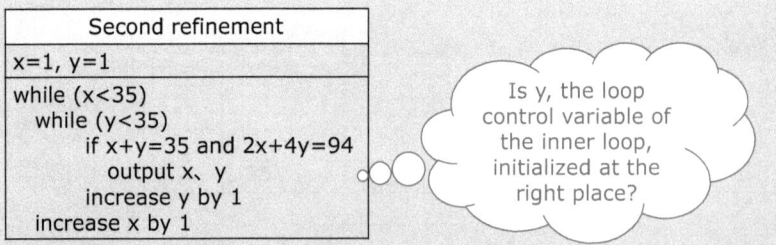

Second refinement
x=1, y=1

while (x<35)
   while (y<35)
       if x+y=35 and 2x+4y=94
           output x、y
           increase y by 1
   increase x by 1

Is y, the loop control variable of the inner loop, initialized at the right place?

Figure 5.60: Algorithm description of chickens and rabbits in the same cage problem.

2. Program implementation

```
01 //Chickens and rabbits in the same cage
02 #include<stdio.h>
03 int main(void)
04 {
05     int x,y;
06     x=1;
07     while (x<35 )
08     {
09         y=1;
10         while (y<35 )
11         {
12             if (x+y==35 && 2*x+4*y==94)
13                 printf("There are %d chickens, %d rabbits\n",x,y);
14             y++;
15         }
16         x++;
17     }
18     return 0;
19 }
```

Execution result:

There are 23 chickens, 12 rabbits

Program reading exercise

Fibonacci's rabbit mating problem

Over 700 years ago, the famous Italian mathematician, Fibonacci, wrote a problem of rabbit mating in his book *Liber Abaci*. Suppose that a pair of rabbits can produce another pair of rabbits each month, and each newly born pair produce another pair in the third month after their birth. If we have one pair of rabbits at first, how many pairs of rabbits are there after one year?

[Analysis]

We can solve this problem by listing the number of pairs in each month:

In the first month, the original pair give birth to another pair so we have 2 (1 + 1 = 2) pairs of rabbits.

In the second month, the original pair give birth to another pair so we have 3 (1 + 2 = 3) pairs.

In the third month, the original pair produce another pair, whereas the pair born in the first month also produce one pair of rabbits. Hence, we now have 5 (2 + 3–5) pairs of rabbits.

Following this pattern, we can list the number of pairs in each month as shown in Figure 5.61. It is clear that we have 377 pairs of rabbits after one year if we start from one pair.

Month	1	2	3	4	5	6	7	8	9	10	11	12
Number of pairs of existing rabbits	1	2	3	5	8	13	21	34	55	89	144	233
Number of pairs of new-born rabbits	1	1	2	3	5	8	13	21	34	55	89	144
Total number of pairs of rabbits	2	3	5	8	13	21	34	55	89	144	233	377

Figure 5.61: Fibonacci sequence.

Fibonacci analyzed the number in each month and wrote the following recurrence formula, where n represents the index of the sequence.

$$\text{fib}(n) = \text{fib}(n-2) + \text{fib}(n-1) \ (n >= 3)$$

Using this formula, we can solve the problem with a loop.

```
01 #include<stdio.h>
02 int main(void)
03 {
04     int n,i,fibn1,fibn2,fibn;
05
06     printf("Enter number of generations n>3: ");
07     scanf("%d",&n);
08
09     fibn1=fibn2=1;
10     printf("Increasing rate starting from gen 1\n",n);
11     printf("1\t1\t");
12     i=3;             //Initial value
13     while (i<=n)  //Loop condition
14     {
15        fibn=fibn1+fibn2; //Find the nth item using recurrence formula
16         printf(i%5? "%d\t" : "%d\n", fibn);//Print 5 items on each line
17        fibn2=fibn1;        //Update the value
18        fibn1=fibn;
19        i++;                //Loop increment
20     }
21     printf("\n");
22     return 0;
23 }
```

The execution result:

```
Enter number of generations n>3: 20
Increasing rate starting from gen 1
1    1    2    3    5
8    13   21   34   55
89   144  233  377  610
987  1597 2584 4181 6765
```

5.5.4 Methods of loop controlling

We have seen two types of loops in examples above: loops with a known number of iterations and loops without a known number of iterations. As shown in Figure 5.62, they are controlled in different ways: one is controlled using counter, whereas the other is controlled using a mark.

Case	Method of loop controlling
Known number of iterations	Controlled by counter
Unknown number of iterations	Controlled by mark

Figure 5.62: Method of loop controlling.

5.6 Do-while loops

5.6.1 Syntax of do-while loops

The syntax and flow of do-while loops in C are shown in Figure 5.63. One can learn the execution process of do-while loops by observing the flowchart: the body gets executed first, then the expression is evaluated. The loop continues if the result is true and terminates otherwise.

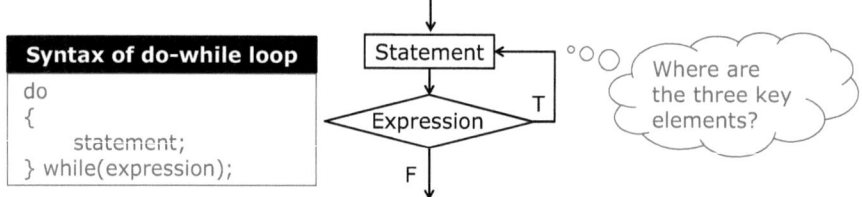

Syntax of do-while loop
| ```
do
{
    statement;
} while(expression);
``` |

Figure 5.63: Do-while loop.
```

Similar to while loops, the three key elements are not clearly shown in the syntax of do-while loops. Programmers should complete the missing parts in their programs based on characteristics of loops used in practical problems.

Example 5.18 Print numbers with a pattern
Use do-while to print numbers with a pattern: 2, 4, 6, 8, 10.

[Analysis]
1. Algorithm description
This problem has been introduced in the section of while loops, where we have done data and key elements analysis. The initial value is 1, execution condition is i < 6 and the increment is i increases by 1 in each iteration.
    With respect to the syntax of do-while loops and the three key elements of loops, we can use our experience of while loops to figure out the flowchart and pseudo code as shown in Figure 5.64.

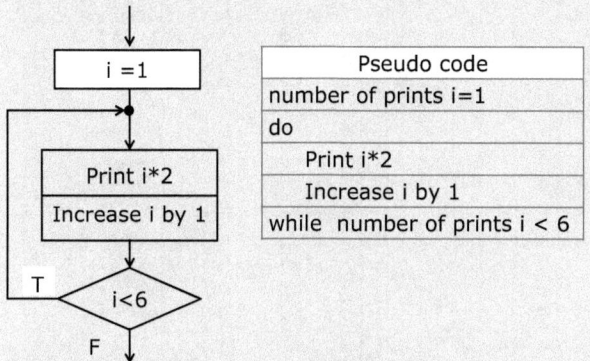

Figure 5.64: Algorithm and flow of printing numbers with a pattern using do-while.

2. Execution analysis
First, the number of prints is initialized to be 1. Then the program enters the do-while loop and prints 2; i becomes 2 after increment, the execution condition i < 6 is true, and the program enters the next iteration. We can use a table to record changes of i, the execution condition and the output as shown in Figure 5.65.

Pseudo code	Program
number of prints i=1	i=1;
do	do
Print i*2	{ printf("%d",2*i);
Increase i by 1	i++;
while number of prints i < 6	} while (i<6);

i	1	2	3	4	5
printf	2	4	6	8	10
i++	2	3	4	5	6
i<6	T	T	T	T	T

Figure 5.65: Analysis of the program of printing numbers with a pattern using do-while.

3. Comparison of implementations using while and do-while

Putting implementations using while and do-while side by side as shown in Figure 5.66, it is clear that they share the same key elements, with the only difference being execution order of loop body and condition checking.

Initial value	i=1
Execution condition	i<6
Increment	i++

while implementation	do-while implementation
i=1;	i=1;
while (i<6)	do
{ printf("%d" ,2*i);	{ printf("%d" ,2*i);
i++;	i++;
}	} while (i<6);

While checks and executes

do while executes and checks

Figure 5.66: Comparison of while and do-while loops.

Example 5.19 Repeated input problem with unknown number of data

Given integer inputs from keyboard, the program should repeatedly read them into variable num and output the value. It terminates and outputs the total number of inputs when the input is larger than a preset value N.

[Analysis]

1. Data analysis and algorithm description

According to the problem description, the three key elements are as follows:

The initial value is the first input of num.

The execution condition is num ≤ N.

The increment is the subsequent input of num.

Using these key elements with the addition of statements that provide required functionalities, we can write the pseudo code of the algorithm as shown in Figure 5.67. Inside the loop body, the input is read into num, which is immediately output. The counter is then increased by 1. When num is less than or equal to the preset value N, the loop continues. Note that "enter an integer" here includes both initial value and loop increment.

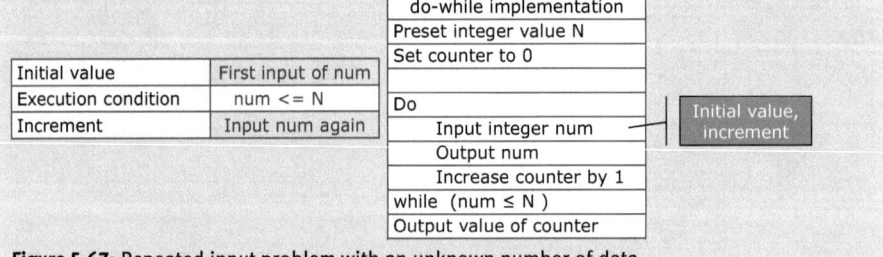

		do-while implementation
		Preset integer value N
		Set counter to 0
Initial value	First input of num	
Execution condition	num <= N	Do
Increment	Input num again	Input integer num
		Output num
		Increase counter by 1
		while (num ≤ N)
		Output value of counter

Initial value, increment

Figure 5.67: Repeated input problem with an unknown number of data.

2. Comparison of solutions using while and do-while

Figure 5.68 shows two solutions side by side. They share the same loop body while the execution order of the loop body and condition check is different.

We shall test them using a special case, where the first input num is larger than the preset value N.

The do-while loop outputs the value and terminates with the counter value being 1.

The loop body of the while loop is not executed as the condition is not met; therefore, there is no output of num and the value of the counter is 0. According to the problem description, the value should be output even if it is greater than N. Thus, the logic of while loop is not suitable here.

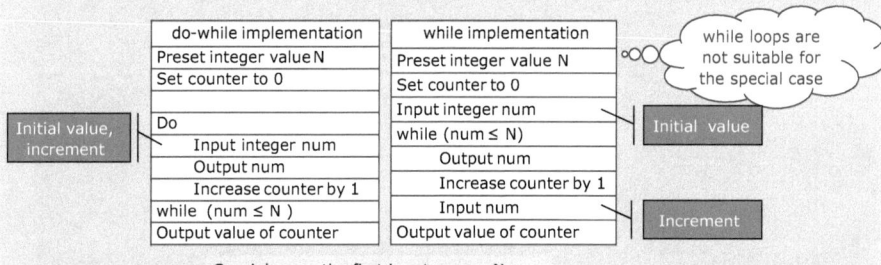

Special case: the first input num > N

Figure 5.68: Comparison of programs using two types of loops.

To solve the problem that the first input does not satisfy the condition of the while loop, we need three additional lines before the loop as shown in Figure 5.69. However, this makes the algorithm more complicated.

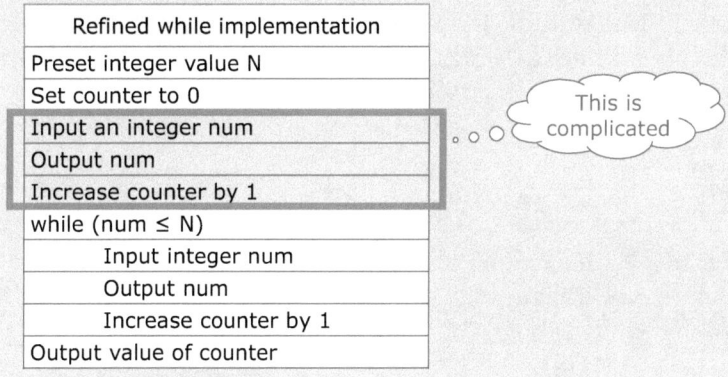

Figure 5.69: Refined while loop.

3. Program implementation
We can write the program based on the pseudo code above.

```c
#include <stdio.h>
#define N 25
int main(void)
{
    int i=0;
    int num;

    do
    {
        scanf("%d",&num);
        i=i+1;
        printf("number=%d\n",num);
    } while ( num <= N );
    printf("total=%d\n", i );
    return 0;
}
```

In addition to data to be output, we can also add some helper text to increase the readability of the result.

5.6.2 Use case of do-while

From what we have seen above, we can conclude that while loops should be considered first when the loop structure is needed for solving the problem. However, if the loop body must be executed at least once regardless of the execution condition, it is more convenient to use do-while as shown in Figure 5.70.

Conclusion

If the loop body must be executed at least once regardless of the execution condition, it is more convenient to use do-while than to use while

Figure 5.70: Use case of do-while.

5.6.3 Example of do-while loops

Program reading exercise
Undefeated general
There are 21 chess pieces and two players take away pieces in turn. Each player can only take away one to four pieces each turn. The player who takes away the last piece loses the game. Please write a computer program against which humans can play the game, where the human

player should take away pieces first and the computer player should be an "undefeated general" (can always win).

[Analysis]

As the computer is the second player to take away pieces, we need to find a strategy to make the program an "undefeated general." Because $21\%5 = 1$, the first player is guaranteed to get the last piece as long as the number of pieces taken away by the first player and the corresponding number of the second player always add up to 5. The program is as follows.

```
01 //A game of 21 chess pieces
02 #include<stdio.h>
03 int main(void)
04 {
05    int num=21,i;
06    printf("Game start\n");
07    while (num>0)
08    {
09       do
10       {
11          printf("Number of pieces (between 1 and %d) you want to take away",
                 num>4?4:num);
12          scanf("%d",&i);
13       }
14       while (i>4||i<1||i>num);   //Read valid input
15       if (num-i>0) printf(" There are %d pieces left\n",num-i);
16       if ((num-i)<=0)
17       {
18          printf(" You took away the last piece.\n");
19          printf(" You lost. Game over.\n"); //Output winning message
20          break;
21       }
22       else
23             printf(" The computer took away %d pieces.\n",5-i);
24       //Output number of pieces taken away by computer
25       num-=5;
26       printf("There are %d pieces left\n",num);
27    }
28    return 0;
29 }
```

If we change the number of chess pieces in this problem, the second player is no longer guaranteed to win. In fact, the second player may be guaranteed to lose. In this case, whether the second player wins is related to the initial number of pieces and the maximum number of pieces allowed to take away each turn. Interested readers can try to write a program to solve this problem.

5.7 Alternative form of while loops

5.7.1 Syntax of for loops

There is another form of loops in C, which is the for loop. Its syntax and processing flow are shown in Figure 5.71. There are three expressions after for, after which are other statements. The execution flow is as follows:

Step 1: Evaluate expression 1.

Step 2: Evaluate expression 2. If the result is false, the loop terminates; otherwise, the statements are executed.

Step 3: Evaluate expression 3 and then go back to step 2.

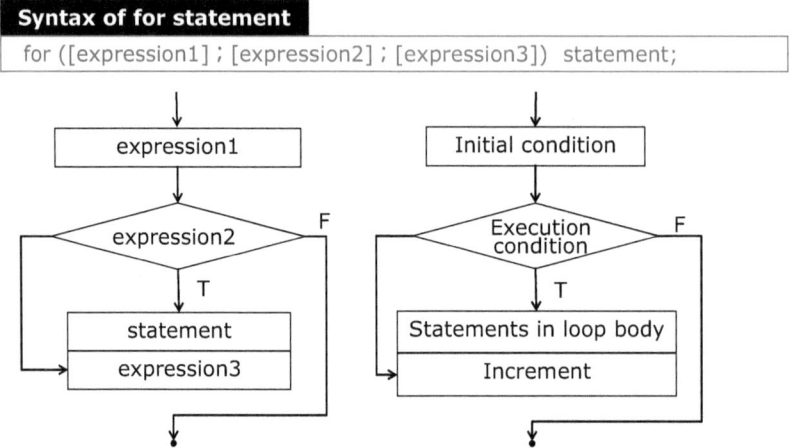

Figure 5.71: Syntax and execution flow of for loops.

Comparing this execution flow with flows of normal loops, it is clear that the logic is equivalent. Hence, the three expressions in the syntax of for are exactly the three key elements of loops. As a result, we can easily write out for statements by extracting the key elements from the problem that involves using loops as shown in Figure 5.72. Note that contents wrapped by [] in the syntax are optional.

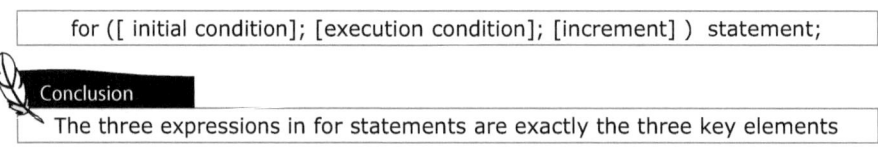

Figure 5.72: For statement and its three key elements.

5.7.2 Example of for statements

Example 5.20 Use for statement to print numbers with a pattern: 2, 4, 6, 8, 10

[Analysis]
1. Problem analysis
Due to the characteristics of for loops, we can simply extract the key elements from the problem and write them as the three expressions in the for statement as shown in Figure 5.73. It is easy and simple to use for statement to implement loops although the program is not very intuitive. We shall analyze the execution flow defined by the grammar.

Initial value	i=1
Execution condition	i<6
Increment	i++

```
for ( i=1;  i<6;  i++ ) printf( "%d", 2*i );
```

Figure 5.73: The key elements of the for loop that prints numbers with a pattern.

2. Flow analysis
First, let us label each step in the flow in the order of execution as shown in Figure 5.74. According to the execution flow, step 1 is executed first, where i is initialized to be 1. Then step 2 is executed, where the execution condition i < 6 is checked. Step 3 is the execution of the loop body. As i = 1, 2 is printed according to the analysis table. Variable i is then increased by 1 in step 4, after which the program returns to step 2 and rechecks the execution condition. The next iteration gets executed if the check yields true; otherwise, the loop is terminated.

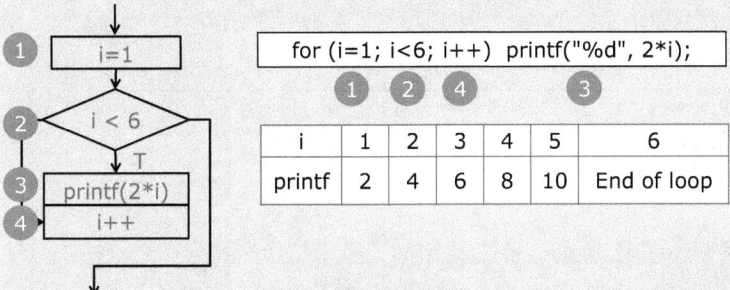

Figure 5.74: Execution process analysis of the for loop that prints numbers with a pattern.

Think and discuss A question on body of for loop
Compare the following two program segments, what are their output?
(1) for (i = 1; i < 6; i++) printf("%d ", 2*i)
(2) for (i = 1; i < 6; i++);   printf("%d ", 2*i)

Discussion: The output of the first segment is 2 4 6 8 10. The loop body in the second segment has nothing but a semicolon inside, so it is an empty statement. In other words, the loop body does nothing during the execution. The value of i after the loop terminates is 6, thus the final output is merely a 12.

Preventing program error
Putting semicolon right next to a for statement makes the loop body an empty statement, which is usually a logic error.

Program reading exercise
Read the following program, describe its functionality and output:

```
1 #include<stdio.h>
2 int main(void)
3 {
4    int sum, i;
5    sum=0;
6
7    for (i=1; i<=100; i++)
8    {
9       sum=sum+i;
10   }
11   printf("%d", sum);
12   return 0;
13 }
```

[Analysis]
We can list the loop variable i and the sum in a table as shown in Figure 5.75. According to the pattern of how sum changes over iterations, it can be shown that sum=1+2+3+...+100 = 5050.

i	1	2	3	...	101
sum	0+1	1+2	1+2+3	...	End

Figure 5.75: Program reading analysis table.

Program reading experience
Tabular method
(1) List variables changed in the loop
When reading a program, we can use the tabular method to list key variables in the program. If a loop exists, we also list changes of the loop control variable and increment in each step. In fact, they are also examined when using debuggers to trace a program step by step. By record-ing the dynamic process of variable changes into a table, we take a "snapshot" of each step to carefully analyze characteristics and patterns of program execution, which makes it easier to obtain results of the program.

(2) List computation method
When there are too many iterations, we do not have to list the values in each iteration. Instead, we write out the computation method to find the relation between the final result and each iteration.

Example 5.21 Chickens and rabbits in the same cage
We have solved this problem using while before and now we are going to solve it using for as shown in Figure 5.76. It is clear that this implementation is simpler. As the code implementation is trivial, we will omit it here.

Second refinement	Third refinement
int x=1, y=1	int x=1, y=1

while (x<35)
　　while (y<35)
　　　　if x+y=35 and 2x+4y=94
　　　　　　Output x and y
　　　　increase y by 1
increase x by 1
	for(x=1; x<35; x++)
	for(y=1; y<35; y++)
	{
	　　if (x+y=35&&2x+4y=94)
	　　printf("%d%d",x,y);
	}

Figure 5.76: Solving chickens and rabbits in the same cage problem using for loops.

Think and discuss The value of the loop control variable after the loop terminates
What are the values of x and y after the for loop ends?

Discuss: Each for loop is executed 34 times; therefore, both values are 35 after the loop terminates. In other words, the program did some meaningless work after obtaining the result. How should we enhance this? Can we use the interruption mechanism in multiple branch structure here to jump out of the loop in time? We will cover how to jump out of the loop after certain conditions are met in the section "Interruption of Loops."

5.8 Infinite loops

5.8.1 Infinite loops in practice

In the example of printing numbers with a pattern using while loop, the loop body will be executed many times if we did not initialize the loop control variable. Similarly, can a loop be executed forever without being terminated?

We have also seen the problem called "things whose number is unknown" in the chapter "Algorithms." It was solved using loops, so the three key elements should exist as shown in Figure 5.77. Based on the pseudo code, it is trivial to find out the initial value and the increment. Nonetheless, the problem did not restrict the number of solutions and there might be multiple solutions. Following this logic, the loop should be executed forever. In this case, what is the execution condition of the loop?

In this problem, the execution condition is "always true" and the number of iterations is unlimited. Hence, there should be a mechanism to enable such infinite loops.

Number of iterations in problem "things whose number is unknown"

Top-level pseudo code	First refinement	Initial value	x=1
x starts from 1	Let x=1	Execution condition	?
	Do the following repeatedly	Increment	x++
Find result that satisfies requirements Output the result	Output result if x satisfies following conditions		
	"2 remains when divided by 3, 3 remains when divided by 5, 2 remains when divided by 7"	When should we terminate the loop?	
	Increase x by 1		

Figure 5.77: Execution condition of the loop in problem "things whose number is unknown".

5.8.2 Infinite loops using while statement

C uses nonzero value, 1 in most cases, to represent true, while(1) then represents a loop whose execution condition is always evaluated to true as shown in Figure 5.78.

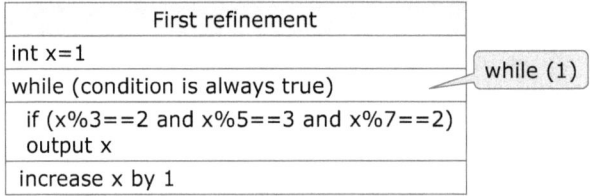

First refinement
int x=1
while (condition is always true)
if (x%3==2 and x%5==3 and x%7==2) output x
increase x by 1

while (1)

Figure 5.78: Representation of an "always true" loop execution condition.

! **Think and discuss** How is a while loop executed when the execution condition expression is 1? Discussion: The grammar flow of while answers the question. The expression is 1, so it is evaluated to "true." It is clear from the flowchart shown in Figure 5.79 that the loop runs forever if the expression is always true.

while (1) statement;

expression   F

T

statement

When "expression" is 1, it always evaluates to true

Figure 5.79: Representation of infinite loop using while.

Similar to how the sun repeatedly rises in the east and sets in the west, infinite loops are very important and are widely used inside computer systems. A program can only be executed once and terminated without infinite loops, so an infinite loop mechanism is necessary to keep the system running repeatedly and normally.

5.8.3 Infinite loops using for statement

For loops are equivalent to while loops in essence, so they can be used to implement infinite loops as well.

The C grammar defines that a for loop always enters the true branch if the execution condition is omitted. In other words, a for loop without execution condition is equivalent to while(1) as shown in Figure 5.80.

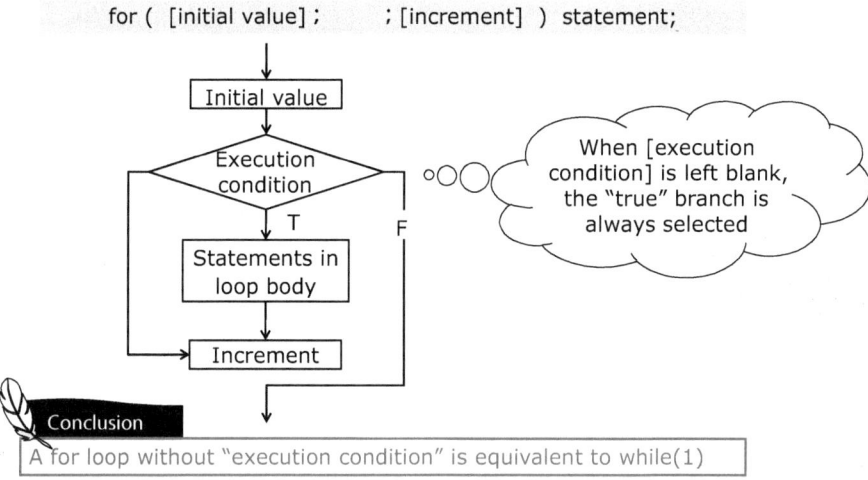

Figure 5.80: Representation of infinite loop using for statement.

There are some special cases of for statement in C as shown in Figure 5.81. Note that all three expressions in for statement are wrapped by square brackets, meaning that they are all optional. However, the semicolons cannot be omitted. In the extreme case, all three expressions are omitted.

for ([expression1]; [expression2]; [expression3]) statement;

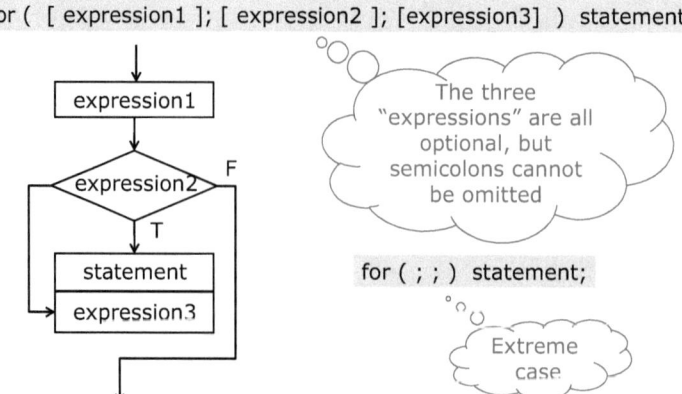

Figure 5.81: Special cases of for statement.

Program reading exercise
A moving smiley face
Note: The library function gotoxy(x, y) moves the cursor to the xth row and yth column. For example, gotoxy(0, 0) moves the cursor to the top left corner of the screen. Note that the loop on line 15 is infinite so that the program keeps running.

```
01 #include <stdio.h>
02 #include <windows.h>
03 void gotoxy(int x, int y) //Move cursor to the x-th row and y-th column
04 {
05     COORD pos;
06     pos.X = x - 1;
07     pos.Y = y - 1;
08     SetConsoleCursorPosition(GetStdHandle(STD_OUTPUT_HANDLE),pos);
09 }
10
11 int main(void)
12 {
13     int x=0,y=0;      //The top left corner of screen
14     int xv=1,yv=1; //Move speed is one char at a time
15     while (1)    //Keep running until Ctrl+Z pressed
16     {
17         gotoxy(x,y); //Move cursor to specified coordinate
18
19         //Move the object at specified speed:
20         x += xv;  //Horizontal speed is xv
21         y += yv;  //Vertical speed is yv
22         gotoxy(x, y);
23
24         //Print smiley face
25         putchar (2) ;   //ASCII value of smiley face is 2
```

```
26        system("cls"); //Clear the screen
27
28        //Bounce the object back at edge
29        if (x >= 80 || x <= 0) xv = -xv;  //Width of screen is 80
30        if (y >= 25 || y <= 0) yv = -yv;  //Height of screen is 25
31   }
32   return 0;
33 }
```

5.9 Interruption of loops

5.9.1 Interruption of loops in practice

5.9.1.1 Example of interruption of loops

Case study 1 Variant of "things whose number is unknown": jumping out of the loop
We consider a variant of the problem "things whose number is unknown," where we are asked to find the maximum number within 2000 that has remainder 2 when divided by 3, has remainder 3 when divided by 5, and has remainder 2 when divided by 7.

In this case, we set the initial value to be 2000, test a number starting from 2000, and decrease it by 1 in each iteration. The loop increment is thus a negative number. There is no restriction on the execution condition as shown in Figure 5.82. Inside the loop body of the infinite loop while(1) in the code implementation, we start from 2000 and repeatedly test whether the value of x satisfies the given condition. If not, we decrease x by 1. Once we have found an x that satisfies the condition, the operation should be terminated after printing the value. In other words, the program should jump out of the infinite loop.

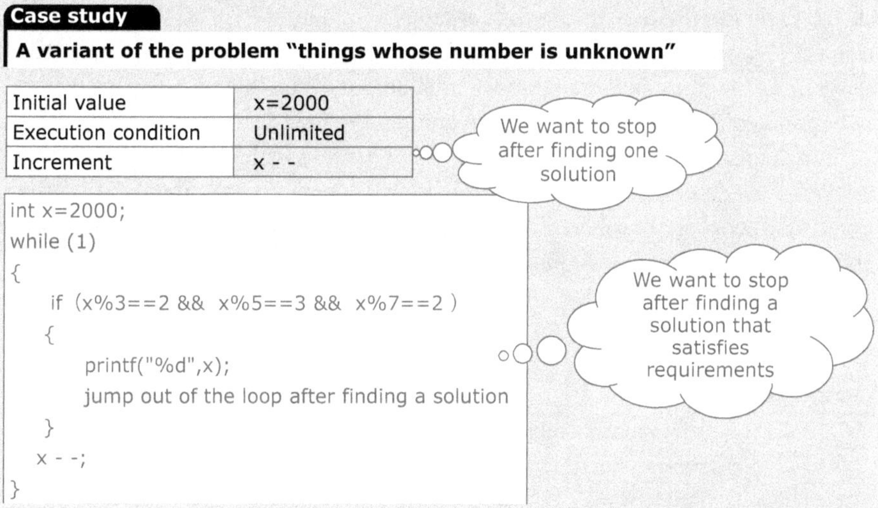

Case study

| A variant of the problem "things whose number is unknown" |

Initial value	x=2000
Execution condition	Unlimited
Increment	x - -

> We want to stop after finding one solution

```
int x=2000;
while (1)
{
    if (x%3==2 && x%5==3 && x%7==2 )
    {
        printf("%d",x);
        jump out of the loop after finding a solution
    }
    x - -;
}
```

> We want to stop after finding a solution that satisfies requirements

Figure 5.82: A variant of the problem "things whose number is unknown".

Case study 2 "Partial sum": jumping inside a loop
Type in 10 integers, compute the sum of positive integers among them.

[Analysis]
Suppose we read the input into x, store the sum in sum and use i as the counter. After determining the three key elements, it is not hard to obtain the processing flow shown in Figure 5.83. With respect to the restriction counter i < 10, we check whether input integer x is positive: if it is, we add it to sum; otherwise, we skip this iteration of the loop body, add 1 to the counter and enter the next iteration.

  If we implement this flow using a for loop, it is clear that the flow should jump to "loop increment" when x < 0, which is a jump within the loop. How do we write such jumps in for loops?

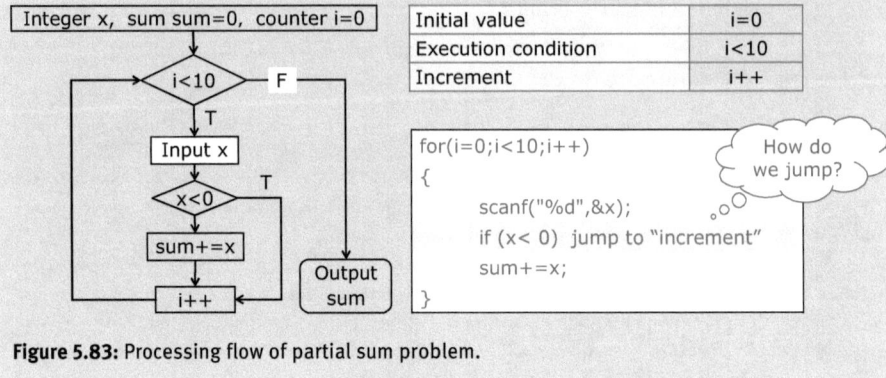

Figure 5.83: Processing flow of partial sum problem.

5.9.1.2 Early termination mechanism of loops

In the problems above, it was necessary to terminate the loop in advance during processing. The difference was that we jumped out of the loop in one case, whereas we terminated the current iteration and entered the next in the other case.

  In response to the needs of jumping when dealing with problems in practice, C provides two statements for early termination of loops: break statement and continue statement as shown in Figure 5.84. The break statement must be used in loops or switch statements to jump out of the loop or switch structure. The continue statement must be used in loop statements to terminate the current iteration of the

Statement	Use case	Role	
break	Loop statements	Jump out of loop	Jump to the end of the loop body containing continue
	switch statement	Jump out of switch structure	
continue	Loop statements	End current iteration	

Figure 5.84: Two statements for early termination of loops.

loop. In other words, the remaining statements in the current iteration are skipped and the next iteration is started.

5.9.2 Jumping out of loops with break statement

To address requirements in practice, C provides the break statement for terminating loops in advance as shown in Figure 5.85. In fact, we have seen its usage when introducing the syntax of switch statement.

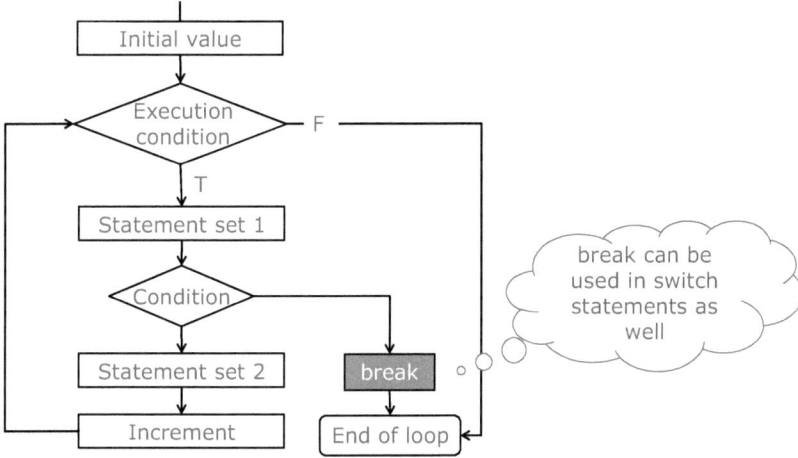

Figure 5.85: Break statement in loop structure.

Example 5.22 Analysis of the variant of "things whose number is unknown" problem
What is the maximum number within 2000 that has remainder 2 when divided by 3, has remainder 3 when divided by 5, and has remainder 2 when divided by 7? Write two programs using while and for, respectively.

[Analysis]
1. Algorithm design
This is a process of repeatedly testing whether an integer satisfies the given condition: we start from $x = 2000$, test whether x satisfies the condition and decrease its value repeatedly. Once we have found the solution, we jump out of the loop. As we do not know what range the solution lies in, we cannot determine the execution condition. Figure 5.86 shows the key elements of the loop and pseudo code.

Initial condition	x=2000
Execution condition	Unknown
Increment	x--

Top-level pseudo code	Refinement
x starts from 2000	x= 2000
while ()	while ()
decrease x by 1 if x doesn't satisfy given conditions. Repeat until finding an x that satisfies the conditions	if (x%3==2 and x%5==3 an x%7==5) break ; x-- ;
Output x	Output x

Figure 5.86: Algorithm description of things whose number is unknown problem.

Note that the loop increment in this example is decreasing. Increment here means the change of the loop control variable, which can be increasing or decreasing, regular or irregular. We should use it with flexibility.

2. Program implementation

```
1 //Things whose number is unknown, solved using for loop
2 #include<stdio.h>
3 int main(void )
4 {
5   int i;
6
7   for ( i=1;  ; i++) //Execution condition is unknown, so we leave it blank
8   {
9     if ( i%3==2 && i%5==3 && i%7==5)
10    {
11      printf("%d\n",i);
12      break;
13    }
14  }
15  return 0;
16 }
```

The program implementation using a while loop is shown in Figure 5.87. With the break statement on line 10, the program produces only one result, 1913. If we delete this break statement and change the execution condition on line 5 to x > 1, we can obtain all solutions that are less than 2000.

```
01 #include <stdio.h>
02 int main(void)
03 {
04    int x=2000;
05    while (1)
06    {
07       if ( x%3==2 && x%5==3 &&x%7==2)
08       {
09          printf("%d",x);
10          break;
11       }
12       x--;
13    }
14    return 0;
15 }
```

Jump out of loop

Solutions within 2000

1913	1808	1703	1598	1493
1388	1283	1178	1073	968
863	758	653	548	443
338	233	128	23	

Result: 1913

Figure 5.87: Program implementation of the variant of things whose number is unknown problem.

Example 5.23 Find the largest number that satisfies given condition
Find the largest number within 100 that can be divided by 19 exactly. Write the program using for loop and trace it in a debugger.

[Analysis]
1. Algorithm design
The pseudo code is shown in Figure 5.88.

Pseudo code	Refinement
i starts from 100	i starts from 100
while i < 100	while i > 1
decrease i by 1 if it is not a multiple of 19,	jump out of the loop if I is a multiple of 19
jump out of the loop when finding an i that satisfies given condition	decrease i by 1
output i	output i

Figure 5.88: Algorithm of finding the largest number that satisfies given condition.

2. Program implementation
```
1 //Find the largest number that satisfies given condition
2 #include<stdio.h>
3 int main(void)
4 {
5    int i;
6    for ( i=100; i>18 ; i--)
7    {
8       if ((i%19)==0) break;
9    }
```

```
10  printf("%d\n",i);
11  return 0;
12 }
```

3. Program tracing

As we do not know the value of i that satisfies condition i%19 = = 0, we need to press hotkey of stepwise tracing F10 again and again to trace step by step, which is inefficient.

In the for loop shown in Figure 5.89, we are interested in the value of i when i%19 = = 0, but the program will jump to the printf statement at that moment due to the functionality of break statement. How can we directly examine the value of i that satisfies given condition starting from I = 100?

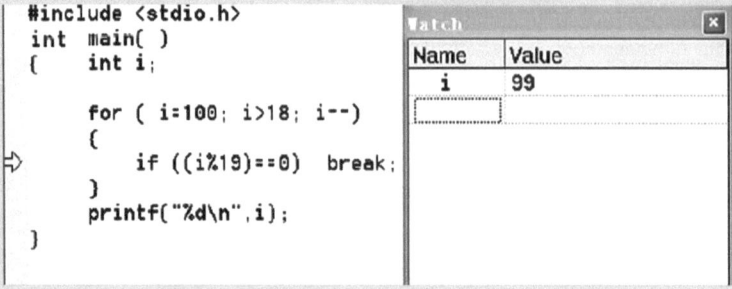

Figure 5.89: Debugging program that finds the largest number that satisfies given condition step 1.

There are two ways of fast tracing:

(1) Use run to cursor command to jump

In Figure 5.90, we move the cursor to the left of the printf statement (by clicking on the left of printf using the mouse). After we have seen the flickering vertical bar, select "Run to cursor" in the "Debug" menu.

```
#include <stdio.h>
int main( )
{    int i;

    for ( i=100; i>18; i--)
    {
        if ((i%19)==0)  break;
    }
    printf("%d\n",i);
}
```

Name	Value
i	99

Figure 5.90: Debugging program that finds the largest number that satisfies given condition step 2.

When i%19 == 0 is met, the program jumps to the printf statement and stops as shown in Figure 5.91. Note that the printf is not yet executed, and the value of i is 95. After executing printf, "95" will be output.

```c
#include <stdio.h>
int main( )
{      int i;

       for ( i=100; i>18; i--)
       {
             if ((i%19)==0)  break;
       }
       printf("%d\n",i);
}
```

Watch	
Name	Value
i	95

Figure 5.91: Debugging program that finds the largest number that satisfies given condition step 3.

(2) Using breakpoints
We can add a breakpoint before printf statement and then execute the Go command (by pressing hotkey F5) as shown in Figure 5.92. F5 is used for debugging one step with the breakpoint. With a single press of F5, the program runs until the breakpoint is met as shown in Figure 5.93.

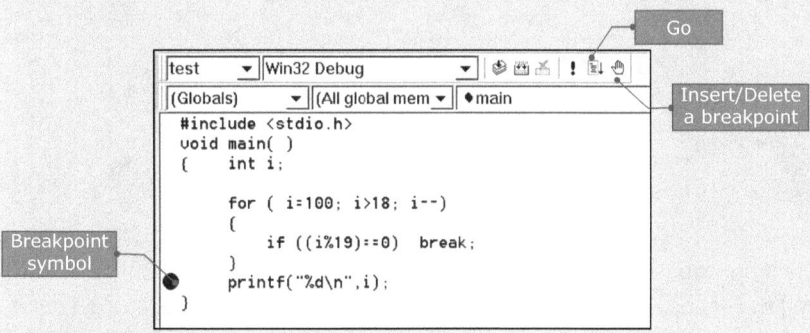

Figure 5.92: Debugging program that finds the largest number that satisfies given condition step 4.

5.9.3 Jumping inside loops with continue statement

5.9.3.1 Functionality of continue statement

In the case study, "partial sum" in section 5.9.1, our brief analysis indicated that we need to find a way for the flow to jump to "loop increment."

```
#include <stdio.h>
int main( )
{     int i;

      for ( i=100; i>18; i--)
      {
           if ((i%19)==0)  break;
      }
      printf("%d\n",i);
}
```

Watch	☒
Name	Value
i	95

Figure 5.93: Debugging program that finds the largest number that satisfies given condition step 5.

As it is tricky to observe the execution order of key elements in a for loop, we shall write out the corresponding while statement and compare these two as shown in Figure 5.94.

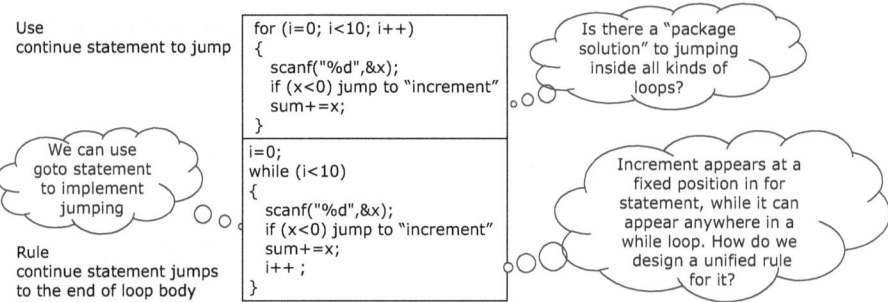

Figure 5.94: Internal jumps in for and while structure.

In a while loop, the jump can be completed by using a goto statement (which will be covered in Section 5.10). However, it is usually not recommended to use the goto statement in programming. Is there a "package solution" to jumping inside all kinds of loops?

C provides a special statement, which is continue, to implement such jumps. Nevertheless, the loop increment appears at a fixed position in for statement while it can appear anywhere in a while loop. How do we design a unified rule for it?

C defines that a continue statement always jumps to the end of the loop, which is the last bracket of loop statement for all three kinds of loops (Figure 5.95).

5.9.3.2 Role of continue in different loops
Now we discuss the role of continue in different loops from the perspective of grammar.

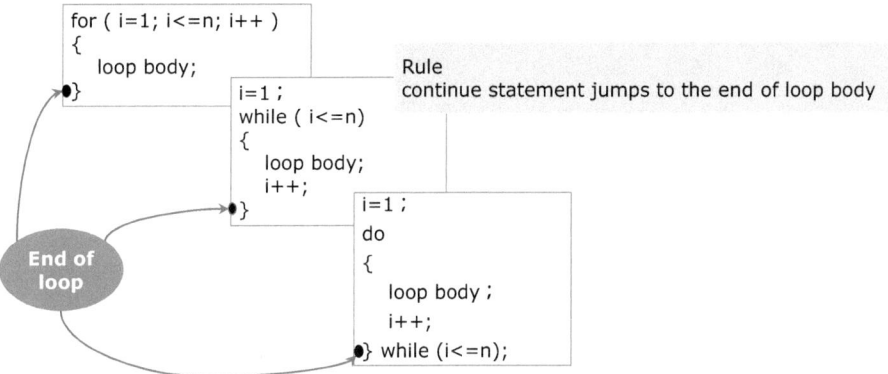

Figure 5.95: The end of a loop.

The "loop increment" is not inside the loop body of a for loop; thus, the increment statement is executed after the jump made by continue statement. See Figure 5.96 for the flowchart.

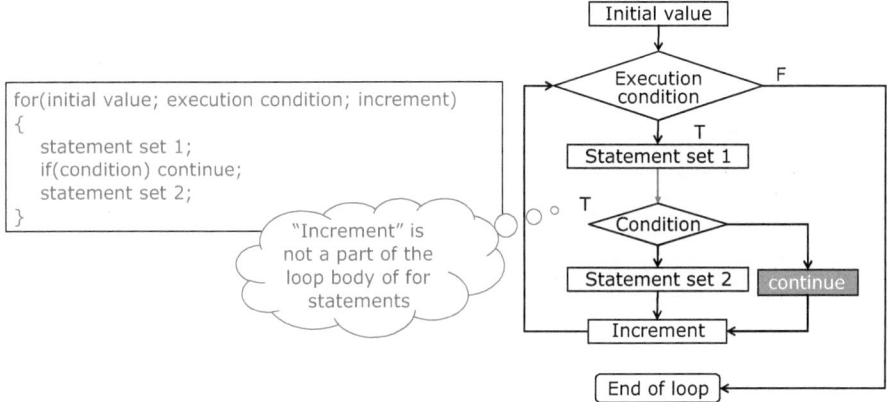

Figure 5.96: Continue in for loops.

Based on the requirements, the "loop increment" of a while loop can be put anywhere inside its loop body. In this case, the continue statement jumps to the condition checking statement as shown in Figure 5.97. In other words, the grammar does not specify whether loop increment is done before the jump, so programmers need to handle it based on the logic of concrete problems.

A continue statement in a do-while loop works in the same way as in a while loop. As shown in Figure 5.98, it is not specified in the grammar whether the loop increment is done before the jump.

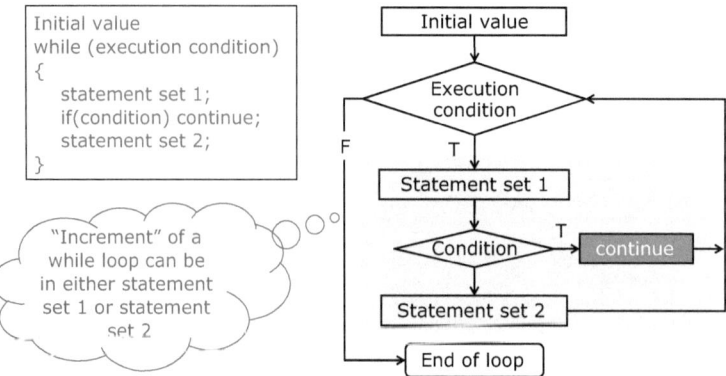

Figure 5.97: Continue in while loops.

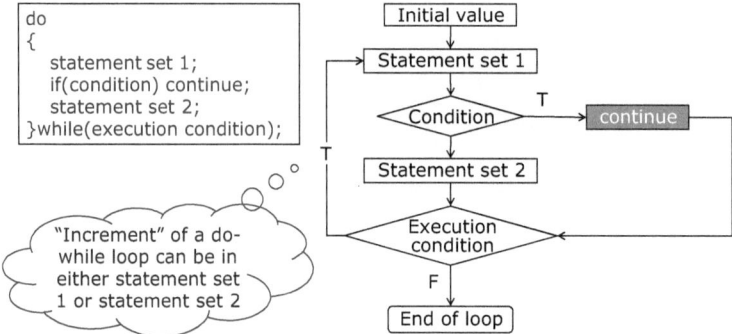

Figure 5.98: Continue in do-while loops.

Now let us take a look at the program implementation of the "partial sum" problem shown in Figure 5.99. It is worth noting that i++ is executed in a for loop regardless of execution of continue. However, whether continue is executed affects the execution of i++ in a while loop, thus the logic of the program is imperfect. Readers can try to revise it on their own.

5.10 Free jump mechanism

5.10.1 Concept of free jump

One day, Mr. Brown went to the new campus of another university in his city for the first time to join a seminar. He drove across the gate and was about to ask a passerby about the direction. However, a sign showing the route was on the roadside.

```
int main(void)
{
    int i,x,sum=0;

    for (i=0; i<10; i++)
    {
        scanf("%d",&x);
        if (x<0) continue;
        sum+=x;
    }
    printf("sum=%d",sum);
    return 0;
}
```

```
int main(void)
{
    int i=0,x,sum=0;
    while (i<10)
    {
        scanf("%d",&x);
        if (x<0) continue;
        sum+=x;
        i++;
    }
    printf("sum=%d",sum);
    return 0;
}
```

These two programs have different logic

Figure 5.99: Program implementations of "partial sum" problem.

He then drove to the destination successfully following the guidance on road signs specially set for the meeting.

There are similar "guiding" statements in C as well. We have discussed jump statements break and continue, which are designed for interrupting loops, in previous sections. The destination they can jump to is strictly restricted by the grammar of C. Meanwhile, there exists a more flexible jump statement in many programming languages. This is the unconditional jump statement, namely the goto statement.

5.10.2 Syntax of unconditional jump statement

The schematic and syntax of the unconditional jump statement are shown in Figure 5.100.

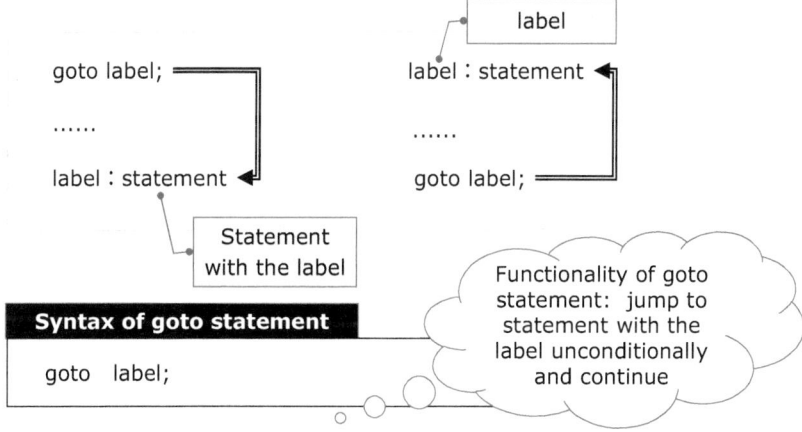

Figure 5.100: Unconditional jump statement.

The "label" here is the "road sign," whereas goto is an instruction of turning. A goto statement can jump either backward or forward. A label is a sign written following the rule of identifiers. It is named in the same way as a variable, but we do not need to allocate memory space or declare it in advance. A label is put in front of a line followed by a colon. It is used to identify a statement and to pair with goto statements. For example:

```
label: i++;
    while(i<7) goto label;
```

C does not restrict the number of labels used in a program, but they must be uniquely named. The goto statement changes the execution path of a program so that the program jumps to the statement marked by the label.

5.10.3 Example of unconditional jump statement

Example 5.24 Print numbers with a pattern
Use goto statement to print numbers with a pattern: 2, 4, 6, 8, 10.

[Analysis]
1. Algorithm implementation
As the goto statement can jump either backward or forward, there are two options for implementation. The processing flows and pseudo code are given in Figs. 5.101, 5.102 and 5.103, respectively.

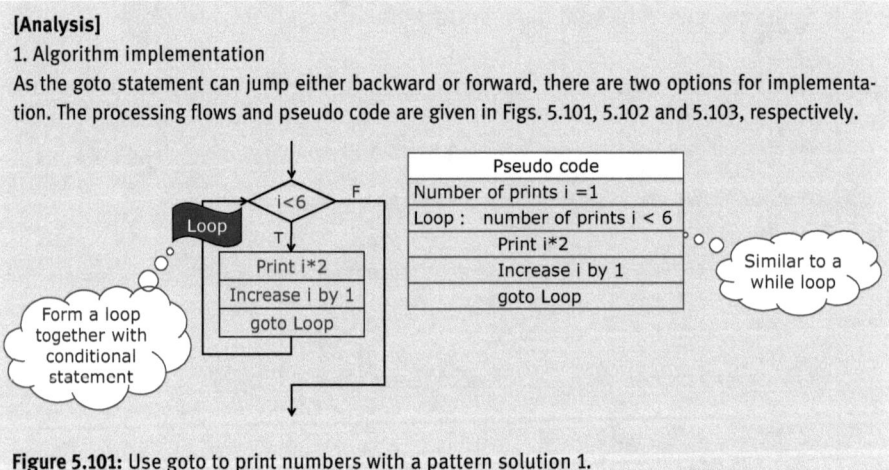

Figure 5.101: Use goto to print numbers with a pattern solution 1.

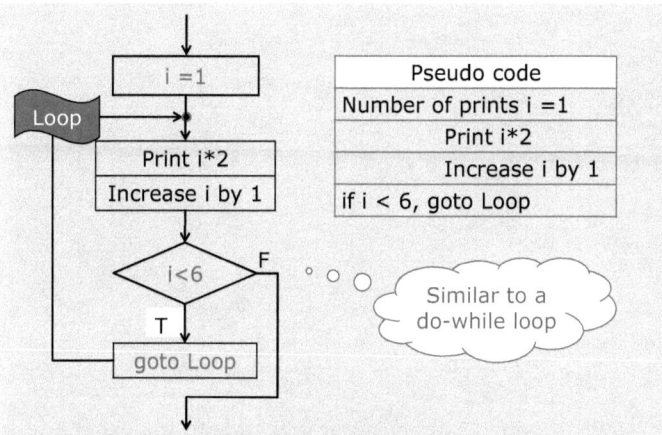

Figure 5.102: Use goto to print numbers with a pattern solution 2.

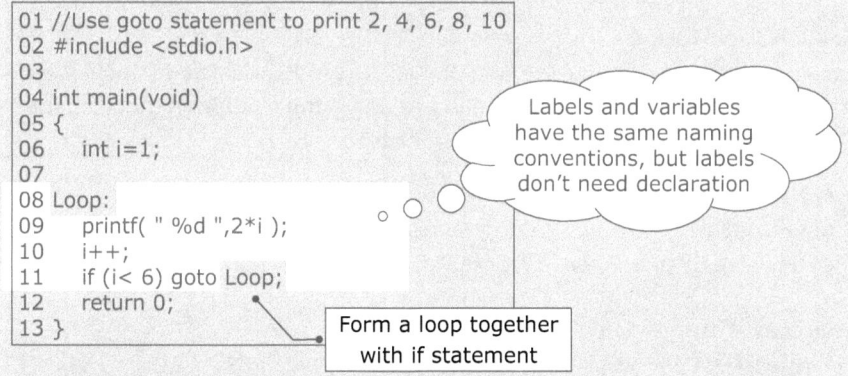

Figure 5.103: Use goto to print numbers with a pattern solution 2.

It is not hard to notice that the backward jump solution is similar to a while loop and the forward jump is similar to a do-while loop. The jumps here are more intuitive than the flow of loop statements as they directly demonstrate the low-level implementation of loops.

2. Program implementation
Goto statements and conditional statements are often used together to provide functions like conditional jumps, loops, and jumping out of a loop.

5.10.4 Characteristics of goto statements

5.10.4.1 Jumping out of a nested loop directly
The most important feature of goto statements is jumping out of nested loops directly. As shown in Figure 5.104, we need one break operation in each for loop if we

```
// Code using break
int flag=false;
// Used as a termination mark
for (int i=1; i<100; ++i)
{
  for (int j=1; j<100; ++j)
  if (i*j==128)
  {
    flag=true;   break;
  }
  ...
  if (flag) break;
}
```

Equivalent

```
// Code using goto
for (int i=1; i<100; ++i)
{
  for (int j=1; j<100; ++j)
    if (i*j==128)  goto End;
    ...
}
End: ...
```

goto statements can jump out of a nested loop directly

Figure 5.104: Jumping out of a nested loop.

want to jump out of a nested for loop of two layers, whereas a goto statement jumps out of the nested loop directly and smoothly.

5.10.4.2 Flexible jumps
Although it is easy to compute the sum of integers 1 to 100 using a for loop, it is also possible to complete the task in a complicated way using multiple goto statements.

5.10.4.3 Note on using goto statements
It is not recommended to use goto statements in modern structured programming. Advice on using them is given in Figure 5.106.

i **Knowledge ABC** Necessity of goto statements
In 1974, Donald E Knuth gave a thorough and fair assessment of goto statements. He claimed that unrestricted use of goto statements, especially backward goto, made it difficult to understand the structure of programs. Thus we should avoid using goto in such cases. In other cases, however, he believed that limited use of goto statements was necessary to increase program efficiency without affecting good program structures. Jumping out of nested loops was one such example (Figure 5.105).

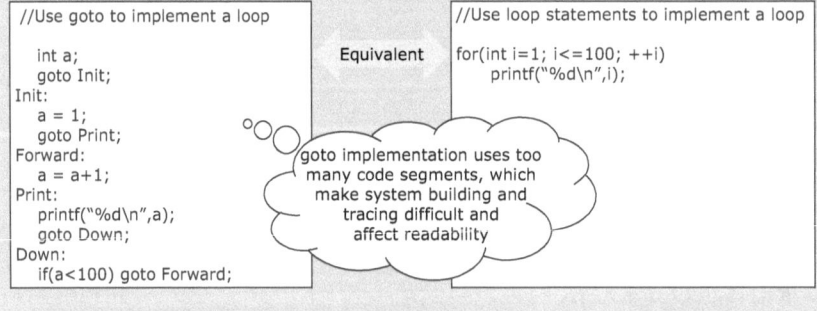

```
//Use goto to implement a loop

  int a;
  goto Init;
Init:
  a = 1;
  goto Print;
Forward:
  a = a+1;
Print:
  printf("%d\n",a);
  goto Down;
Down:
  if(a<100) goto Forward;
```

Equivalent

```
//Use loop statements to implement a loop
for(int i=1; i<=100; ++i)
  printf("%d\n",i);
```

goto implementation uses too many code segments, which make system building and tracing difficult and affect readability

Figure 5.105: Different implementations of loop.

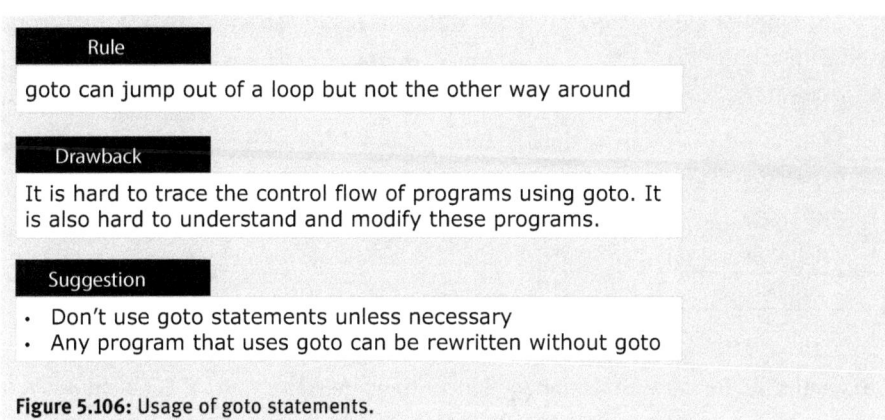

Figure 5.106: Usage of goto statements.

5.11 Summary

The main concepts and relations between them are shown in Figure 5.107.

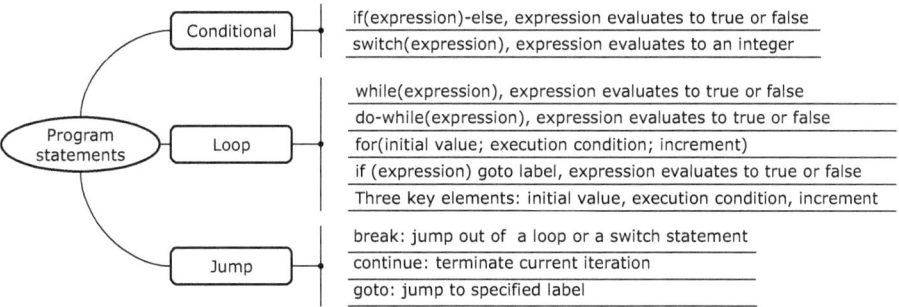

Figure 5.107: Concepts related to program statements.

When using statements, sometimes expressions are required by the grammar. We need to pay attention to the types of these expressions. As shown in Figure 5.108,

Statement	Type of expression in the statement		
if	Conditional/Logical		
switch	Arithmetic		
while	Conditional/Logical		
do-while	Conditional/Logical		
for	Expression 1	Expression 2	Expression 3
	Assignment	Conditional/Logical	Arithmetic

Figure 5.108: Type of expressions in different statements.

different types of expressions can yield different results so they are not to be confused with each other.

Types of results of different expressions are shown in Figure 5.109.

Expression	Result
Arithmetic	Numeral
Relational	True/False
Logical	True/False

Figure 5.109: Type of result of expressions.

Grammatically, for statements are equivalent to while statements. It is recommended to use for statements because they have a more straightforward and clearer form. Do-while is more convenient than for and while if the loop body needs to be executed at least once.

Program statements are instructions that drive computers.

There are three kinds of branch statements and four kinds of loop statements, each with its own syntax.

If statement is used in single- and double-branch structures, whereas it is better to use "switch" for multiple branches.

If we check a condition and select one branch from two based on the result, we should use if statement.

If we compute a value, which may be one of many cases, the path corresponding to the correct constant is chosen.

Default is used to handle exceptions not included in any case.

Doing things repeatedly and tirelessly is the merit of computers, do-while is straightforward, thus it executes the loop body first anyways; while is smart, thus it checks the condition to determine whether the loop body should be executed; for is an alternative form of while that has a more straightforward form.

Using goto to implement loops is tricky; using it with care is the advice from those masterminds.

Loops can be interrupted in special cases,

Where continue skips remaining statements in the current iteration and jumps to condition checking;

While break terminates the loop immediately without hesitation.

These four kinds of loops provide the same functionality,

So they should share some common attributes.

The initial value, execution condition and loop increment are the three key elements,

We should extract them from the problem if a loop is needed.

5.12 Exercises

5.12.1 Multiple-choice questions

(1) [Exception in if]
int x = 0x13;
if (x = 0x12) printf("True");
printf("False\n");
What is the output of the program above? ( )
A) True   B) TrueFalse   C) False   D) TrueFalseTrue

(2) [While]
Which of the following is not an infinite loop? ( )
A) for(y=0,x=1; x>++y; x=i++) i=x;
B) for(; ; x++=i);
C) while(1){x++;}
D) for(i=10; ;i--) sum+=i;

(3) [Do-while]
We want to compute s=1 + 2*2 + 3*3 + ... + n*n +... until s>1000 with the following program.
int s=1,n=1;
  do
  { n=n+1;
      s=s+n*n;
    } while(s>1000);
printf("s=%d\n",s);
After executing the program, we find that the result is wrong. Which of the following changes make the program correct? ( )
A) Change while(s>1000) to while(s<=1000)
B) Change s=1 to s=0
C) Change n=1 to n=0
D) Change n=n+1 to n=n*n

(4) [Break and continue]
Suppose x and y are both int variables. What is the value of y after executing
the following loop? ( )

```
for(y=1,x=1;y<=50;y++){
      if(x>=10) break;
      if(x%2==1) {x+=5;continue;}
      x-=3;
}
```

A) 2   B) 4   C) 6   D) 8

(5) [Switch]

Suppose we have the following definitions: float x=1.5; int a=1, b=3, c=2;
Which of the following switch statements is correct? ( )

A) switch(a+b)
   { case 1:   printf("*");
      case 2+1: printf("**"); }

B) switch((int)x);
   { case 1:   printf("*");
      case 2:   printf("**"); }

C) switch(x)
   { case 1.0:   printf("*");
      case 2.0:   printf("**"); }

D) switch(a+b)
   { case 1:   printf("*");
      case c: printf("**"); }

(6) [While and switch]

```
int main(void)
{ int s;
  scanf("%d", &s);
  while( s>0 )
  { switch(s)
    { case 1: printf("%d", s+5);
      case 2: printf("%d", s+4); break;
      case 3: printf("%d", s+3);
      default: printf("%d", s+1); break;
    }
    scanf("%d", &s);
  }
  return 0;
}
```

Suppose the input is 1 2 3 4 5 0<Return>. What is the output of the program above? ( )

A) 66656   B) 6566456   C) 66666   D) 6666656

(7) [For]

What is the output of the program below? ( )

```
int x=10,y=10,i;
for(i=0;x>8;y=++i) printf("%d,%d; ",x--,y);
```

A) 10,1; 9, 2;   B) 9, 8; 7, 6;   C) 10, 9; 9, 0;   D) 10,10; 9, 1;

5.12.2 Fill in the tables

(1) [Continue]

Fill in the table in Figure 5.110.

```
int main(void)
{
    int i=0;
    while(i<100)
    {
        i++;
        if(i%2==0 ||i%3==0)continue;
        printf("%d ",i);
    }
    return 0;
}
```

i	1	2	3	4	5	6	7	...	99	100
Is a multiple of 2 or 3	False	True								
Output	1	None								
The functionality of the program										

Figure 5.110: Program statements: Fill in the tables question 1.

(2) [Nested loop]
Fill in the table in Figure 5.111.

```
main()
{
    int n,s,sum=0;
    scanf("%d",&n);
    for(int i=1;i<=n;i++)
    {
        s=0;
        for(int j=1;j<=i;j++) s+=j;
        sum+=s;
    }
    printf("%d",sum);
}
```

i	1	2	3	...	n
j	1	1~2		...	
s	1	1+2		...	
sum	1	1+(1+2)		...	

Figure 5.111: Program statements: Fill in the tables question 2.

(3) [Nested for]
Fill in the table in Figure 5.112.

```
int main(void)
{
    for (int i=0; i<2; i++)
    for (int j=3; j>0; j--) printf("*");
    return 0;
}
```

i	0	1	2
j			
Out-put			

Figure 5.112: Program statements: Fill in the tables question 3.

5.12.3 Programming exercises

(1) The elevation of Mount Everest is 8844 m. Suppose we have a piece of paper of infinite size. Its thickness is 0.05 mm. We want to repeatedly fold the paper in half so that the total thickness exceeds the elevation of Mount Everest. How many folds are needed?

(2) Given n numbers, find a pair of two numbers whose difference has the smallest absolute value and output it.

(3) Suppose that abc + cba = n. a, b, and c are all one-digit numbers. n is in the range (1000, 2000). Write a program that finds all possible combinations of a, b, and c.

(4) The Renminbi has banknotes for 100, 50, 20, 10, 5, and 1 yuan. Given an integer price, please figure out a way to pay exactly with as few notes as possible.

(5) Given two integers and an operator (+, -, *, /, %), compute and output the result. Note that the divisor cannot be 0 in division and remainder operations.

(6) Write a program that outputs lowercase English letters in the alphabetical order and the reversed alphabetical order.

(7) Given a natural number N (N < 10), use a double for loop to compute N! and ΣN!.

(8) Given two integers a and b, compute their greatest common divisor. (Hint: when one of the numbers is 0, the gcd is the one that is not 0. When two numbers are relatively prime, their gcd is 1.)

(9) A ball falls off from the height of 100 m. Upon falling on the ground, it bounces back to half of the original height. It then falls off and bounces back again. Compute the total distance the ball has traveled when it falls on the ground the n-th time and the bouncing height. Suppose $5 \leq n \leq 15$.

(10) Compute the value of π using the following sequence.

$$\pi = 4 - \frac{4}{3} + \frac{4}{5} + \frac{4}{7} + \frac{4}{9} + \frac{4}{11} + \dots$$

(11) Write a program that outputs the following pattern.

```
AAAAAAAAAAAA
BBBBBBBBBB
CCCCCCCC
DDDDDD
EEEE
FF
```

(12) Write a program that converts a line of input into an integer. The input consists of digits separated by spaces (each digit, excluding the first and the last, is prefixed and suffixed with a space) and ends with EOF (by pressing Ctrl + Z). For example, the program should output "2483" given the input "2 4 8 3."

6 Preprocessing: work before compilation

Main contents
- Definition and characteristics of preprocessing;
- Definition and usage of macros;
- Meaning and usage of file inclusion;
- Rule and usage of conditional compilation;

Learning objectives !
- Know usage and effect of file inclusion, be able to develop a program using multiple files through #include
- Be able to use #define to create ordinary macros
- Understand conditional compilation

6.1 Introduction

The midterm exam of the programming course Mr. Brown taught was conducted using an online judge system. There were five problems in the exam and the score was calculated as follows: the five scores of each problem were sorted in descending order, and their weighted sum was computed with weight 0.3, 0.25, 0.2, 0.15, and 0.1. Mr. Brown wrote a program to compute the grades of students. After calculating all the grades, he found that the passing rate was low. As a result, he updated the calculation rule in the final exam, where a student passed the exam as long as he/she was able to solve two problems. To achieve this goal, he changed the weight to 0.3, 0.3, 0.15, 0.15, and 0.1 However, when he modified his program at the end of the semester, he forgot which scores corresponded to 0.25, 0.2, and 0.15 because it had been a long time since he wrote the program. It was quite tedious to review and verify. Moreover, these weights were used multiple times in the program; therefore, it was possible that he wrongly updated a weight or even forgot to update at all.

There was a large crossborder bank that asked Mr. Brown to write a piece of software to manage multiple currencies. One of the desired functions was to display the exchange rates in multiple languages. After analyzing this requirement, Mr. Brown realized that although the exchange rate should be displayed in multiple languages, the business logic of reading, calculating, and displaying exchange rates remained unchanged. In other words, displaying exchange rates in multiple languages was only a matter of user interface support. There would be two problems if he created a separate project for each language: first, a large amount of repeated work would be done as the same business logic needed to be implemented in all the projects; and second, Mr. Brown must modify the same code in all projects if he needed to change a logic processing flow.

https://doi.org/10.1515/9783110692327-006

Solutions above may not be preferable for many of us, but is there a better way to meet these requirements in C? The answer is affirmative: the preprocessing directives we are going to introduce now are the solution.

6.1.1 Preprocessing

What is preprocessing? It is a simple concept. When we write C programs, we can include some compilation instructions in the source code to tell the compiler how the program should be compiled. When the program is compiled, these instructions are executed before the compilation of source code. Hence, these instructions are also called preprocessing directives. See Figure 6.1 for the process of compiling source code into executable files.

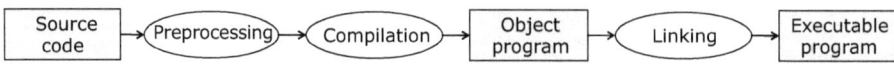

Figure 6.1: The process of compiling source code into executable files.

6.1.2 Preprocessing directives

Preprocessing directives are defined by the American National Standards Institute C standard. They include macro definition, file inclusion, and conditional compilation. See Figure 6.2 for their corresponding keywords.

Type	Keyword
Macro definition	#define, #undef
File inclusion	#include
Conditional compilation	#if, #ifdef, #else, #elif, #endif

- Start with "#"
- Each on its own line
- No semicolon at the end

Programming Error

Use semicolons after #define and #include directives

Preprocessing directives are not C statements

Figure 6.2: Preprocessing directives.

We can see from these keywords that all preprocessing directives start with the "#" sign, whose scope extends to the first newline symbol after it. In essence, its scope is one logical line. If a directive is too long in actual programs, the logical line can be divided into multiple physical lines using "\". Compilers can recognize these lines before compilation and process them as a single logical line.

In fact, preprocessing directives are not statements in C, but they enhance the power of C programming. Reasonable use of these makes programs we write more natural to read, modify, port, and debug. For example, we can adopt the modularization approach to divide a system into multiple relatively independent function modules based on user requirements and functionality design by using file inclusion directives. Besides, conditional compilation directives enable us to compile programs into different versions to accommodate different requirements without modifying the rudimentary code. The code reuse rate is thus increased.

Note that we should not add semicolons after preprocessing directives. Preprocessing directives are not C statements and the end of their scope is the end of a logical line.

6.2 Macro definition

6.2.1 Simple macro definition

Sometimes, a constant is used in multiple places in a program and its value may be updated as needed during testing. A convenient way to do this is to use a dedicated symbol and assign a value to the symbol when defining it. In this way, we can simply modify the value in the definition to update all occurrences of the constant rather than updating every one of them. This prevents us from leaving one occurrence unchanged and creating a bug in the program. What we have just described is the purpose of using macro definitions. In short, the macro definition is essentially text replacement of the source code done before compilation.

The syntax of a simple macro definition is shown in Figure 6.3, where define is the keyword of macro definition directive, < macro name > is an identifier, and < string > can be a constant, an expression or a formatted sequence.

Syntax of simple macro definition

```
#define  <macro name>  <string>
```

Figure 6.3: Syntax of simple macro definition.

Note:
(1) The source code will be checked before compilation. Whenever a macro name is encountered, it is replaced with the string specified in the macro definition. The compilation will not be started until all replacements are completed.
(2) The replacement process is called macro replacement in American National Standards Institute C.

(3) We usually use capital letters for macro names in C programs. This helps us find macro replacements when reading a program and avoids confusion between macro names and normal identifiers. Because a macro name is essentially an identifier, spaces are not allowed in it. Moreover, it must be a combination of letters, numbers, and underscores with the exception that the first character must not be a number.

(4) Good programming habits are beneficial. It is recommended to put shared macro definitions at the beginning of a header file and use them through file inclusion with #include directive. Modularization makes functionalities of files clearer. It is also easier to modify these definitions later.

Let us take a look at how macro definitions work through some examples.

Example 6.1 Example of macros 1
Use a sequence to replace the identifier.

```
#define MAX 128
int main(void)
{
  int max_value =MAX;
  return 0 ;
}
```

[Analysis]
A macro MAX is defined in this example, which corresponds to 128. When the compiler processes this program, it replaces the MAX in the source code with 128. In other words, the actual code that will be compiled is "int max_value = 128." It is worth noting that this is merely text replacement. No variable assignment is done in this process. Variable MAX never existed in the program. It is equivalent to use "Find-> Replace" to find MAX and replace it with 128 in a text editor.

Example 6.2 Example of macros 2
Use a sequence to replace the identifier.

```
#define TRUE  1
#define FALSE 0
printf("%d %d %d", FALSE, TRUE, TRUE+1);
Output:
0 1 2
```

[Analysis]
The output arguments of printf are provided by macro replacement. The printf statement is replaced with printf("%d %d %d", 0, 1, 1 + 1) before compilation. It is worth noting that preprocessing directives does not generate code or participate in code execution. They are simply a porter of code. Hence, the actual computation of 1 + 1 will be done in the compilation phase.

Example 6.3 Example of macros 3
The most common use of macros is defining names for constants.

[Analysis]
The following code uses macro MAX_SIZE as the length of the array.

```
#define MAX_SIZE 100
float balance[MAX_SIZE];
```

Example 6.4 Example of macros 4
Macro replacement is only done for identifiers. Values in strings are not replaced.

[Analysis]
A macro definition and statements using it are shown in Figure 6.4.

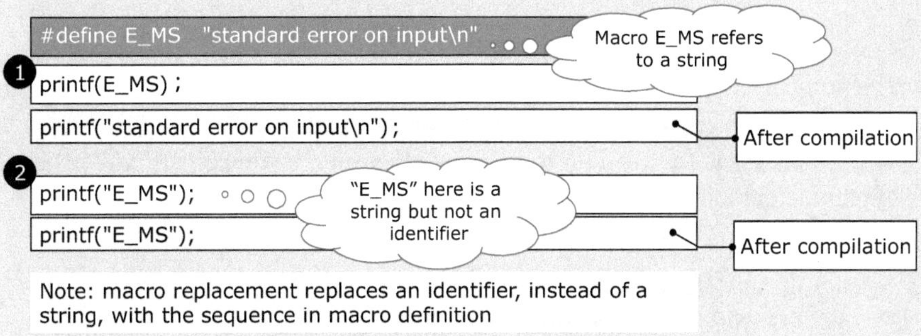

Figure 6.4: Macro in a string.

The first E_MS is an identifier, thus it is replaced with the corresponding string. The second E_MS is a string wrapped in double quotation marks, thus it is not replaced. Beginners may feel confused about the difference and make a mistake when programming.

Furthermore, this example shows that we can define a macro if an output statement with the same format, like the first E_MS above, is used multiple times in a program. In this way, we do not have to write the same code again and again; therefore, it is less likely to output incorrectly formatted contents due to typos. Also, we only need to update the definition if we want to change the format.

6.2.2 Macro definitions with parameters

We have introduced simple macro definitions, which are merely text replacement. Now we are going to learn a more complex macro definition: macro definition with parameters.

Macro definitions with parameters are more abstract and universal. We can define parameters in macros in a way similar to how we use parameters in function definitions and pass arguments when calling functions. The syntax of macro definition with parameters is shown in Fig. 6.5, where:

Syntax of macro with parameters

```
#define <macro name>(parameter list) <macro body>
```

Fig. 6.5: Syntax of macro with parameters.

(1) < macro name > is again an identifier.
(2) The number of parameters in the parameter list can be one or more. When there are multiple parameters, they are separated by commas.
(3) < macro body > is the string for replacement, which is an expression consisting of parameters in the parameter list.

Example 6.5 Example of macros 5
Figure 6.6 shows a macro definition with parameters.

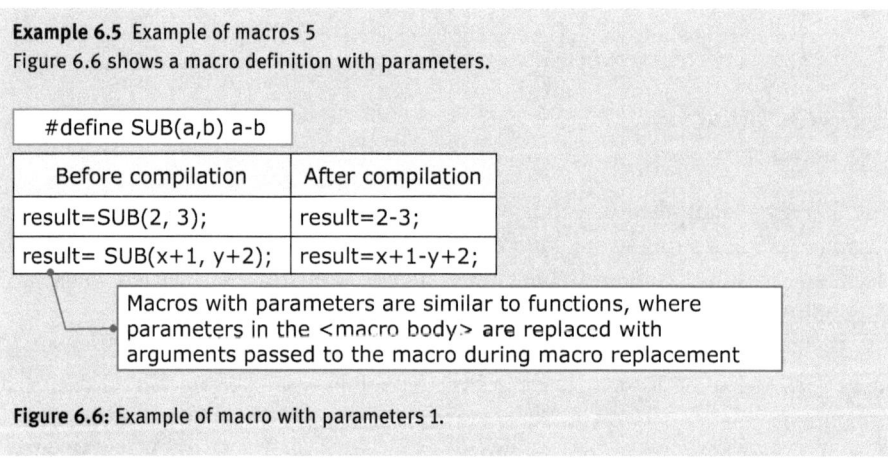

Figure 6.6: Example of macro with parameters 1.

The definition of macro SUB was an abstraction of subtraction a-b. In contrast to macro definitions without parameters, a and b here are used as parameters, because we do not know their values when defining the macro. Arguments we pass in when doing text replacement determine their values. Isn't it similar to function definitions? In macro replacement, we are replacing parameters in the < macro body > with arguments. Let us see an example now (Figure 6.7).

Example 6.6 Example of macros 6

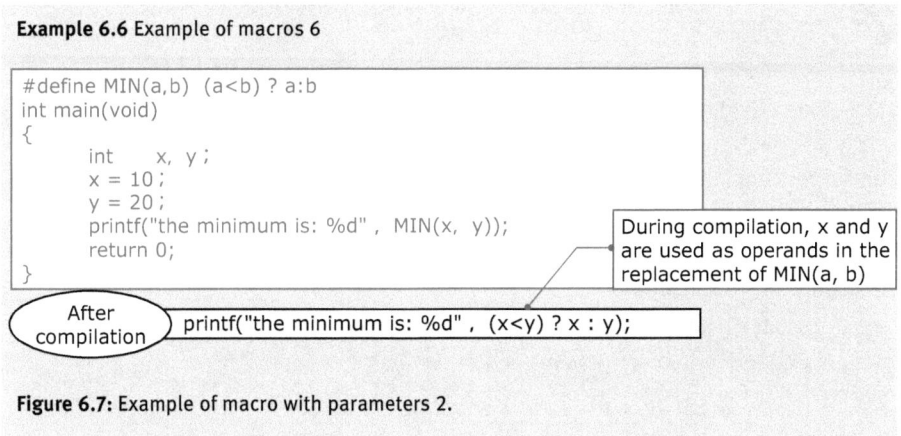

Figure 6.7: Example of macro with parameters 2.

It seems that MIN(x,y) in this example is a function call. However, expression MIN(x,y), which is defined by macro MIN(a,b), will be replaced during compilation. Arguments x and y will be used to replace a and b.

The merit of using macros instead of functions is that we use something in the form of functions without the overhead of function call. The source code is written in a similar style, but it can be executed faster as there is no expense of function call. On the other hand, since macros are text replacements, we are still using redundant code in our program, which in fact increases the length of our program despite higher execution speed.

Although macros with parameters are similar to functions with parameters, they are different things in essence. Figure 6.8 shows the differences between them.

	Macro with parameter	Function
Processing phase	Preprocessing	Runtime
Parameter type	No type issues	Need to define types of parameters and arguments
Processing process	No memory allocated, simple text replacement	Memory allocated, compute values of arguments and pass to function
Program length	Increased	Unchanged
Execution speed	No extra expense	Function call and return take time

Figure 6.8: Differences between macros and functions.

> **Good programming habit**
> When there are many references to a relatively long variable (usually member of a structure) in the function, we can use an equivalent macro to replace it. This improves programming efficiency and readability.

6.2.3 Side effects of macros

Careful readers may have noticed that the macro replacement in Example 6.5 is logically wrong. Let us sift through this example. The code is as follows.

```
#define SUB(a,b) a-b
result=SUB(2, 3);  //Replaced with: result=2-3;
result= SUB(x+1, y+2);  //Replaced with: result=x+1-y+2;
```

We wanted to implement the subtraction of a and b with the macro. The replacement of SUB(2-3) with result = 2-3 is correct and consistent with our design. However, SUB(x + 1, y + 2) is replaced with result = x + 1-y + 2, instead of desired result result = x + 1-y-2.

The reason behind this error is that macro replacement is merely text replacement. Neither concrete computation nor precedence of operators is actually involved. Hence, we should design macros carefully to avoid side effects.

Is it impossible to use macros in such examples, then? The answer is no: we can still use macros, but we need parentheses in their definitions. For example, if we change the macro definition in Example 6.5 into #define SUB(a, b) (a)-(b), the subtraction will be correctly implemented no matter what a and b we are using.

Now we are almost done with macros, but there is one more thing to remember: we should avoid increment or decrement operators in macros. For example, if we want to compute SUB(++x, ++x) in Example 6.5, it will be replaced with (++x)-(++x) according to the macro definition. Although the subtraction logic is correct, the system has no rule regarding which operand of "−" is read first. The value of this expression is compiler-dependent during execution.

6.3 File inclusion

When writing a large-scale program, we often divide the system into different modules based on modularization principles. Each module is implemented by one or more files and provides an interface for other modules to use. To use these interfaces or variables in a module, we often need to define the same variable (or function interface) in multiple files. For example, when calculating the area of circle, annulus, and surface of a sphere, we implement area calculations of different geometric objects in multiple

C files for better extensibility because each shape has its own area formula. All these formulas need the value of r^2; therefore, we use a function pow to compute it as a practice of code reuse. However, we also know that if we define it in every C file, we are making the mistake of repeatedly defining the same identifier in the same scope, which is usually resolved by a single definition and multiple declarations. Even so, it is still tedious to declare it multiple times. Is there a more convenient solution?

In this section, we are going to learn how file inclusion directives are used to solve this problem.

A file inclusion directive inserts the specified file at its location so that the file is linked with the current one to form a single source file. It can be considered as enhanced text replacement.

The syntax of the file inclusion directive is shown in Figure 6.9.

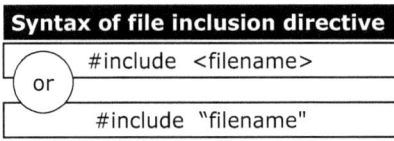

Figure 6.9: Syntax of file inclusion directive.

In the syntax:
(1) include is the keyword.
(2) the filename is the full name of the file to be included. If it is in angle brackets, the compiler will search for the header file in a directory specified by the system (e.g., one or multiple standard system directories in UNIX systems); if it is in double quotation marks, the compiler will first look for the header file in the current directory and go to the system specified directory if none is found.

In the preprocessing phase, the preprocessing directive is replaced with the content of the file to be included. The file after inclusion is then treated as a single source file during compilation. The processing flow is shown in Figure 6.10.

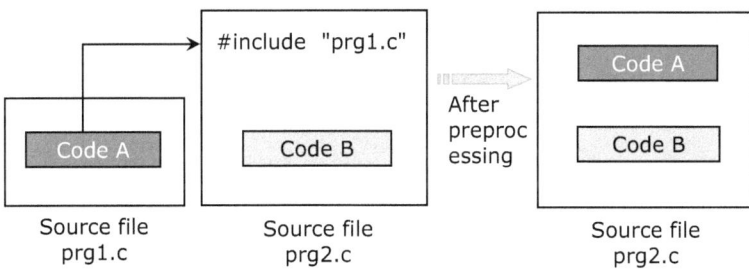

Figure 6.10: Process of file inclusion.

File inclusion is beneficial in programming. When a program is too long, we can split it into multiple shorter programs whose functions are independent of each other so that multiple programmers can work on them simultaneously. Shared information in these programs can be extracted and put into a separate file, which is used by other files by adding include directive at the beginning. For example, constants and definitions of functions can be put into a separate .h file (file with extension .h, also called a header file). Then we use include directive at the beginning of other files to include this header file so that we do not have to write these shared items in every file. Doing so reduces development time and mistakes. We can either write our own header files or use those provided by the system. For example, stdio.h is a header file related to input/output operations provided by the system.

Note that there is no limit on file types. As shown in Figure 6.10, we can include a ".c" file in another ".c" file. However, this is merely an example showing that there is no limit on file types. In practice, we usually include ".h" files only for a better coding style and avoiding redundant definitions.

Example 6.7 Edit a file and include it in another file
In Figure 6.11, the definition of function fun is done in file fun.c, whereas its declaration is made in file fun.h; the file main.c obtains declaration of function fun by including fun.h and calls the function in main function.

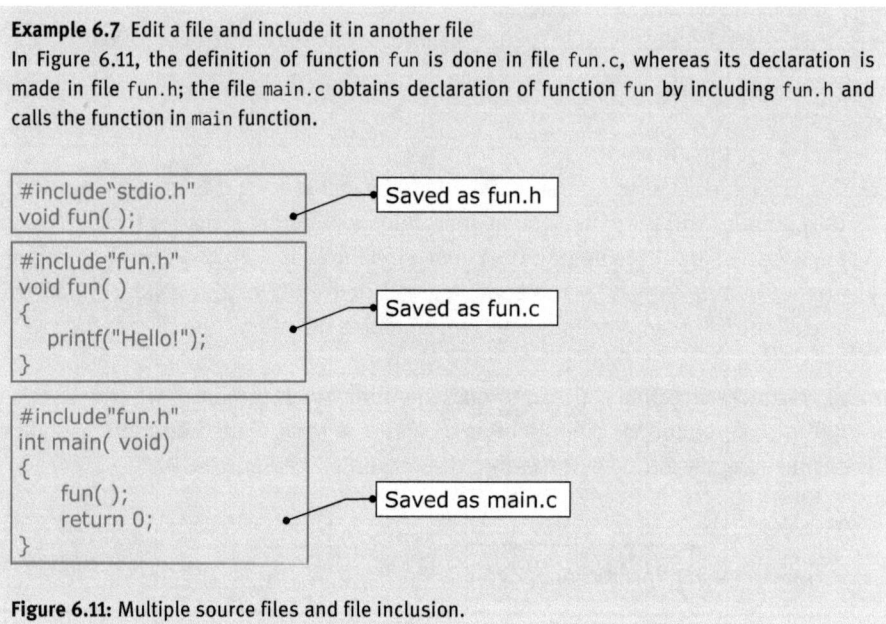

Figure 6.11: Multiple source files and file inclusion.

6.4 Conditional compilation

Conditional compilation directives instruct the compiler to compile different parts of the program under different conditions to produce different object code files as shown in Figure 6.12. In other words, we can set conditions by using conditional

Conditional compilation

Conditional compilation directives instruct the compiler to compile different parts of the program under different conditions in order to produce different object code files

Figure 6.12: Definition of conditional compilation.

compilation directives so that some parts of the program would not be compiled unless the conditions are met.

6.4.1 Format of conditional compilation 1

There are three commonly used formats of conditional compilation. Figure 6.13 shows the first one, where ifdef, else, and endif are the keywords. Code segments 1 and 2 consist of preprocessing directives and statements. The conditional compilation here works as follows: if a #define directive has defined the identifier, then code segment 1 will be compiled; otherwise, segment 2 will be compiled.

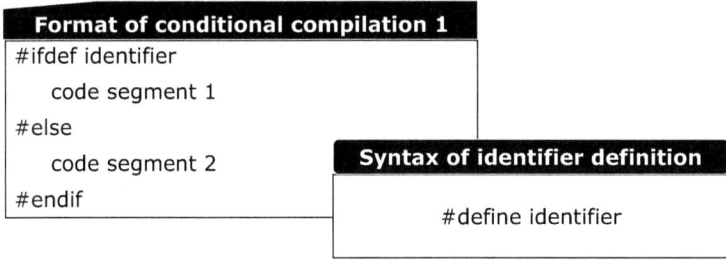

Format of conditional compilation 1

```
#ifdef identifier
      code segment 1
#else
      code segment 2
#endif
```

Syntax of identifier definition

```
#define identifier
```

Figure 6.13: Format of conditional compilation 1.

The #else in this form is optional, so it can also be written as follows:

```
#ifdef identifier
        Code segment
#endif
```

Example 6.8 Example of conditional compilation 1

```
1 #include <stdio.h>
2 #define TIME
3 int main(void)
4 {
5 //#undef TIME Uncomment this line when we want to cancel the definition
6 #ifdef TIME
7  printf("Now begin to work\n");
```

```
8 #else
9  printf("You can have a rest\n");
10 #endif
11 return 0;
12 }
```

Because conditional compilation directives are used in this example, which printf is compiled depends on whether TIME has been defined using #define. If it has been defined, "printf("Now begin to work\n");" on line 6 will be compiled, otherwise "printf("You can have a rest\n");" on line 8 will be compiled. TIME has been defined in this example, so the output is:

```
Now begin to work
```

If we want to change the compilation condition, for example, we want to output "You can have a rest," we do not have to write the program again. It can be done by commenting line 2, which makes TIME undefined, or by canceling the definition of TIME using #undef as shown on line 5 of the program.

6.4.2 Format of conditional compilation 2

The second format is shown in Figure 6.14.

Format of conditional compilation 2
#ifndef identifier
Code segment 1
#else
Code segment 2
#endif

Figure 6.14: Format of conditional compilation 2.

The only difference between the first two formats is that the keyword ifdef is replaced with ifndef. In this case, code segment 1 is compiled if a #define directive has not defined the identifier; otherwise, segment 2 is compiled. This is exactly the opposite of the first form. Let us take a look at the following example:

```
#ifndef NULL
#define NULL ((void *)0)
#endif
```

This segment of code ensures that symbol NULL is defined as ((void *) 0) exactly once. It works as follows: when the compiler first processes this directive, NULL has not been defined. The compiler proceeds to the macro definition of NULL since the condition of #ifndef is met. If the same directive appears again, the macro definition will not be processed because NULL has already been defined. Hence, NULL is guaranteed to be defined only once.

Good programming habit

In practice, especially in large-scale programs, nested inclusion can often be found as the inclusion relations between source files are complicated. For example, file1.h may include file2.h and file3.h, whereas file2.h also includes file3.h. If there is no guarding mechanism, file3.h will be included twice in file1.h, which leads to code redundancy in source files. Moreover, if there are definitions of identifiers in file3.h, repeated definition errors will occur. Hence, we should use a guarding mechanism like the second format introduced above and put everything before #endif when defining header files. This effectively prevents double inclusions.

6.4.3 Format of conditional compilation 3

The third format is shown in Figure 6.15.

Format of conditional compilation 3
#if constant expression
Code segment 1
#else
Code segment 2
#endif

Figure 6.15: Format of conditional compilation 3.

if, else and endif are the keywords of this format. Code segments 1 and 2 consist of preprocessing directives and statements. It works as follows: if the constant expression evaluates to true, code segment 1 will be compiled; otherwise, segment 2 will be compiled. With this directive, our program can complete different tasks under different conditions.

Example 6.9 Example of conditional compilation 2

```
1  #include <stdio.h>
2  #define R 1
3  int main(void)
```

```
4 {
5   float c,s;
6   printf("input a number: ");
7   scanf("%f",&c);
8   #if R
9   s=3.14*c*c;
10  printf("area of round is:%f\n",s);
11  #else
12  s=c*c;
13  printf("area of square is%f\n",s);
14  #endif
15  return 0;
16 }
```

In this example, lines 9 and 10 are compiled if the expression R evaluates to true.

```
s=3.14159*c*c;
printf("area of round is:%f\n",s);
```

Otherwise, lines 12 and 13 are compiled.

```
s=c*c;
printf("area of square is%f\n",s);
```

6.4.4 Nested conditional compilation directives

We can only implement a double-branch structure with #if and #else, thus C also provides #elif directive, which means "else if." It can be used with #if and #else to form an if-else-if structure for multiple branch cases. Its syntax is shown in Figure 6.16.

```
Format of nested conditional compilation
#if constant expression 1
    code segment 1
#elif constant expression 2
    code segment 2
#elif constant expression 3
    code segment 3
......
#else
    code segment n+1
#endif
```

Figure 6.16: Format of nested conditional compilation.

Example 6.10 Use cases of conditional compilation
There are two merits of using conditional compilation: easier debugging and better portability. When a program has multiple versions, we can use the code segment shown in Figure 6.17 to make porting easier. If the program is to be compiled and executed in the Borland C environment, we can add #define BORLAND_C at the beginning.

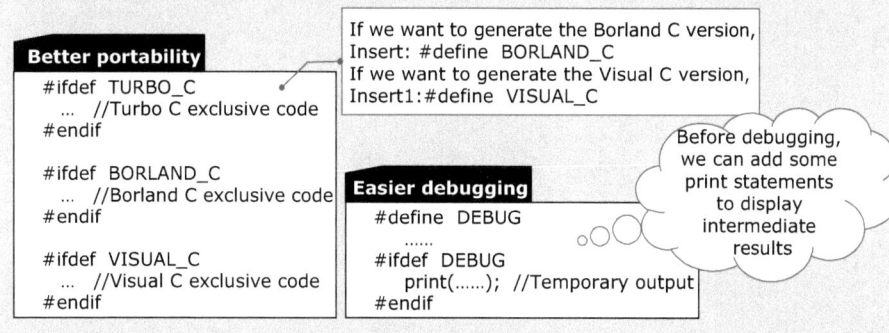

Figure 6.17: Use cases of conditional compilation.

When debugging, we can use some print statements to display intermediate results. After debugging is done, we can remove #define DEBUG so that these statements would not be compiled.

Good programming habit
Instead of maintaining a release version and a debug version of source files simultaneously, it is better to use a debug switch to switch between them, which makes maintenance easier.

6.5 Summary

The main concepts and their relations are given in Figure 6.18.
Compilation translates statements into machine code.
Preprocessing is work done before compilation.
File inclusion allows us to use existing files.
The macro definition does replacement, which makes code editing easier.
Conditional compilation compiles code as needed, which makes debugging more convenient and enhances the flexibility of code.

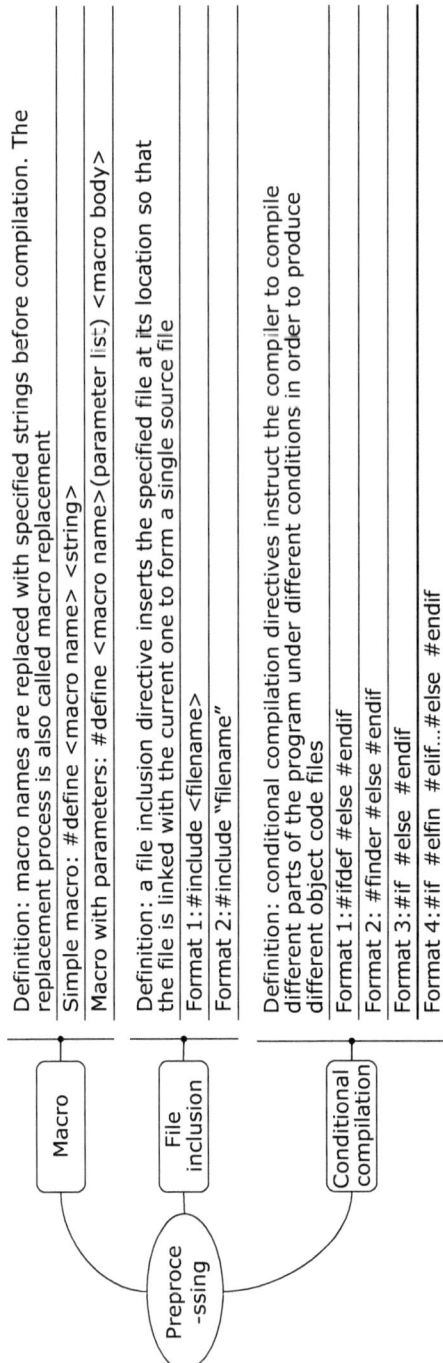

Macro

Definition: macro names are replaced with specified strings before compilation. The replacement process is also called macro replacement

Simple macro: #define <macro name> <string>

Macro with parameters: #define <macro name>(parameter list) <macro body>

File inclusion

Definition: a file inclusion directive inserts the specified file at its location so that the file is linked with the current one to form a single source file

Format 1: #include <filename>

Format 2: #include "filename"

Conditional compilation

Definition: conditional compilation directives instruct the compiler to compile different parts of the program under different conditions in order to produce different object code files

Format 1: #ifdef #else #endif

Format 2: #finder #else #endif

Format 3: #if #else #endif

Format 4: #if #elfin #elif...#else #endif

Preproce-ssing

Figure 6.18: Relations between concepts related to preprocessing.

6.6 Exercises

6.6.1 Multiple-choice questions

(1) [Simple macros]

In macro definition #define A 3.897678, A represents a ( )

A) float number   B) double number   C) constant   D) string

(2) [Define]

Which of the following statements is correct? ( )

A) Preprocessing commands must be at the beginning of a file.

B) We can have multiple preprocessing commands on a single line.

C) Macro names must be capital letters.

D) Macro replacements are not done during program execution.

(3) [define]

C compilers process macros ( )

A) at runtime

B) during linking

C) at the same time as they compile other statements

D) before they compile other statements

(4) [Define]

Which of the following statements is correct? ( )

A) #define and printf are both C statements.

B) #define is a C statement, but printf is not.

C) printf is a C statement but #define is not.

D) Neither #define nor printf is a C statement.

(5) [Macro with parameters]

What is the output of the following program?

```
#include<stdio.h>#define PT 5.5
#define S(x) PT*x*x
int main(void)
{
    int a=1, b=2;
    printf("%4.1f\n", S(a+b));
    return 0;
}
```

A) 49.5   B) 9.5   C) 22.0   D) 45.0

(6) [File inclusion]

In file inclusion preprocessing directives, how is the file searched when the filename is inside "< >"? ( )

A) It is searched in the system-defined directory.

B) It is searched in the directory of the source file first and then in the system-defined directory.

C) It is only searched in the directory of the source file.

D) It is only searched in the current directory.

6.6.2 Fill in the tables

(1) [Define]

Suppose we have the following macro definitions. Fill in the table in Figure 6.19 with the statements after macro replacement.

```
#define MAX 10;  #define PI 3.1415926
#define area(r) (PI*r*r)
#define A 3+2
#define INPUT "Please input your name.\n"
```

Original statement	Compilation result
int array[MAX];	
printf(INPUT);	
int temp = A/5;	
printf("INPUT A");	
float temp = area(5.5);	

Figure 6.19: Preprocessing: fill in the tables questions 1.

(2) [Conditional compilation]

Fill in the table in Figure 6.20.

```
#include <stdio.h>#define CHINESE //————①
int main( void)
{
char name[MAX];
int age;
#ifdef CHINESE
```

```
printf("输入您的姓名和年龄：\n");
#else
printf("Please enter your name and age:\n");
#endif
scanf("%s %d",name, &age);
#ifdef CHINESE
printf("%s，您好！您已经%d岁了，欢迎加入C语言的学习大军！\n",name,
age);
#else
  printf("Hello %s! You are %d years old!Welcom to C language!\n",name,
age);
#endif
return 0;
}
```

Input	Output
张三 23	
(Comment statement ①)	
Bob 28	

Figure 6.20: Preprocessing: fill in the tables questions 2.

(3) [Conditional compilation]
    Fill in the table in Figure 6.21.

```
#include <stdio.h>
#define DEBUG //————①
int swap(int *p, int *q)
{
int temp=0;//————②
#ifdef DEBUG
printf("debug: *p= %d, *q=%d \n",*p, *q);

#endif
if (*p > *q)
{
  temp = *p;
  *p = *q;
  *q = temp;
  temp=1;
}
```

```
#ifdef DEBUG
  printf("debug: *p= %d, *q=    %d \n",*p, *q);
#endif
return temp; //————③
}
int main()
{
  int a = 5, b = 4;
  int c = swap( &a, &b );
  printf(" a= %d, b= %d, c= %d \n",a, b,c); //————④
  return 0;
}
```

Variable	a	b	*p	*q
Value after statement ②				
Value after statement ③				
Value after statement ④				
Program output				
Program output after commenting statement ①				
Functionality of statement ①				

Figure 6.21: Preprocessing: fill in the tables questions 3.

6.6.3 Programming exercises

(1) Find the maximum of three numbers using functions and macros, respectively.
(2) Use conditional compilation to complete the following task:
    Given a line of telegram text, output it in one of the two formats:
    1) output as-is
    2) convert each character to its next character in the alphabet, that is, "a" is output as "b", . . ., "z" is output as "a"
    The program should use #define to control which format is used:
    #define CHANGE 1 // Output encrypted text
    #define CHANGE 0 // Output as-is

(3) Write a program that converts letter inputs to uppercase or lowercase, depending on the conditional compilation command.

(4) Given keyboard input y, use a macro to evaluate the value of the following expression:

$$3(y^2 + 3y) + 4(y^2 + 3y) + y(y^2 + 3y)$$

(5) Define a macro with parameters that swap its two parameters. Use the macro to swap two input numbers and output the new values.

7 Execution of programs

Main contents
- Introduction of VC6.0
- Debugging methods
- Testing methods

Learning objectives !
- Know the typical process of software development, can follow this process to write programs
- Know the purpose and meaning of compilation and linking
- Know basic approaches of debugging
- Know testing methods

> Even a genius can't guarantee that his code is completely correct from the beginning. Every single, if not all, program that is not trivial is written after debugging and modifying again and again. –Experience of debugging

Debugging is the process of finding and correcting errors in programs. It is the most fundamental skill that a programmer should possess. It is more important to learn to debug than to learn a programming language. A programmer cannot write good software without knowing how to debug, even if he/she knows a programming language well.

Few, if not none, codes are correct when first being written. It is nearly impossible to debug by reading source code for programs of a reasonable scale. The most efficient way of debugging is to use debugging tools.

Debugging helps programmers to learn the actual execution process of their programs. It also allows programmers to check whether their design works as expected, which improves the development efficiency in return. Mastering debugging techniques enables programmers to write codes that are easier to debug. They can gain better perception and control of code.

Debugger tools can help us learn the computer system and other knowledge of software and hardware. We can quickly learn modules, architecture, and working flow of software or a system through debugging.

7.1 Runtime environment of programs

From being written to being executed and outputting a result, a program needs to go through several processing phases as shown in Figure 7.1. The functions of each phase are as follows.
- Edit: Type in source code and save it to generate C source file, whose extension is .c (or .cpp in VC6.0 environment).

https://doi.org/10.1515/9783110692327-007

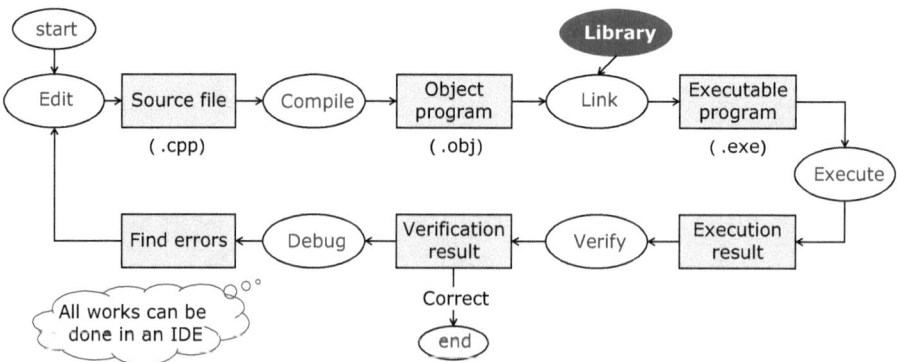

Figure 7.1: Process of program execution.

- Compile: Execute compile command. The compiler scans the source code for syntax errors. If none is found, it generates code in machine language, which is called the object program and has extension .obj. If syntax error exists, programmers should modify the code based on the warning or error message given by the compiler until the program is successfully compiled.
- Link: Execute the link command. The system links obj files, which can be written by programmers or library functions used in the program, together to generate an executable file, whose extension is .exe.
- Run: Execute run command. The program is executed to produce a result.
- Verify: Programmers check the output at the specified output location, for example, a specified window or file, and compare it with the expected result to determine whether the program is correct.
- Debug: If the result is wrong, programmers need to use various debugging techniques to find the error and modify the code. The above steps are repeated until a correct result is obtained.

All these activities related to program execution, including editing, compiling, linking, executing, and debugging, can be done in an integrated development environment (IDE).

An IDE is an application that provides a program development environment. As shown in Figure 7.2, it is a software that provides integrated services like code editing, compilation, debugging, and so on. All software that has such features can be called IDEs.

The basic idea and general methods of debugging are applicable in all debugging environments, thus it is important to master this fundamental knowledge. Visual C++ 6.0 (VC 6.0 for short) is a small but robust IDE. It provides powerful debugging tools and is compatible with multiple versions of the Windows operation system; therefore, it is recommended for beginners of C. Moreover, it enables a

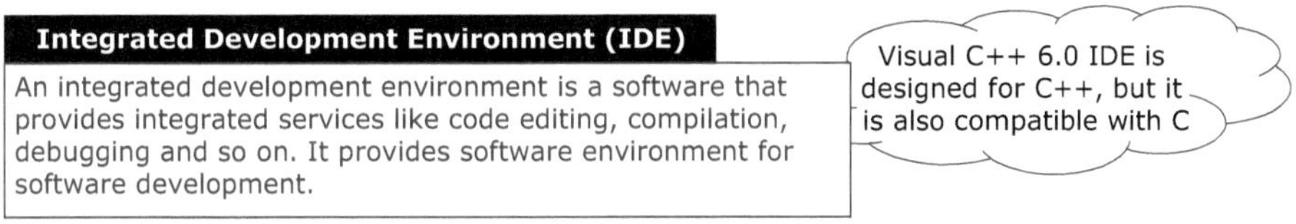

| **Integrated Development Environment (IDE)** |
| An integrated development environment is a software that provides integrated services like code editing, compilation, debugging and so on. It provides software environment for software development. |

Visual C++ 6.0 IDE is designed for C++, but it is also compatible with C

Figure 7.2: Definition of IDE.

smooth transition to the Visual Studio IDE, which provides similar functionalities but has a more complex user interface. After comparing multiple IDEs, we shall use VC6.0 to run and debug programs in this book.

Knowledge ABC Visual C++ 6.0 IDE

Microsoft visual C++ 6.0 (abbreviated as Visual C++, MSVC, VC++, or VC) is an application development environment used to develop C++ programs created by Microsoft. It integrates tools like code editor, compiler, debugger, and graphical user interface. VC 6.0 has been widely used due to its good interface and usability.

Using the console operation provided by Visual C++ 6.0, we can create C applications. Win32 console applications are a type of Windows program, which communicates with users through a standard console without using a complex graphical user interface. We shall introduce how to use Visual C++ 6.0 to write simple C programs from the seven perspectives shown in Figure 7.3.

1	Main screen of integrated environment
2	Create a project
3	Create a source file
4	Edit a source file
5	Compile a source file
6	Link a program
7	Execute a program

Figure 7.3: Steps of using VC6.0 integrated environment.

7.1.1 Main screen of integrated environment

After installing Visual C++ 6.0, we can start the application through the "Start" menu or desktop shortcut. The IDE is shown in Figure 7.4. Similar to most Windows applications, the menu bar and toolbar are on the top, and the three areas below are workspace, editor panel, and output panel.

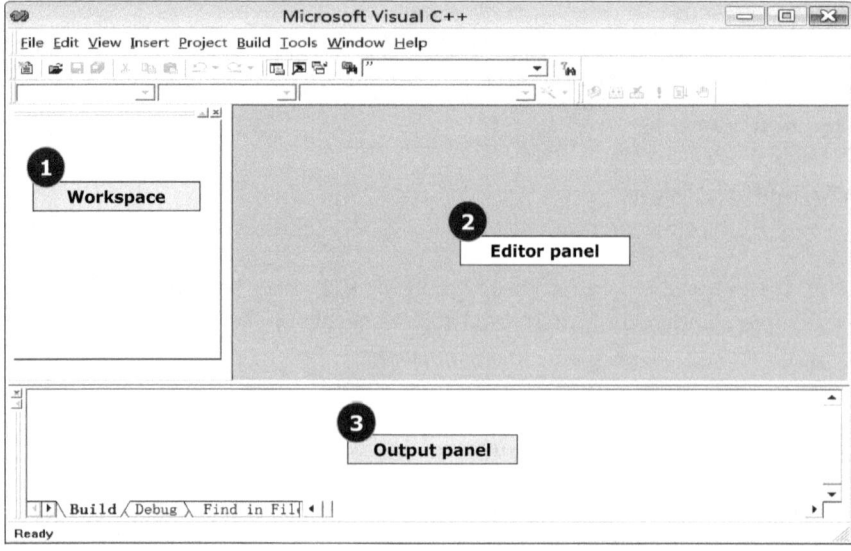

Figure 7.4: Main screen of the Visual C + + 6.0 integrated environment.

The workspace records the status of users' work and will be automatically saved when VC6.0 is closed; code files are edited in the editor panel; the output panel shows messages, errors, or results generated during the creation and debugging of the program.

7.1.2 Create a project

The execution of a program is a systematic project, which is similar to a theater play. Actors cannot start performing until every environment setting is done, including stage, setting, light, sound, and so on.

IDE is such a "stage." VC6.0 puts every environment resource needed for a play into a "project." Programs are like actors of the play. As shown in Figure 7.5, a project is a series of correlated activities that are done following a set of rules with given time and resources to achieve a specific goal.

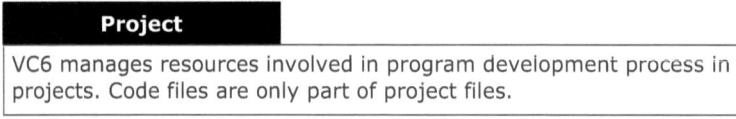

Figure 7.5: Definition of project.

Knowledge ABC Projects

In Visual C++ IDE, a project is the set of correlated C++ source files that implement required functionalities, resource files, and classes that support these files. Projects are the basic unit of program development in Visual C++ IDE. They are used to manage all elements that construct an application and eventually generate the application.

The steps of creating an application are as follows.

Figure 7.6 shows the screen of steps 1 and 2, where we select the "File" menu in step 1 and the "New" sub-menu in step 2.

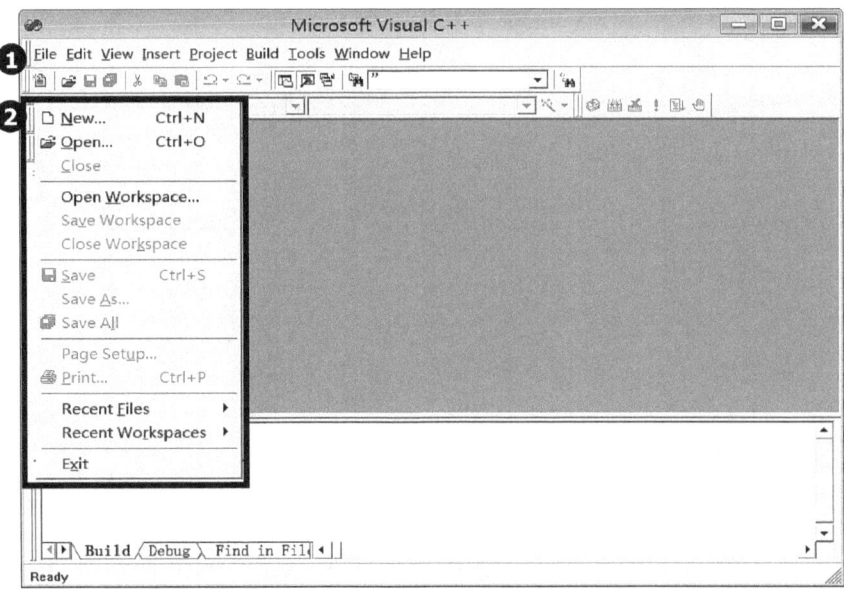

Figure 7.6: Project creation step 1 and 2.

Figure 7.7 shows the screen of steps 3–7. In step 3, we switch to the "Project" tab; in step 4, we select the "Win32 Console Application" as our project type, which is a console application working in a 32-bit Windows environment (it is a character program without a graphical interface); in step 5, we type in project name; in step 6, we specify the save location; and in step 7, we click "OK" to confirm.

Figure 7.8 shows the screen of steps 8–10. After we confirm our input in the Project tab, a wizard, which is shown on the left of Figure 7.8, pops up and guides users to generate the framework of the program. In step 8, we select "An Empty Project"; in step 9, we click "Finish" to close the wizard, then the dialog box on the right pops up; and in step 10, we click "OK" to confirm.

Note that beginners often make the mistake of selecting project types other than console application, which leads to linking errors later as shown in Figure 7.9.

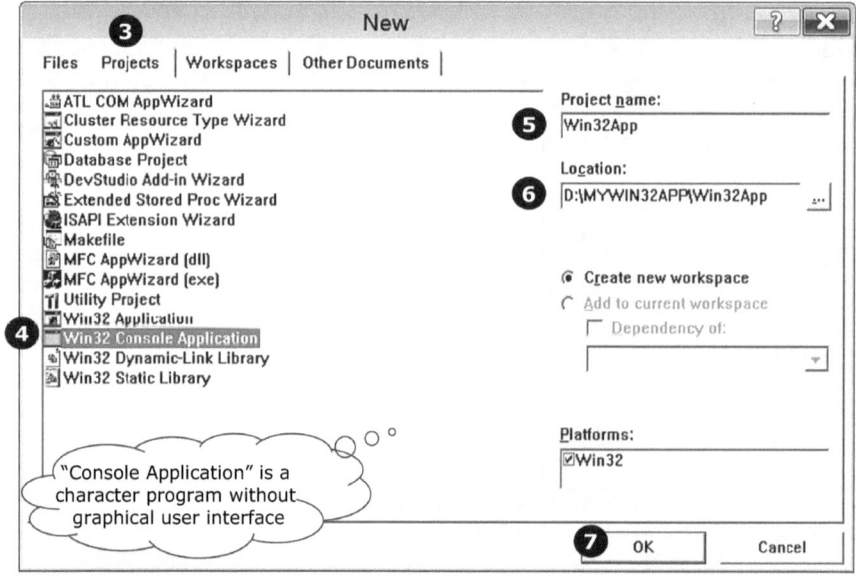

Figure 7.7: Project creation steps 3–7.

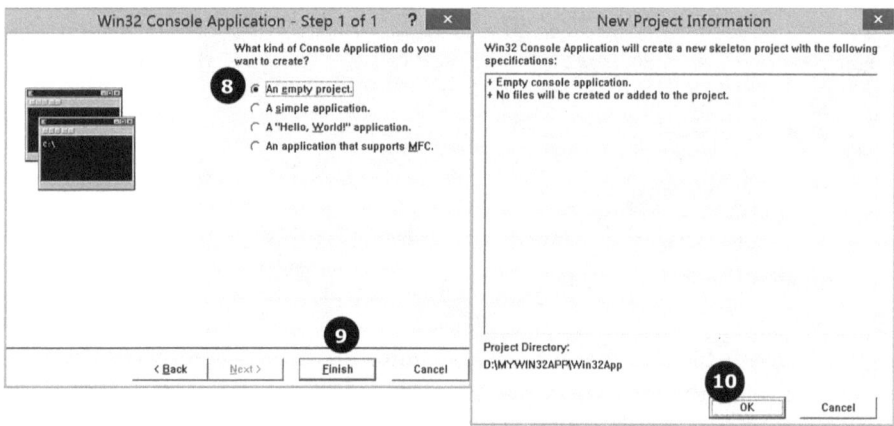

Figure 7.8: Project creation step 8 to 10.

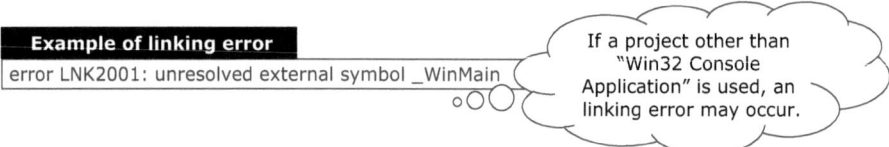

Figure 7.9: Example of linking error.

7.1.3 Create a source file

The steps of creating a source file are as follows.

Steps 1 and 2 are shown in Figure 7.10. In step 1, we select the File menu; in step 2, we select the "New" sub-menu.

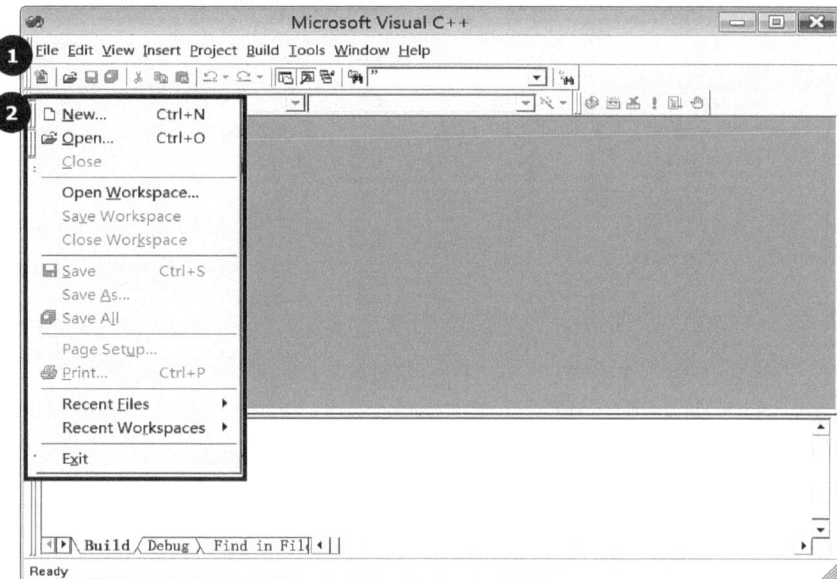

Figure 7.10: Source file creation step 1 and 2.

Steps 3–7 are shown in Figure 7.11. In step 3, we choose the "File" tab; in step 4, we select *C++* Source File as the file type; in step 5, we check the box "Add to project"; in step 6, we type in the file name; and in step 7, we click "OK" to confirm.

Notes on file names: (1) do not include file extensions; (2) use meaningful names for easier management.

7.1.4 Edit a source file

As shown in Figure 7.12, we can perform various editing operations to source files in the editor window, which includes opening and browsing files, input, modification, copy, cut, paste, find, replace, undo, and so on. They can be done either through the menu or through the buttons in the toolbar. In essence, everything is similar to their counterparts in other Windows text editors, for example, Word.

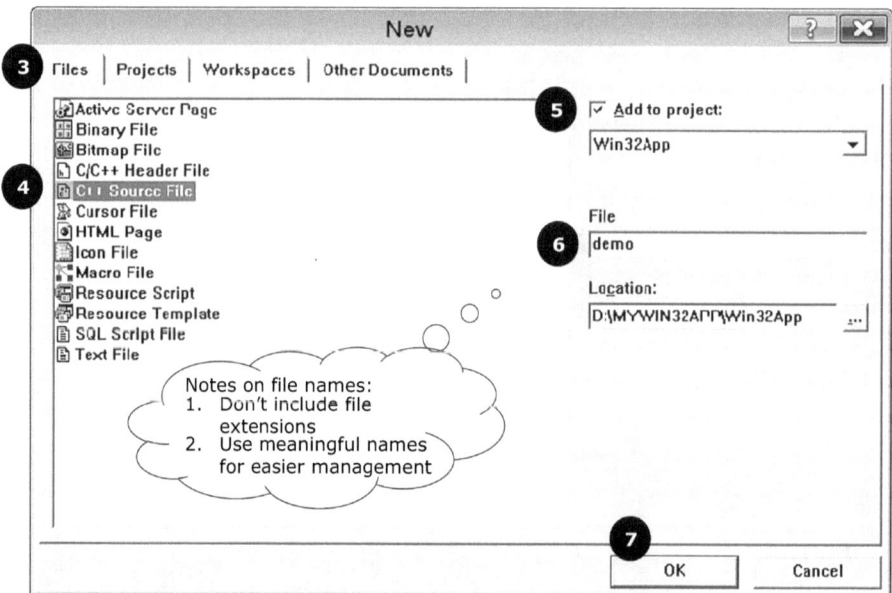

Figure 7.11: Source file creation steps 3–7.

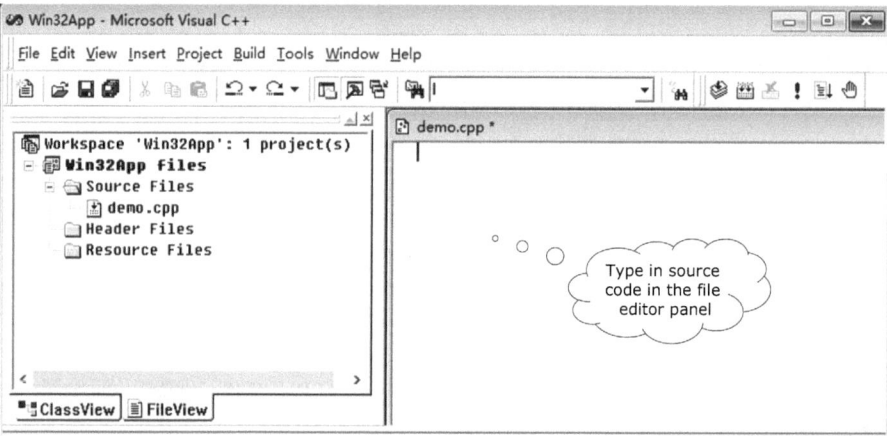

Figure 7.12: File editor panel.

Figure 7.13 shows how to change the code format or font used. If the code is not well formatted, we can format it using "Format Selection" in the menu "Edit-Advanced," whose hotkey is Alt + F8. "Format" here means adjusting the alignment of code as required.

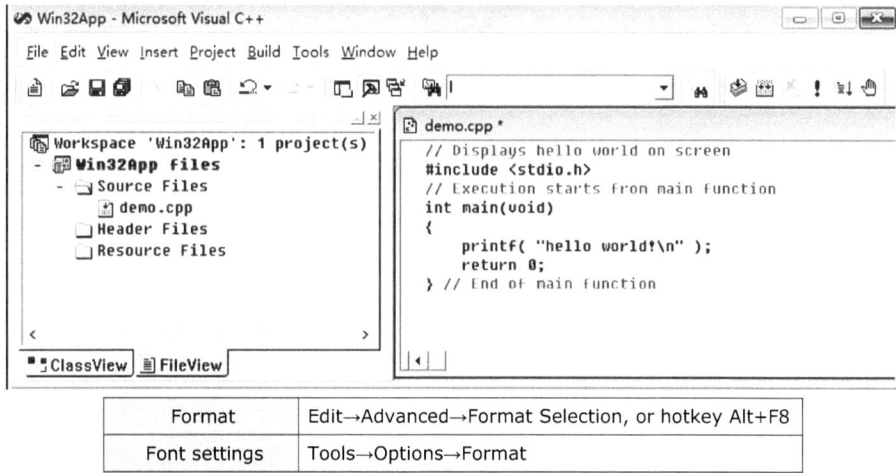

| Format | Edit→Advanced→Format Selection, or hotkey Alt+F8 |
| Font settings | Tools→Options→Format |

Figure 7.13: Edit a file.

If we are not satisfied with the font used in the editor, we can customize it in the "Format" tab of the "Options" dialog box in the "Tools" menu.

Note that Chinese punctuation marks are invalid. Moreover, we should remember to use Ctrl + S to save our code all the time.

> **Good programming habit** Type in parentheses in pairs
> When typing in programs in practice, it is better to enter parentheses in pairs. For example, we should type main()() first and insert statements inside {} later. Doing so prevents us from forgetting the ending parenthesis even if the program is long, which is a common compilation error created by beginners. It often takes a long time to find such errors because the error message is not clear enough.

7.1.5 Compile a source file

The compile command is "Compile" in the menu "Build," whose hotkey is Ctrl + F7. The "Compile" button is located at the first position in the "Build MiniBar" toolbar as shown in Figure 7.14. Users can use any one of these three to compile a source file.

If the compilation is completely successful, "0 error(s), 0 warning(s)" will be shown in the message panel at the bottom.

Object file with extension .obj will be generated after a successful compilation as shown in Figure 7.15.

If an error occurs during compilation or linking as shown in Figure 7.16, there will be an error message indicating the line on which the error exists and the type of error in the message panel at the bottom. For example, there is a message indicating

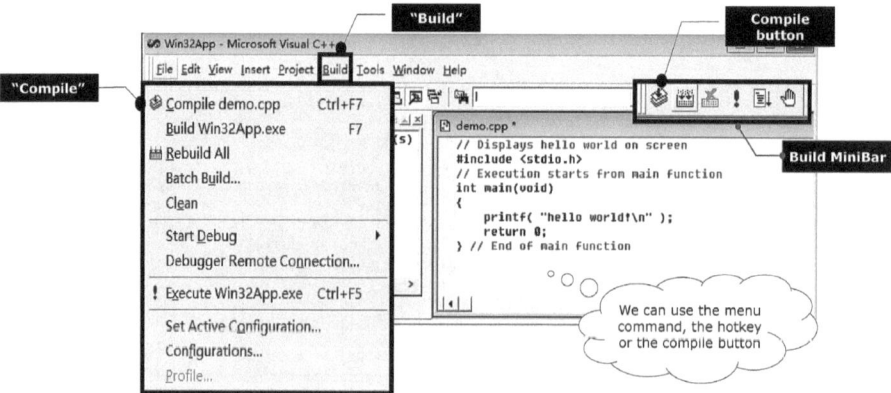

Figure 7.14: Compile command.

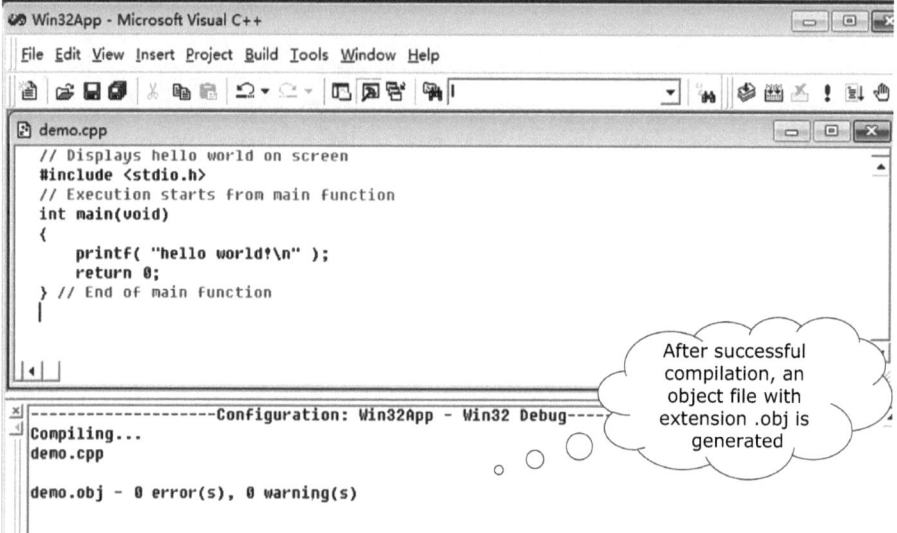

Figure 7.15: Successful compilation.

a "syntax error" on line 7. If we double click on the error message, a blue arrow appears in the editor pointing at the line of the error, so we can check the corresponding code. The cause of the error here is the comma after return 0, which should be a semicolon as defined by the grammar of C. We can modify the code, restart the compilation and linking process, and repeat until there is no syntax error in the program.

Error messages are displayed during compilation to help programmers to find the error and correct it. It is worth noting that the error location showed in the error message may not be correct. If we cannot find an error in the line indicated by the

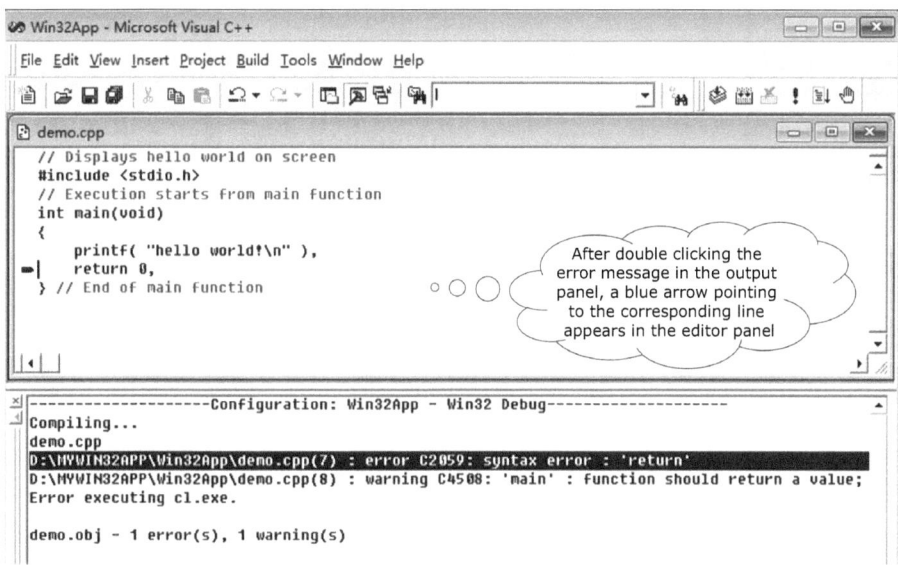

Figure 7.16: Error in compilation.

error message, we should look for it in the lines above. Sometimes the error type is not correct either because errors may occur in many cases or they are often correlated. We need to analyze the code carefully to find the true error, instead of spending time looking for the exact error indicated by the error message.

Method of finding and correcting syntax errors: pay attention to the number of errors and warnings in the message window; find and correct errors before warnings; and correct errors in order. Do not try to find the next error before correcting the current one.

7.1.6 Link programs

There are two types of linking commands: Build and Rebuild All in the "Build" menu. Both of them are used to generate executable .exe files. Because a program can consist of multiple files, these files need to be compiled separately to generate the corresponding object file. The purpose of linking is to link these obj files and other library files used in the program together to construct a single exe file, which can be executed in the operating system.

In Figure 7.17, the difference between Build and Rebuild All is that the former compiles the source file that is modified most recently and does linking while the latter compiles all source files and links them regardless of their modification time. If linking is completed successfully, an executable file will be generated. Note that the file name is projectname.exe as shown in Figure 7.18.

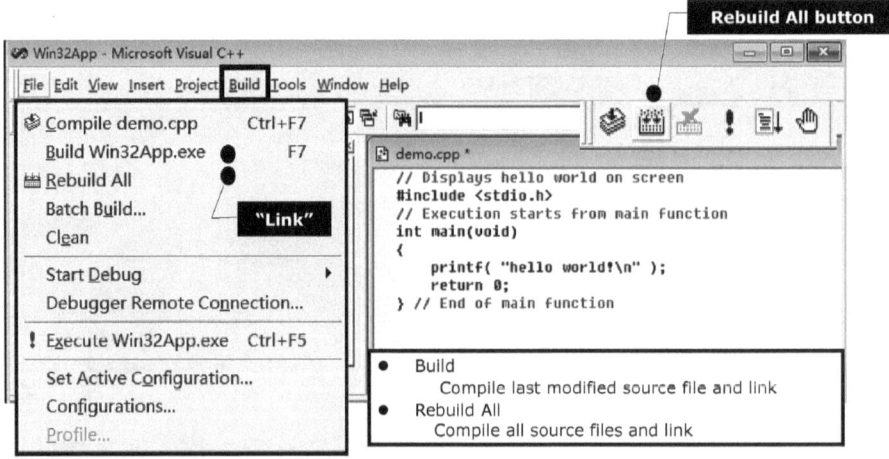

Figure 7.17: Link command.

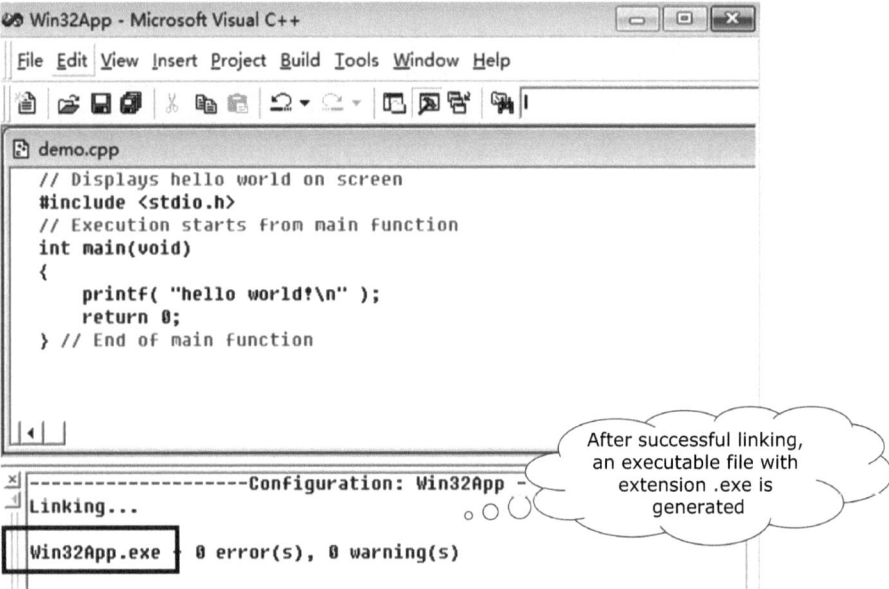

Figure 7.18: Successful linking.

7.1.7 Execute program

To execute a program, we use the Execute command in the "Build" menu as shown in Figure 7.19. The Execute button is the one with an exclamation mark in the Build

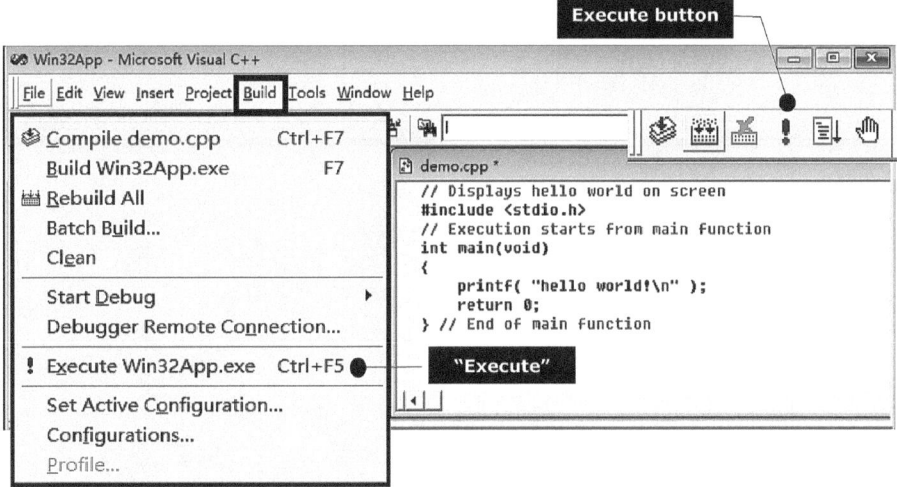

Figure 7.19: Execute command.

Minibar. There is another execution button on its right, which is called the Go command. They can both be used to execute a program, with the only difference being that the console disappears after execution when Go is used while it remains active in the case of Execute. It is easier to view the program result in the console using Execute.

We can check the execution result in the console. If the program is executed successfully, the result will be output to the screen as shown in Figure 7.20.

7.2 Testing

7.2.1 Introduction

7.2.1.1 Defect in arithmetic question generator

We will first tell a story of Brown's family. Mr. Brown wrote question-generator software for his son Daniel. The program generated random arithmetic problems and checked whether the answer entered was correct.

Daniel did exercises using the software happily, until one day, the program crashed after he typed in two numbers. After asking his son about the input and the operation he has done, Mr. Brown checked the value of a variable in his program using the Watch window of the debugger. As shown in Figure 7.21, the value was 1. #INF. The system suggested that he should check whether there was overflow caused by division by zero. After investigating his code, Mr. Brown then realized that he did not restrict the divisor to be nonzero value in division operations. The program shown here is a simplified version for easier demonstration.

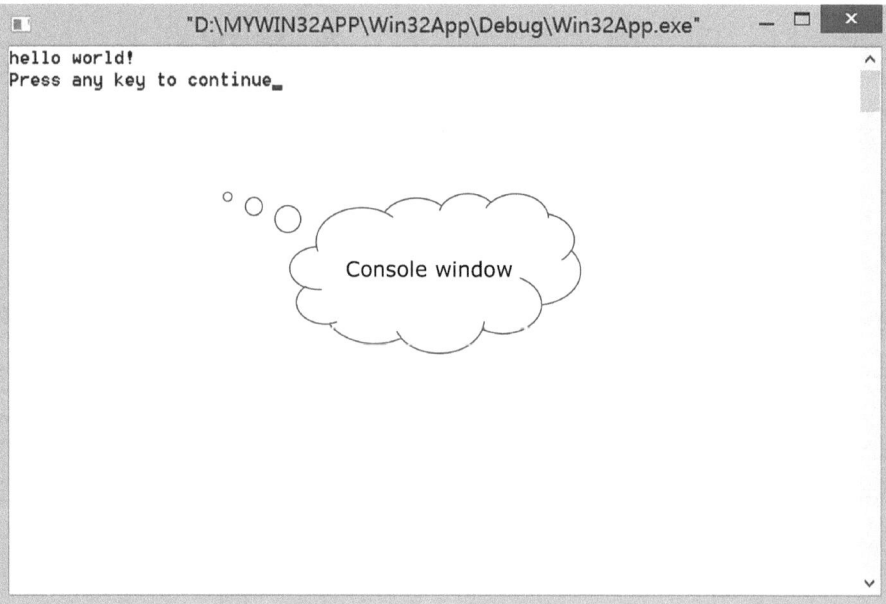

Figure 7.20: Inspection of execution result.

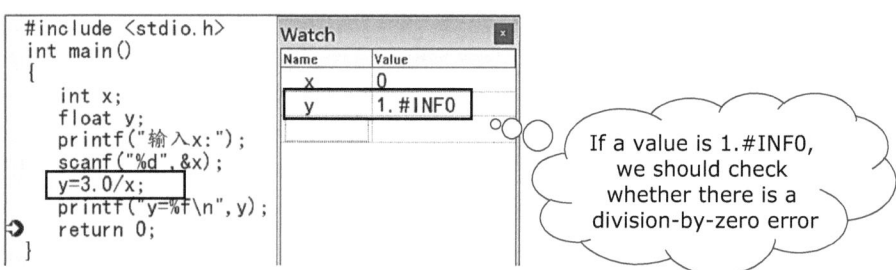

Figure 7.21: Division by zero error.

Division by zero is not just a minor problem. In 1997, the propulsion system of USS Yorktown (CG-48) crashed due to a division-by-zero error. The ship was paralyzed and stuck in the water for nearly three hours. If there were war at that time, the consequences could be disastrous. Of course, modern operating systems can handle such errors elegantly by displaying warnings instead of crashing immediately.

7.2.1.2 Error handling in the n! program

In an example in the section "Comprehensiveness of Algorithms," we found defects in our n! algorithm by testing the program with a special input n = 1, which showed that we should consider cases where "data are out of range" in addition to data within the normal range as shown in Figure 7.22. Consequently, we need a complete and reasonable mechanism of testing to find errors in programs and enhance their quality. Meanwhile, we need to design test data before writing code; therefore, we "have a guideline to follow" when programming. In other words, the test case design should happen before algorithm design.

$$n! = \begin{cases} 1, & \text{when } n = 0 \\ n*(n-1)!\,, & \text{when } n \geq 0 \end{cases}$$

> Ideally, test case design should be done before algorithm design

Test cases

	General case	Edge case	Error case
Input data	n>1 and n is an integer	n=0, n=1	n<0
Expected result	Value of n!	1	Warning

Figure 7.22: Test data of n! algorithm.

Numerous problems arise in the world each year due to software defects. Loss due to software defects is enormous. In 2002, research by the National Institute of Standards and Technology showed that losses incurred by software defects were at $59.5 billion per year. Over one-third of the losses could have been avoided through software testing.

An undetected error in software may bring down the entire system or even lead to disastrous consequences. Hence, testing of software products is of significant importance.

7.2.2 Program testing

7.2.2.1 Errors and warnings
Errors and warnings may occur during compilation and execution as shown in Figure 7.23.

(1) Compile-time error
There are two types of compile-time errors:
- Syntax errors: Arise when we do not use statements in the way specified by the grammar. Wrongly spelled keywords, wrongly defined variable names, incorrect use of punctuation marks, incomplete or unmatched branch and loop structure, and missing or incompatible arguments in function calls are all syntax errors.

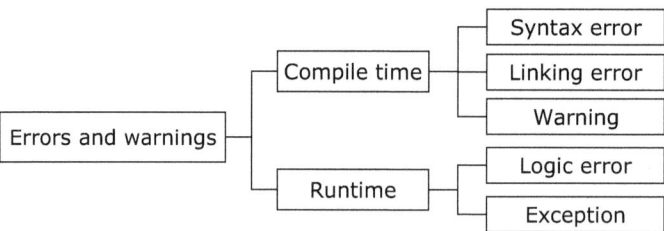

Figure 7.23: Categorization of errors and warnings.

- Linking errors: Found by the linker during the construction of object programs. Wrong library function names, missing files, and wrong path of included files are all linking errors.

(2) Runtime error
There are two types of runtime errors:
- Logic errors: Errors in program design that make the result inconsistent with programmers' expectations. For example, inappropriate execution conditions or number of iterations in loops falls into this category. Such errors cannot be found during compilation or execution, thus it is hard to find and correct them. Programmers have to rely on their proficiency in the language and programming experience to find logic errors.
- Runtime exceptions: These arise when a program attempts to execute an invalid operation during execution. Such operations include division by zero, invalid input format, opening nonexisting files, and not enough space on the disk.

(3) Compilation warnings
When statements in a program are against grammar rules of C, the compiler will show an error message. Sometimes, however, the compiler generates a warning message, which indicates that the code is not technically wrong but unusual. An error may exist in this case. During development, we should consider warnings to be errors as well. Linking can be successfully completed with warnings.

7.2.2.2 Definition of testing
The definitions of testing and test cases are given in Figure 7.24.

7.2.2.3 Purpose of testing
No matter how proficient a programmer is in programming or how well-designed a software product is, it is difficult to ensure high quality in software without testing against adequate and appropriate test cases. The quality of software depends mostly on the number and quality of test cases. Test case design involves complex analysis

> **Testing**
>
> Testing is the process of comparing actual output of a program with its expected output and determining whether the program satisfies design requirements.

> **Test case**
>
> A test case is a combination of certain input data, corresponding execution conditions and expected execution results. It is carefully designed to test or verify whether a program satisfies certain requirements.
>
> Test case design should be done before algorithm design. We should consider as many situations as possible during this process.

Figure 7.24: Definition of testing and test cases.

of problems. Test-driven development is also a challenge for developers. A developer cannot master program development without knowledge of test case design.

7.2.2.4 Principles of test case design

The goal of testing is to find defects in software using as few test cases, as little time and as few people as possible so that quality is ensured. Figure 7.25 shows the principles of selecting test cases. We should do a comprehensive test using a small number of test cases that help us find errors efficiently. In addition to regular input, we also need to consider invalid or abnormal input.

> **Principle of test case design**
>
> We should do a comprehensive test using a small amount of test cases that help us find errors efficiently. In addition to normal input, we also need to consider invalid or abnormal input.

Figure 7.25: Principle of test case design.

The test case design is a complex process. Readers can refer to resources on software engineering for a detailed discussion on this topic.

7.2.2.5 Methods of testing

As shown in Figure 7.26, there are two types of testing, namely white-box and black-box testing. The box here refers to the software being tested.

White-box testing treats the testing object as a transparent box. Testers have full knowledge of the internal logic as well as other information about the program. Test cases are designed to exercise all logic paths of the program.

In black-box tests, on the other hand, testers do not possess or choose to ignore the knowledge of the internal logic and characteristics of the program. They check whether a program works as intended based solely on the requirements and specifications of the program.

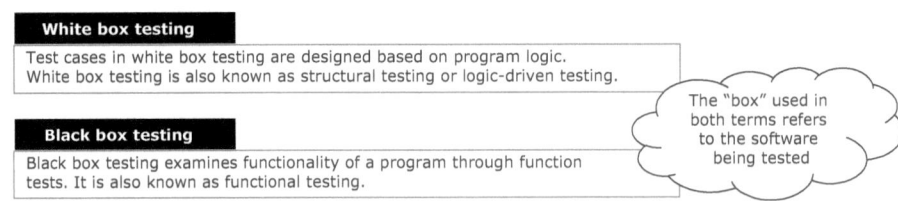

Figure 7.26: Methods of testing.

7.2.2.6 Basic approaches to test case design

As shown in Figure 7.27, white-box and black-box testing use many concrete methods of test case design, where each of them produces a special set of test cases. Using one of these methods cannot test programs comprehensively, so we often use a combination of them to design test cases in real-life projects.

White box testing	Logic coverage	We must test both true and false branches for logical values
	Path testing	All independent paths in each module should be executed at least once
Black box testing	Equivalence partitioning	We divide all possible input data into several classes, and select a few typical data from each class as test cases
	Boundary value analysis	We select valid and invalid boundary values as test cases
	Error guessing	We list all possible errors and special cases that are error-prone of a program, and select test cases based on them
	Cause-effect graph	It is an approach that takes combinations of input cases and constraints between input conditions into consideration

Figure 7.27: Basic techniques of test case design.

7.2.2.7 Order of testing

We can also divide testing methods into two categories based on the order of testing: bottom-up testing and top-down testing as shown in Figure 7.28.

Bottom-up testing	We first test modules on the lowest level, then modules on a higher level, and finally the main module
Top-down testing	We first test the main module, then modules it calls, and finally modules on the lowest level

Figure 7.28: Order of testing.

Example 7.1 Palindrome checking program
A "Palindrome" is a word or sequence of characters that reads the same backward as forward. Write a program that determines whether a sequence of characters is a palindrome.

[Analysis]
A test case consists of input data and expected output. We first consider possible cases of input data; the length of the sequence can be odd or even and the sequence is either a palindrome or not. These cases are obtained by dividing all possible cases into equivalent classes. There is also a special case where the length is zero as shown in Figure 7.29. This case is obtained by corner case analysis.

Finally, we list the expected output: if the sequence is not a palindrome, the program returns 0; otherwise, the program returns 1.

Test case					
	String length is odd		String length is even		Edge case
Test data	Not a palindrome	Palindrome	Not a palindrome	Palindrome	Empty string
Expected result	Return 0 "Not a palindrome"	Return 1 "Palindrome"	Return 0 "Not a palindrome"	Return 1 "Palindrome"	Return 0 "Not a palindrome"

Madam,I'm Adam — Was it a cat I saw — Live on no evil. — deified

Figure 7.29: Test case design of the palindrome checking program.

Example 7.2 Testing a sorting program
Test a program that sorts data.

[Analysis]
In addition to general cases where data are distinct, some corner cases that need special testing are listed below:
- There is no input data.
- There is only one input number.
- Numbers are already sorted in order.
- Numbers are sorted in reverse order.
- There are duplicates.

Figure 7.30 shows the expected results for different input cases.

Test case			
	Number sequence	Special case	Invalid case
Input data	• Distinct • Has duplicates • Already in order	Single number	No data
Expected result	Sorted sequence	Single number	Warning

Figure 7.30: Test case design of sorting program.

7.3 Concept of debugging

Mr. Brown has been working overtime lately. Sometimes he even stayed up all night. Mrs. Brown asked his husband what he was working on. He blinked his eyes and responded, "I have been producing bugs and debugging." However, Mrs. Brown became more confused after knowing the meaning of "bug" and "debug." She asked, "Why do programmers debug all the time? Why cannot you write programs without bugs?"

"Good question!" our professor commended. He then started to think about this question raised by his wife, a complete amateur. Although the question seemed funny, it reflected many problems in the lifespan of software products, from being designed to going live.

7.3.1 Bug and debug

A bug refers to a small insect or a defect. People nowadays refer to defects or problems hidden in computer systems or programs as bugs and the process of finding bugs as debugging.

There is a story behind this. On 9 September 1947, a computer operator found a moth trapped next to relay #70 on the circuit when tracing an error in a Harvard Mark II computer. It was the moth that led to the error. The moth was later removed and taped to the logbook with the caption "First actual case of bug being found" as shown in Figure 7.31. These operators also suggested using the word "debug," thus

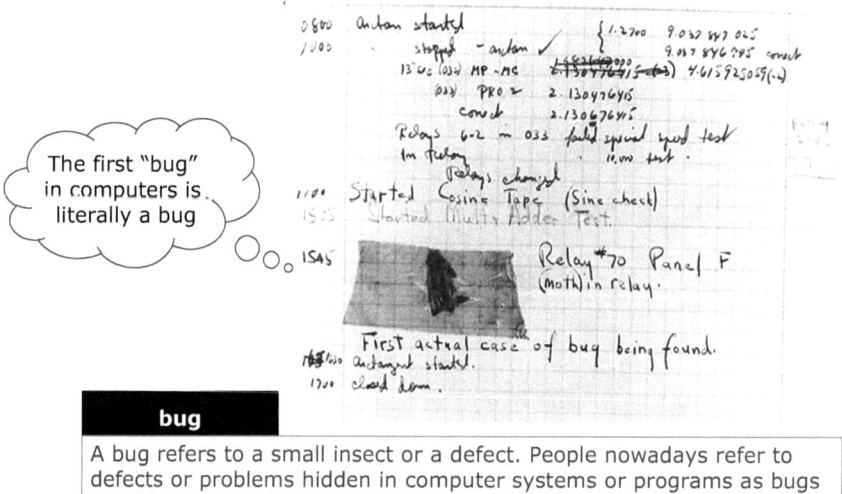

The first "bug" in computers is literally a bug

bug

A bug refers to a small insect or a defect. People nowadays refer to defects or problems hidden in computer systems or programs as bugs

Figure 7.31: Bug in programs.

creating the new term "debugging a computer program." The plain meaning of the word debug is to remove a bug, but in practice it also refers to the process of finding and locating a bug, which is more complicated than removing bugs in most cases.

Since the 1950s, people have been using the word debug to refer to the process of correcting errors, which solve problems in software through reproducing errors and locating bugs. The definition of software testing is shown in Figure 7.32. Debugging is the process of solving problems in software using debugging tools. The goal of debugging is to find the root cause of software defects and figure out a solution to it. In addition to debugging, debuggers are used for other purposes as well. For example, we use debuggers to analyze how the software works, why a system crashes, and to solve problems of the system and hardware.

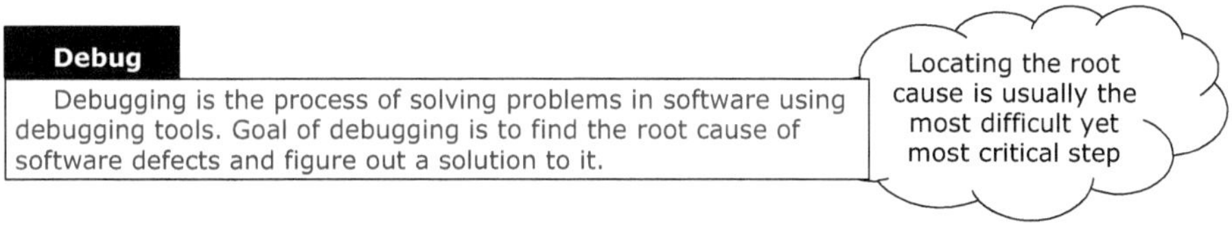

Debug

Debugging is the process of solving problems in software using debugging tools. Goal of debugging is to find the root cause of software defects and figure out a solution to it.

Locating the root cause is usually the most difficult yet most critical step

Figure 7.32: Definition of debugging.

7.3.2 Bugs are everywhere

Figure 7.33 shows the flowchart of program development, which we introduced in the chapter "Introduction to Programs." As shown in the flowchart, software design is not an easy task. Errors may occur in any phase: problem abstraction, data analysis, algorithm design, program design, and so on. Almost every program of a reasonable scale has been debugged and modified again and again. Hence, debugging is an essential skill in programming.

Except for problems that arise in the development, bugs may appear when users are using the software as testing may not find all problems. As a result, the most frequent job of programmers is to modify programs repeatedly, whether it is during or after development. This is why amateurs have the impression that "programmers are always fixing bugs."

7.3.3 Difficulties in debugging

Debugging is a thinking and analysis process of uncovering the root cause of a phenomenon. It requires strong skills, thus it is hard even for experienced programmers. Brian W. Kernighan, one of the creators of the C language, and Yinkui Zhang, author of the book *Software Debugging*, have both commented on the difficulties in debugging as shown in Figure 7.34.

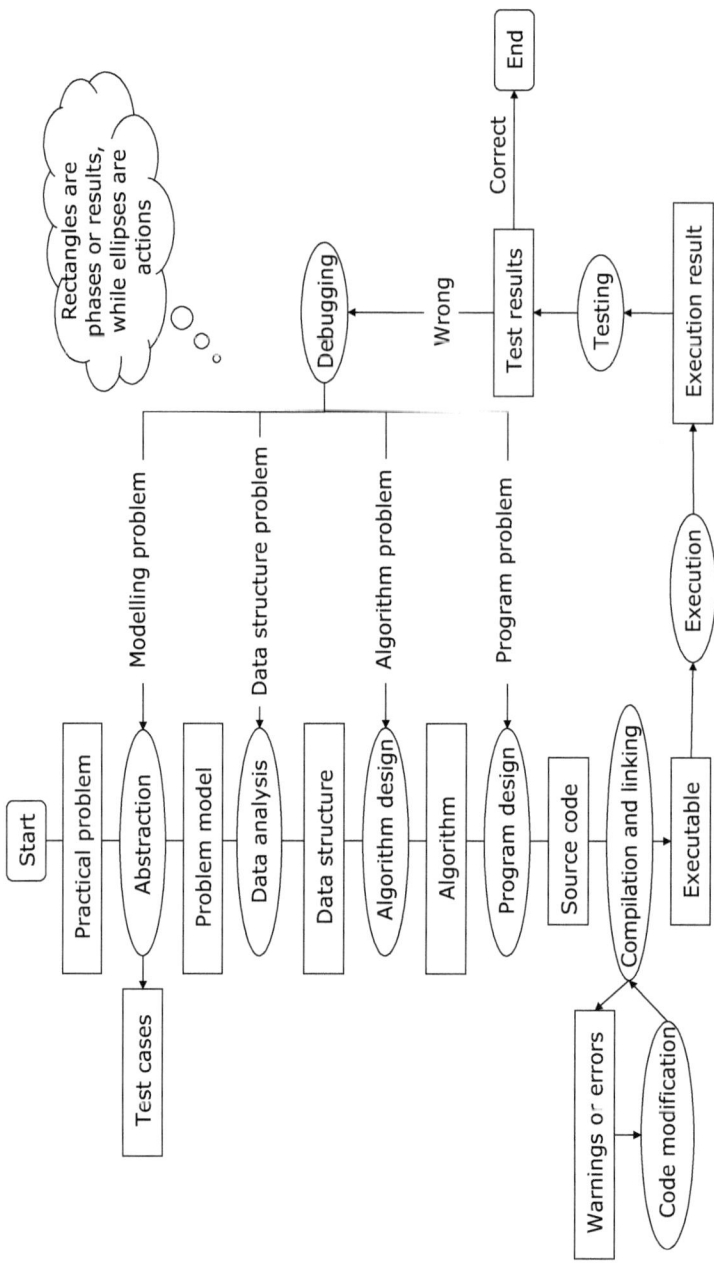

Figure 7.33: Flow of program development.

Debugging is twice as hard as writing the code in the first place. Therefore, if you write the code as cleverly as possible, you are, by definition, not smart enough to debug it.

——Brian W.Kernighan, one of creators of C

Debugging is a task frequently done in software development and maintenance. It is not easy to find software defects in complex computer systems. It often takes more time to debug a program than to write one.
—— *Software Debugging*, Yinkui Zhang

Figure 7.34: Degree of difficulty of debugging.

We shall discuss methods of debugging in the following sections.

7.4 Methodology of debugging

7.4.1 Introduction

7.4.1.1 Finding errors in a domino sequence

Daniel participated in a game of building large-scale domino sequences in the summer camp. "Large scale" here means that it is hard to figure out the global status of the sequence at a glance. After Daniel's team had set up their sequence, they toppled the first domino. However, some dominoes in the sequence did not fall, so there must be something wrong. How could they find the problem? What strategies could they take to find the problem? After a discussion, the team believed that they could walk through the sequence and check the setup of each domino. They could also divide the sequence into segments so that each member was in charge of the inspection of one segment.

Structure and execution of programs are similar to dominoes, where a statement is a domino and a program consisting of multiple statements or functions is the sequence of dominoes. Hence, we can use the strategy of Daniel's team in debugging as well as shown in Figure 7.35.

7.4.1.2 Collapse of the domino sequence

After finding the error in the sequence, Daniel's team started repairing at once. It was late at night when they finished so they decided to test it the second day.

When they entered the stadium the next day, however, they found that some parts of the sequence were completely ruined. The stadium was locked at night, so they wondered what happened. A member then noticed that there was a surveillance

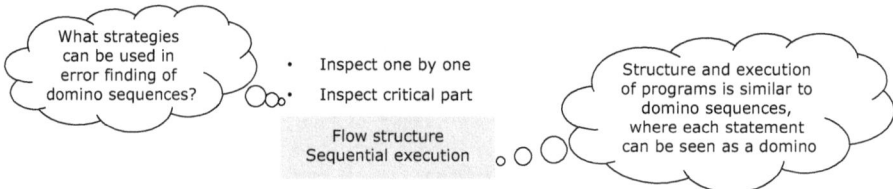

Figure 7.35: Strategy of error finding.

camera on site so she suggested investigating the video recording by rewinding it from the current time.

Similarly, when a program crashes, we can investigate the log to find out which module of the program caused the crash if the execution process of the program has been recorded. By playing the video recording backward, we can also figure out call stacks of each module as shown in Figure 7.36.

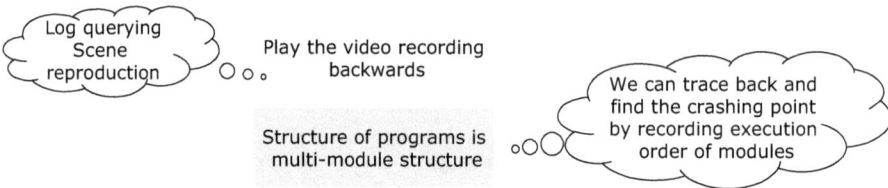

Figure 7.36: Inspection of crash log.

7.4.2 Basic flow of debugging

Daniel's team searched for the error in their sequence on site. The error was a domino that fell down before the sequence was toppled. To find bugs in a program, we need to reproduce the problem to be solved on the system. We will debug by repeating steps that led to the failure, and analyze the root cause of the failure, derive a solution, modify the program, and verify whether the problem has been solved.

As shown in Figure 7.37, a complete debugging process should be a loop with four phases, namely bug reproduction, root cause searching, solution exploration and implementation, and solution verification. In root cause searching, we should use various debugging tools and methods to find the major source of software failure. In solution exploration and implementation, we design and implement a solution based on the root cause found in the last phase and resources we possess. Finally, we test whether the solution is effective in the target environment in the verification phase.

The prerequisite of fixing a bug is finding its root cause. Root cause searching is always the most critical yet most challenging step among all steps of debugging. Finding the root cause is the core of debugging.

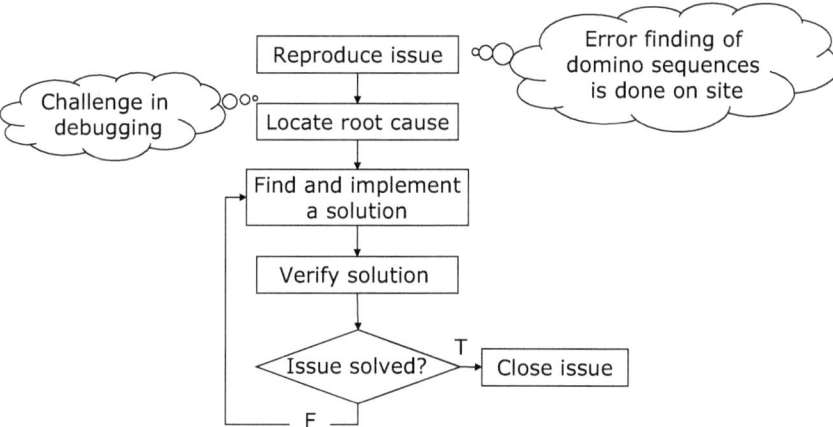

Figure 7.37: Basic flow of debugging.

How do we find errors in the logic of programs? We shall introduce the methodology based on the characteristics of programs and the execution of programs.

7.4.3 Discussion on methods of finding errors in programs

7.4.3.1 Analysis of flow of program execution

In the process of solving problems with programs, data are processed to produce results. If data to be processed are inherent in the problem, that is, the logic structure of data is determined solely by the problem, then there is no error within them. Consequently, we need to focus on "processing" and "result" to find errors. In other words, finding errors is done by tracing how data are processed and inspecting results as shown in Figure 7.38.

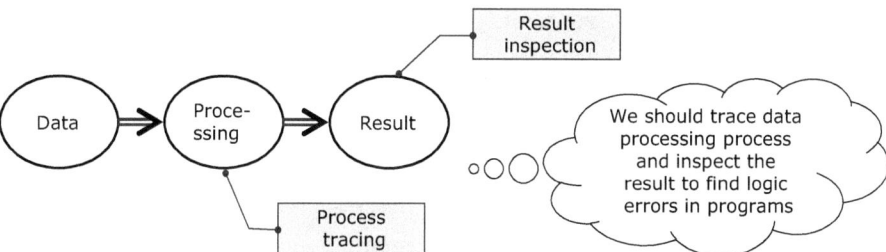

Figure 7.38: Steps in program execution flow to focus on when finding errors.

> **Knowledge ABC** Logic errors in programs
> Logic errors of a program are reflected in the differences between execution results and the expected results of the program. For example, if we forget to add parentheses to make addition evaluated before multiplication in an expression, the result may be wrong. Programs with logic errors can be executed successfully most of the time, so there is no error message indicating the location of the error.

7.4.3.2 Relations between modules

Modules interact with each other through calls. Let us examine a concrete example first. There are three child functions in Figure 7.39, namely function a, b, and c. The calling stack of the main function and these child functions is shown in the figure as numbers: the main function calls function a first, function a calls function b, which later calls function c. Once function c is completed, it returns to the remaining instructions in function b, which later returns to function a when it is done. Finally, function a returns to the main function when the remaining statements in it are completed.

The general rule of nested call of multiple functions is: the last called function returns first.

We can imagine the CPU as a stage. Only one scene can be presented on stage at a time; similarly, only one function can be executed at a time. To call another function, the CPU pauses execution of the current one and switches to the child function called. Due to the way CPU executes programs, some information should be stored so that the CPU can complete what has been left in the calling function after the child function called is done.

The context information of a function call stored by the system includes return address and some variables and parameters. Similar to the order of calling and returning, the process of saving and restoring contexts follows a "Last-In-First-Out" order as shown in Figure 7.40. The numbers in the figure correspond to those in Figure 7.39, where numbers 1–4 refer to the saving order and numbers 6–8 refer to the restoring order. The "Last-In-First-Out" principle followed by memory space management of context information is the classic way stacks work. In this way, the execution path of modules of a program is recorded.

7.4.3.3 Problems involved in error finding

The execution path is constructed based on the structure of the program as shown in Figure 7.41. Programs are constructed by multiple modules, so we should trace errors both inside modules and between modules.

Based on how data are organized and stored, result inspection involves inspecting a single variable, address of a variable and a series of variables with continuous addresses, which is also called an array.

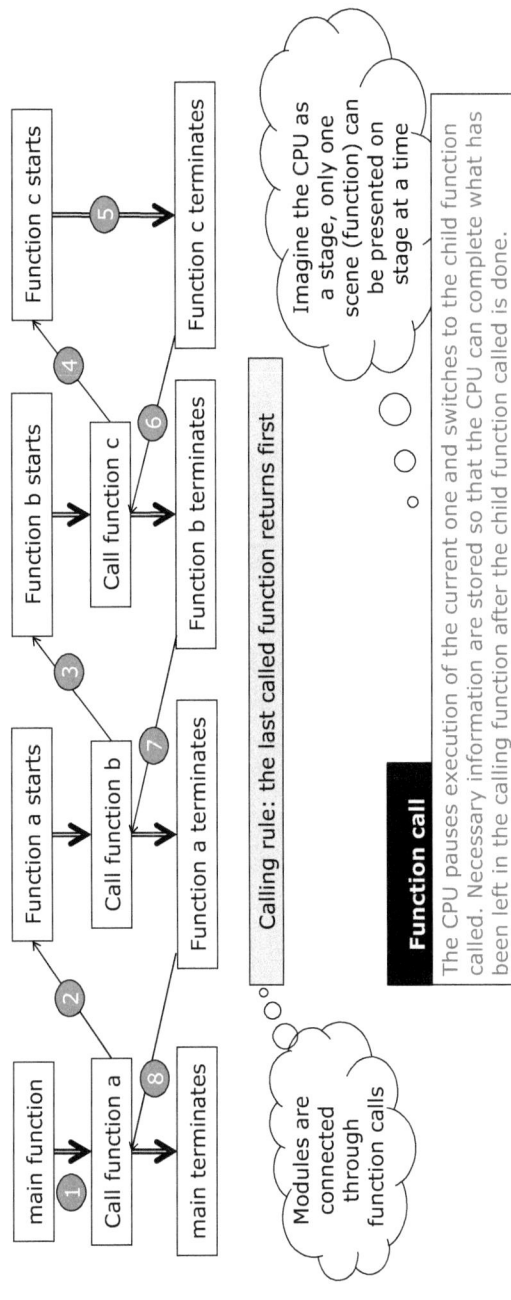

Figure 7.39: Calling relation between modules.

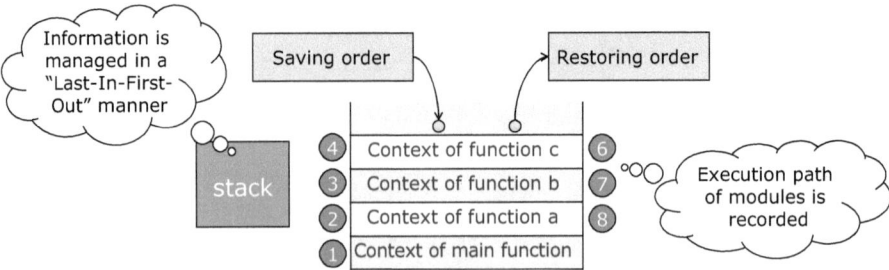

Figure 7.40: Saving and restoring context information.

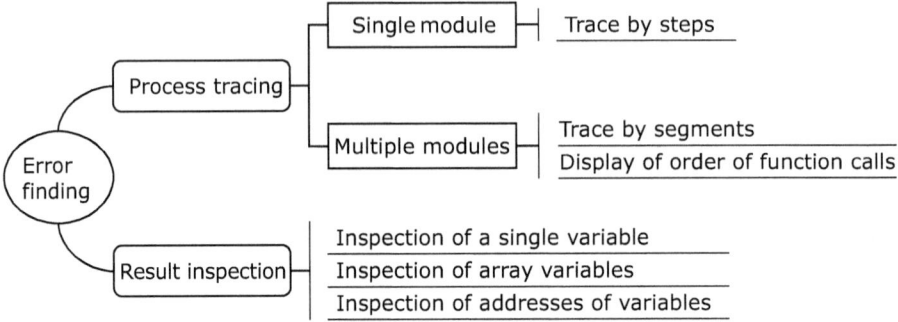

Figure 7.41: Problems involved in error finding.

We shall discuss concrete strategies for tracing programs.

7.4.4 Exploration of tracing methods

To find errors in a domino sequence, we can inspect dominoes one by one or segment by segment, where we randomly check certain places in each segment. We can also combine these two methods together. During the inspection, we can stop, observe, think, and rebuild the sequence at any time. Similarly, we need a pausing mechanism for program execution to inspect data and processing results. Such a mechanism is provided by the IDE, enabling us to use an error finding strategy similar to the one used in dominoes.

7.4.4.1 Trace by statements

7.4.4.1.1 Stepwise tracing
The debugger starts from the main function and executes statements one by one using stepwise commands. A child function is also treated as a statement. For

example, the calling of function func in main function shown in Figure 7.42 is handled as a single statement in stepwise tracing.

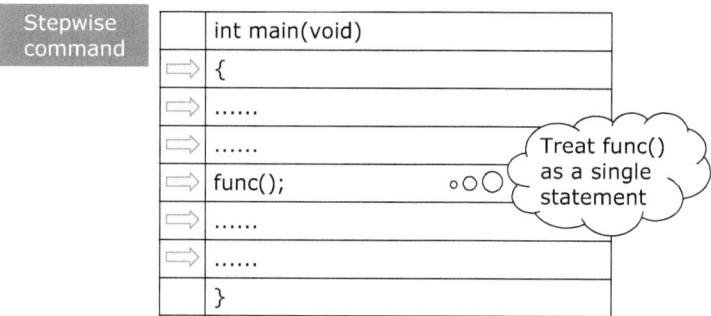

Figure 7.42: Stepwise tracing.

7.4.4.1.2 Statement-wise tracing

The debugger starts from the main function and executes statements one by one using stepwise commands. Upon reaching a child function call, the debugger enters the child function using the step-into command as shown in Figure 7.43. Statements inside the child function are again executed one by one. After the child function is done, the debugger returns to the calling function and executes the remaining statements.

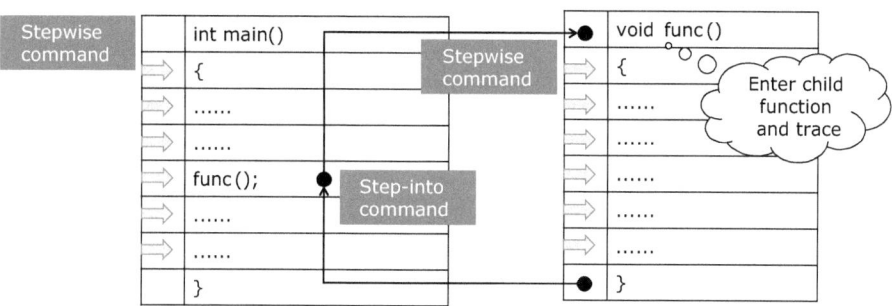

Figure 7.43: Statement-wise tracing.

7.4.4.2 Trace by segments

We first set breakpoints at the line we want to inspect. As shown in Figure 7.44, a dot is inserted to pause the program execution, which is resumed after we have inspected data we would like to see. We can use multiple breakpoints and the jump to breakpoint command so that programs are stopped when a breakpoint is encountered. The

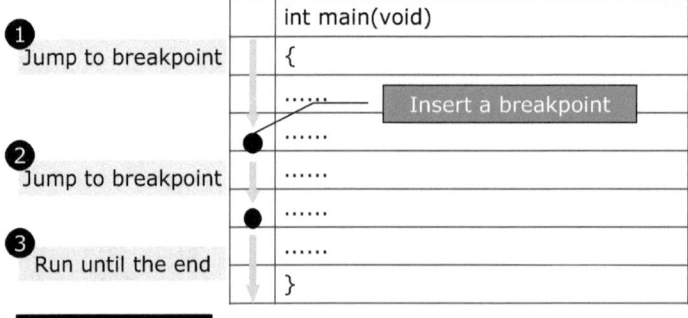

Interruption

When a special event (interrupting event) happens, a computer will pause the task (program) that is currently being executed, switch to another task (interrupt handler) and eventually resume the initial task.

Figure 7.44: Interruption and tracing by segments.

execution is done after multiple jumps. The action of setting breakpoints is called "Break" in computers.

7.4.4.3 Reversed inspection of call stack

The execution path of modules is automatically recorded into a call stack by the system during program execution. When a program crashes, we can quickly locate the function in which error occurs by checking the information in the call stack as shown in Figure 7.45. Which module does the problem arise in? As the program terminates when crashing, the error should be in the function on the top of the stack.

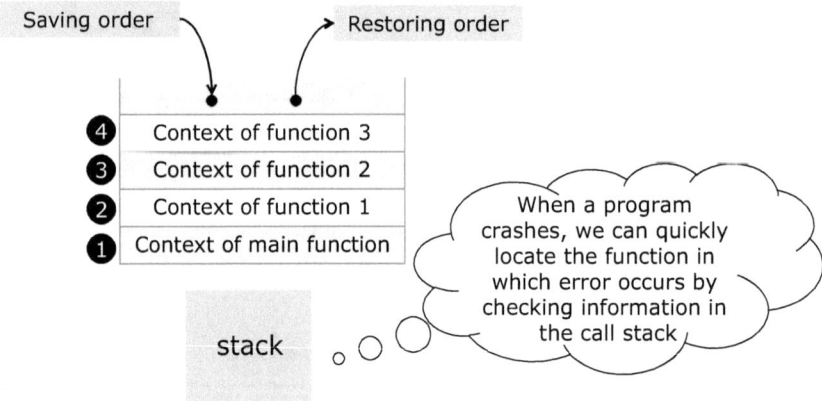

Figure 7.45: Record of execution path.

We have discussed the methodology of debugging and now we are going to introduce how to apply these strategies in an IDE.

7.5 Debugging tools

In the world of software, inspection and repairing tools like corkscrew or multimeter no longer work. They are replaced by debugging tools with the debugger as the core. – Software Debugging, Yinkui Zhang

Debugging is of vital importance to software. Inspection tools in the world of software are debugging tools. Using the right debugging tools properly can largely increase the efficiency of finding bugs.

7.5.1 Functions of debugger in IDE

Typically, programs are executed continuously. However, debugging can control the pace of program execution. Execution of a program can be paused, done step by step, or done with jumps. When a program is paused, we can inspect its status. The debugger provides functions like controlling execution pace or inspecting execution status in IDE. More specifically, it traces and records how the CPU executes a program and takes snapshots of this dynamic process for programmers to inspect and analyze as shown in Figure 7.46.

Function	Meaning	Case
Control execution pace	• Control size of step • Inspect execution path	Stepwise execution: use a statement or a function as a step in program execution
		Jumping execution: run a program to the cursor or a breakpoint set by programmers
Inspect execution status	Inspect internal data when a program is paused	Variable values, memory values, register values, stack, etc.

Figure 7.46: Functions of the debugger in IDE.

By controlling the execution pace, the debugger controls the number of lines executed in each step so that programmers can better observe the execution path. Stepwise execution uses a line of code or a function as a step in program execution. In contrast, jumping execution can run a program to the cursor or breakpoints set by programmers. Stepwise execution is an effective method of diagnosing dynamic characteristics of software. However, stepwise tracing a program or even a module is usually not efficient. A commonly used comprehensive debugging approach is to run the program to the line we are interested in using breakpoint and execute critical code stepwise after that.

With the execution status displayed, the debugger allows us to observe internal data of a computer when the program is paused. We can inspect variable value, memory value, register value, and stack value in the IDE. Observing the status of a program during execution is one of the most critical tasks in debugging. To efficiently debug a program, we need to combine these functions and use them flexibly.

> **Knowledge ABC** Debug version and release version of programs
>
> To enable us to debug a project using debuggers, the information needed for debugging must be stored in the compilation units. Consequently, we need to use the compiler to Insert debugging information into compilation units before using debuggers.
>
> The debug version of a program contains the debugging information. We can debug the program conveniently through stepwise execution and tracing. No optimization is done for the debug version, thus the executable file generated is larger in size and runs slowly.
>
> The release version often contains a series of optimization. The size and execution speed of the generated file is fully optimized for better user experience. However, we cannot use debuggers in the release version.
>
> There is a compilation option for switching between these two versions. It is located in the menu Build->Batch Build, in which we can choose to compile one of them or both as shown in Figure 7.47. The debug version uses a group of compilation options to support debugging. Before a program is published after being written and debugged, we can discard debugging information through the compilation option and generate the release version with efficient code.

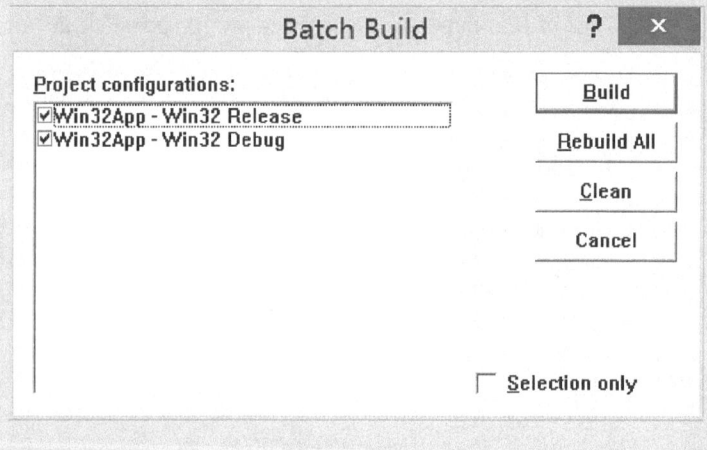

Figure 7.47: Option of debug and release version.

We shall introduce various commands of debuggers and their usage in the following sections.

7.5.2 Debugging commands

7.5.2.1 Enter the debugging environment

As shown in Figure 7.48, we can enter debugging environment through three steps:
(1) select "Build" in the menu bar, (2) select "Start Debug" in the Build menu, and
(3) click "Step Into" command (whose hotkey is F11). Alternatively, we can press
hotkey F10 (Step Over) to enter the debugging environment. After that, we have entered the fourth step, where the Build menu has turned into the Debug menu,
which contains various debugging commands.

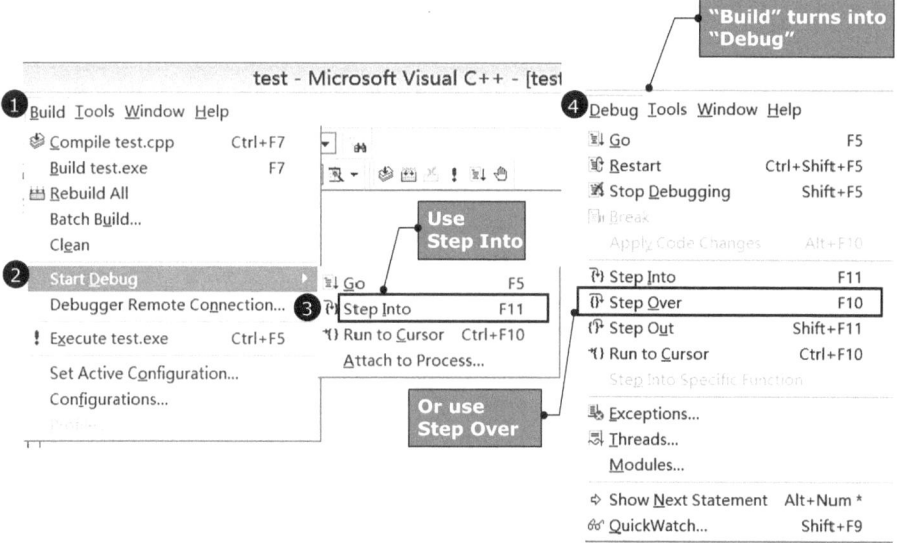

Figure 7.48: Enter debugging environment.

7.5.2.2 Commands controlling program execution

Together with breakpoints and jump command, commands controlling program execution are used to complete stepwise execution and jumping execution. We shall
see their usages in concrete debugging examples. Major commands are shown in
Figure 7.49.

7.5.2.3 Set breakpoints

Setting breakpoints is one of the most commonly used techniques to trace a large
scale program.

A breakpoint is a mark set in programs by the debugger. When the program runs
to a breakpoint, its execution is paused and the program returns to the debugger so

Menu command	Hotkey	Notes
Go	F5	Run to a breakpoint. Used with breakpoints
Step Over	F10	Stepwise command that doesn't enter child functions
Step Into	F11	Stepwise command that enters child functions
Run to Cursor	Ctrl+F10	Run to the cursor. Used with cursor setting
Step Out	Shift +F11	Run to the end of current function and return to the calling function
	F9	Insert/Delete (a location breakpoint)
Stop Debugging	Shift+F5	Exit debugging and return to editing mode

This is the Debug toolbar. Icons in it correspond to commands in the Debug menu. It is recommended to use hotkeys though

Figure 7.49: Major commands of program execution controlling.

that programmers can inspect the code or variable values. After a program is paused, we can further execute it step by step to determine whether it is running as expected.

There are three kinds of breakpoints in the IDE, namely location breakpoints, data breakpoints, and message breakpoints. Only the first two are involved in console applications of C. The configuration screen of breakpoints is shown in Figure 7.50, which can be invoked by selecting the "Breakpoints" sub-menu inside the "Edit"

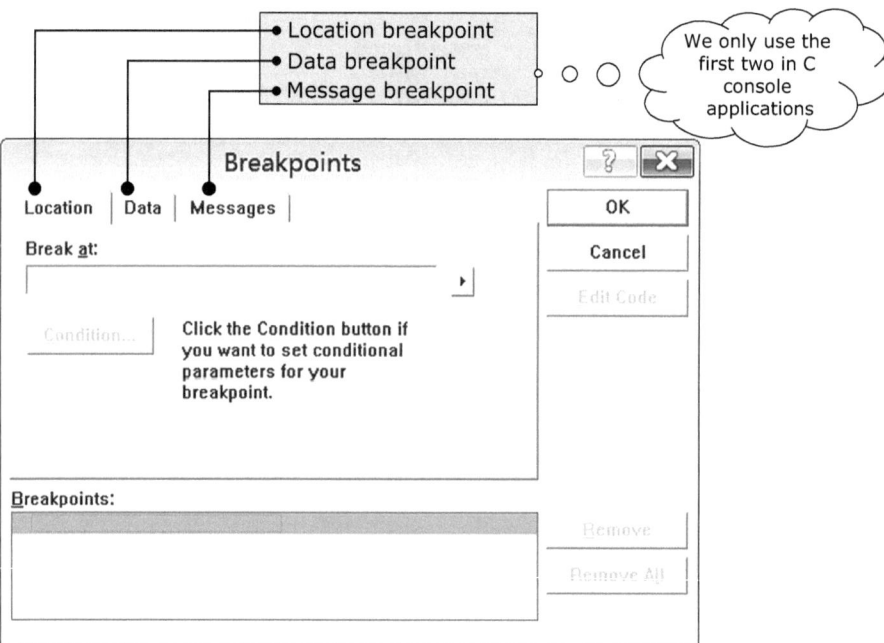

Figure 7.50: Configuration screen of breakpoints.

menu on the main menu bar. There are multiple breakpoint options that we can use as needed.

(1) Location breakpoint

Location breakpoints are the most commonly used breakpoints. They are usually inserted at a specified line of source code, the beginning of a function, or a specified address in memory.

We can insert a breakpoint by selecting "Edit->Breakpoints" in the menu or pressing hotkey F9. To use the shortcut for breakpoints, we first move the cursor to the line where we want to insert the breakpoint, then press hotkey F9 or click the "Hand" shape button in Build Minibar. A dark red dot will appear on the left of that line after insertion. To clear the breakpoint, we press F9 or click the button one more time. The hotkey for clearing all breakpoints is Ctrl + Shift + F9.

(2) Data breakpoint

Data breakpoints are set on variables or expressions. Program execution is paused when the value of the variable or expression is changed.

(3) Message breakpoint

Message breakpoint is set on window function WndProc. Program execution is paused when a particular message is received.

7.5.2.4 Inspect execution status

The most crucial thing in debugging is to inspect the status of the program during execution. It is through this process that we find errors in the program. The status here refers to values of variables and values in registers, memory, and stack. There are windows to view these values in the IDE as shown in Figure 7.51. Inspection of these values must be done during stepwise tracing or pausing upon breakpoints.

The window for execution status inspection and items it can display is shown in Figure 7.52. When debugging, we determine which window we should inspect based on program logic and execution controlling commands. The usage of these windows are introduced in corresponding sections. This section only covers some examples of them. A sample of these windows is shown in Figure 7.53.

(1) Watch window

By typing in the variable or expression we want to inspect in the Watch window, we can obtain its value. During stepwise debugging, we can inspect variable values dynamically inside the Watch window, which helps us determine whether the program is running correctly.

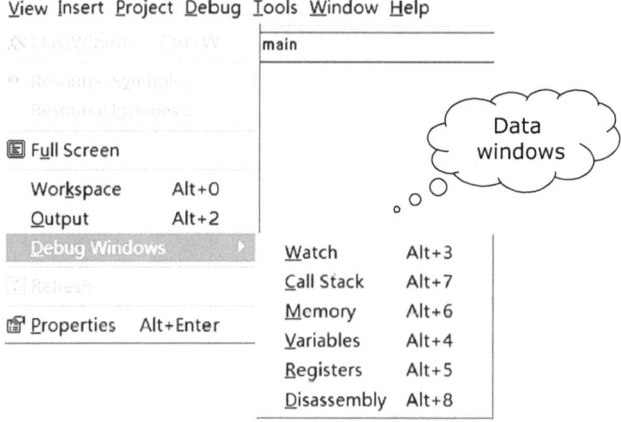

Figure 7.51: Debug windows.

Window	Function
Watch	Display values of expressions or variables we type in the window
Variables	Automatically display values of all variables that are visible in the current context
Memory	Display the memory starting from a specific address
Registers	Display current values of all registers
Call Stack	Display all functions that have been called and not yet terminated in the order of calling

Figure 7.52: Functions of debug windows.

(2) Variables window

It automatically shows the values of variables that are visible in the current context. In particular, the variables involved in the current statement are displayed in red. If there are many local variables, it can be tricky to inspect them in this window. It is recommended to use the Watch window in this case.

(3) Memory window

It shows the memory starting from a specific address, which defaults to 0x00000000. Length of memory we can view in Watch window is limited by sizes of variables, but the Memory window can show memory in a range of continuous addresses. To use this window, we need to type in the starting address, which can be found in the Watch window. The watch window shows the values and addresses of variables.

(4) Register window

Displays the current values of all registers.

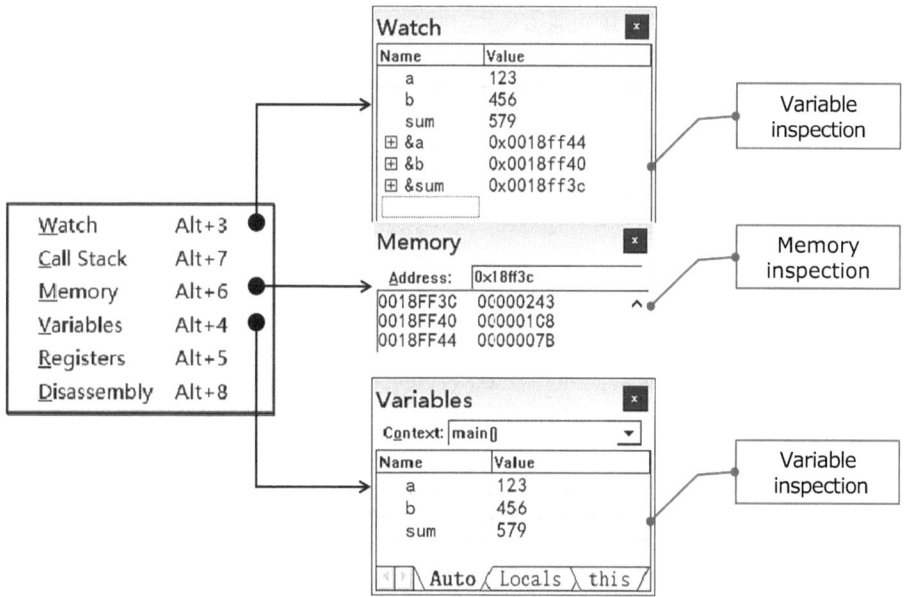

Figure 7.53: Sample of data windows.

(5) Call stack
It reflects which functions called the function currently being paused and how this function is called. The call stack window shows a series of function calls, where the current function stays on top and callers are listed below in the order of calling. We can jump to corresponding functions by clicking on the function names.

(6) Display format of data
Each window provides various display formats of data as shown in Figure 7.54. For example, the Memory window can display addresses in Byte Format or Long Hex Format. We can choose the format we are comfortable with by right-clicking in the Memory window and selecting the corresponding item in the popup menu. Note that memory is displayed in byte in a reversed order when using Byte format. Lower bits are on the left and higher bits are on the right. The values of variables are displayed as decimal numbers by default in the Watch window. We can also choose to display in hexadecimal format.

Knowledge ABC Online help
Visual C++ 6.0 provides detailed help information. Microsoft Developer Network (MSDN) is an information service for software developers provided by Microsoft. Programmers can use MSDN in various ways based on their needs: they can either install it locally or use it online.
　　After installing MSDN locally, we can enter the help system by selecting "Contents" under the "Help" menu. Alternatively, we can enter the help system of Visual C++ 6.0 by moving the

cursor in the editor to a word we want to look up and press F1. Users can obtain almost all technical information on Visual C++ 6.0 through the help system, which is one of the reasons Visual C++ is called a friendly development environment.

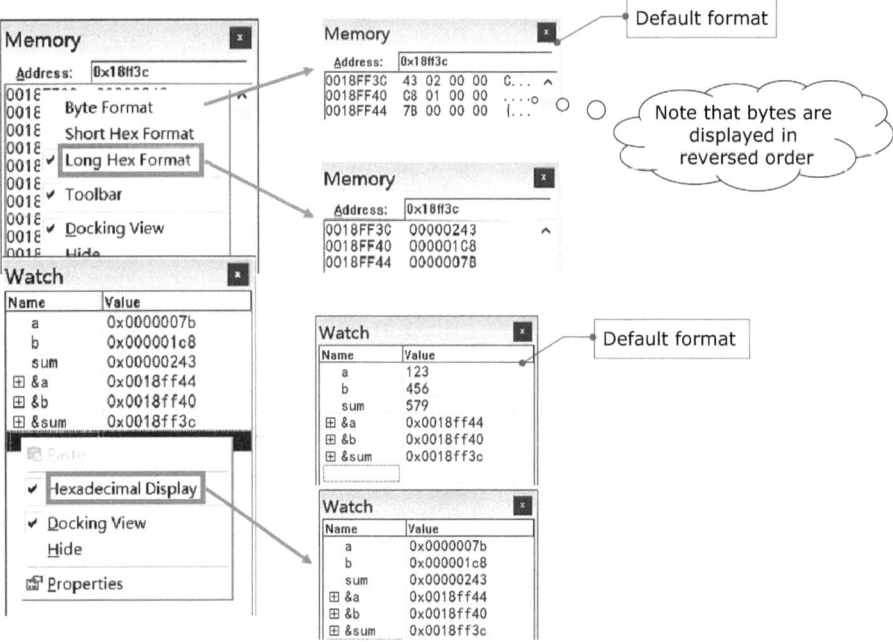

Figure 7.54: Display formats of Debug windows.

7.6 Examples of debugging

7.6.1 Demonstration of basic debugging steps

We shall use a simple program to learn the necessary steps of debugging. The program is given in Figure 7.55.

7.6.1.1 Tracing by setting breakpoints
As shown in Figure 7.55, we insert a breakpoint at the printf statement.

To run the program to breakpoint, we select command Build→Start Debug→Go or press hotkey F5. The program will be executed until the first breakpoint is encountered. A yellow arrow was inserted to the left of the current line by the debugger, indicating the next statement to be executed. Users can inspect data like

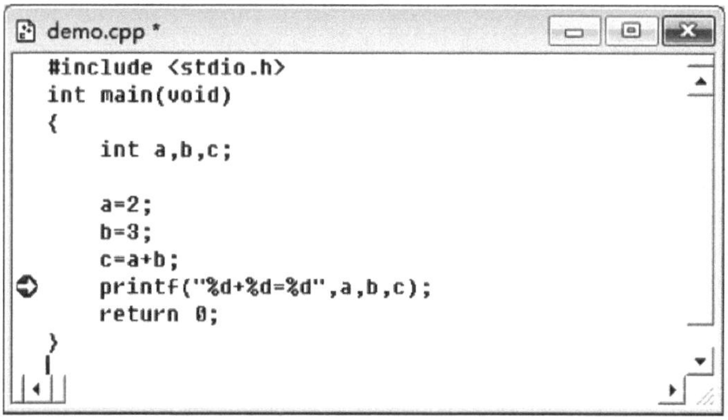

Figure 7.55: Tracing by setting breakpoints step 1.

variables or expressions. By executing the Go command again, we run the program to the next breakpoint or to the end if there is no more breakpoint.

We can use the View→Debug Windows→Watch command to inspect the values of variables. By entering the variable name in the Watch window, we can view its value as shown in Figure 7.56. Column Name lists expressions or variables we are watching, whereas column Value displays their corresponding values. We can observe changes in expression values during program execution in this window.

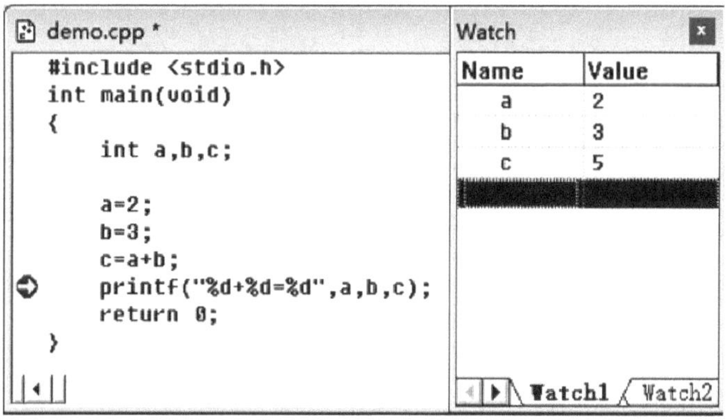

Figure 7.56: Tracing by setting breakpoints step 2.

We resume the execution by pressing F5, and the result is shown in Figure 7.57. The console window will appear for a moment and quickly disappear, leaving no time for us to see the result. To prevent the console from disappearing after the program terminates, we can execute the program using command Build→Execute (Ctrl + F5).

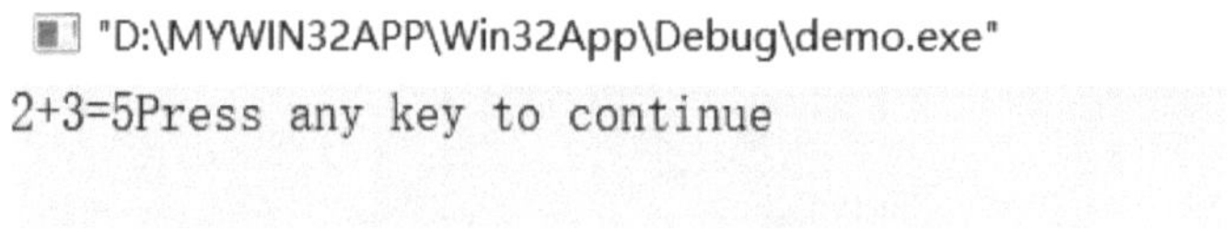

■ "D:\MYWIN32APP\Win32App\Debug\demo.exe"

2+3=5Press any key to continue

Figure 7.57: Tracing by setting breakpoints step 3.

7.6.1.2 Stepwise tracing

We can start stepwise tracing by selecting command Build→Start Debug→Step Into or pressing hotkey F11.

The program will be executed starting from main function as shown in Figure 7.58. Note that the Build menu will turn into the Debug menu.

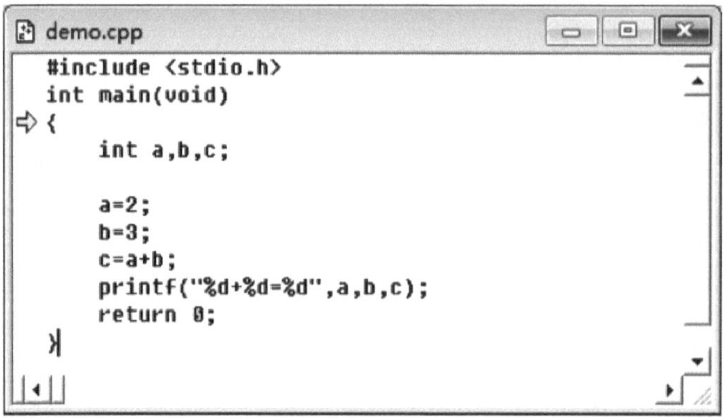

```c
#include <stdio.h>
int main(void)
{
    int a,b,c;

    a=2;
    b=3;
    c=a+b;
    printf("%d+%d=%d",a,b,c);
    return 0;
}
```

Figure 7.58: Stepwise tracing step 1.

To run the program step by step, we can use command Debug→Step Over or press hotkey F10.

Figure 7.59 shows the second step of stepwise tracing. Upon each press of F10, one statement in the program is executed and the statement indication arrow moves to the next line.

The third step is shown in Figure 7.60, where we inspect related variables in the Watch window. Values of variable b and c are different from what are assigned to them in the program because the assignment statements have not been executed at this time. Hence random values are displayed here instead of the assigned values.

To run a program until a specified location, we can use the command Debug→Run to Cursor or press hotkey Ctrl + F10.

Figure 7.61 shows the fourth step of stepwise tracing.

We first move the cursor to a specific location, for example, the return 0 statement, then press Ctrl + F10 to run the program until the statement before return 0.

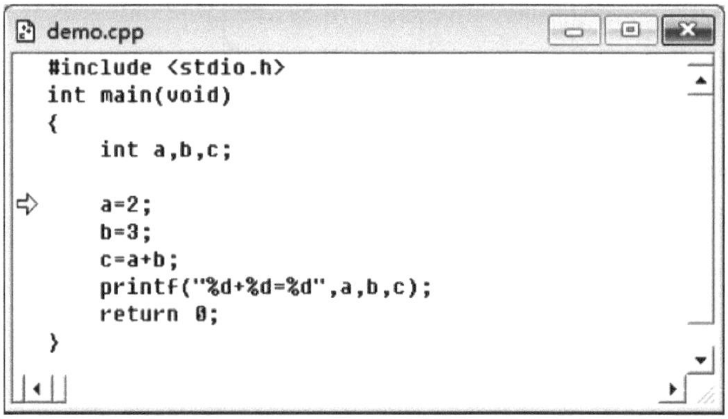

Figure 7.59: Stepwise tracing step 2.

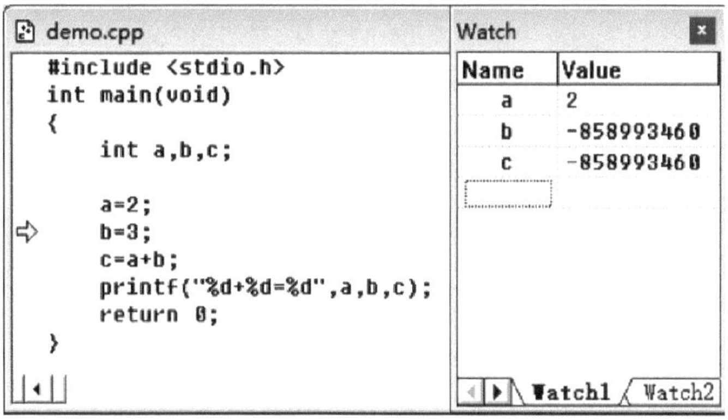

Figure 7.60: Stepwise tracing step 3.

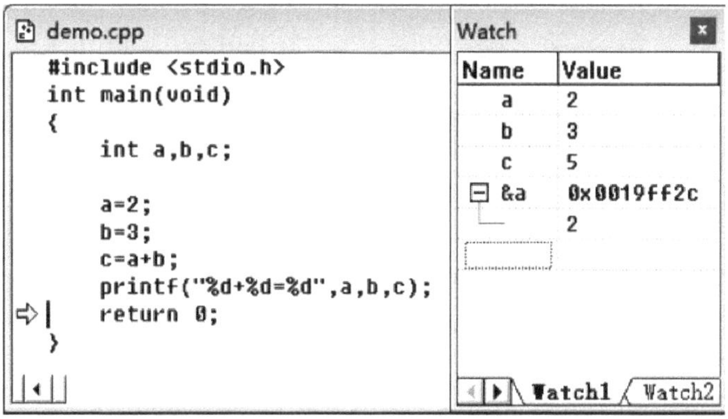

Figure 7.61: Stepwise tracing step 4.

The yellow arrow now points to statement return 0. As shown in Figure 7.62, we can inspect values and addresses (variable names prefixed with & sign) of variables, and the console output.

D:\MYWIN32APP\Win32App\Debug\demo.exe

2+3=5

Figure 7.62: Console window.

Figure 7.63 presents multiple windows. 0x19ff2c is the address of variable a, whose value is shown in Memory window. In Memory window, the leftmost column displays addresses in memory, while the first four columns to its right present contents stored in these addresses in hexadecimal form. The last column is the text representation of memory contents.

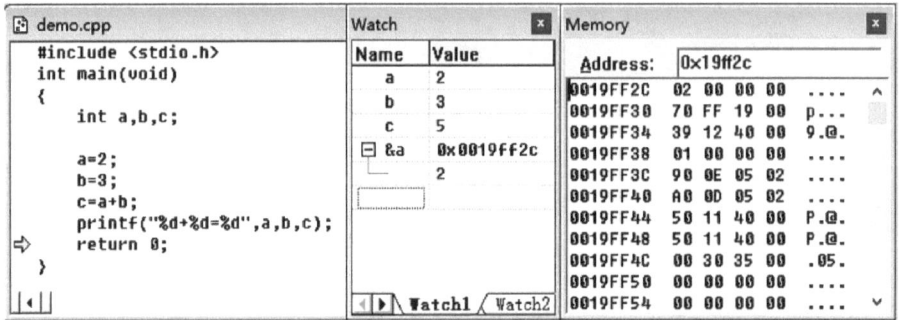

Figure 7.63: Information in multiple windows.

7.6.2 Example of debugging

Mr. Brown prepared an assignment for his students, in which they were asked to read a string that ended with a newline and output the string backward. For instance, "ABCD" should be output as "DCBA."

A student emailed his source code to Mr. Brown for help as shown in Figure 7.64. He claimed there was a bug, but he could not find it.

We shall debug the program using the debugging environment in the IDE. Steps of creating a project and a file were introduced in sections 7.1.1 and 7.1.2. In this example, the project is located at D:\MYWIN32APP\ with the name test and the source file name is Debugdemo.

Problem
Enter a string (less than 80 characters) that ends with newline, output the string backwards. For example, given input "ABCD", the output should be "DCBA".

Main steps of debugging
• Create a project and a source file • Format the code (modify in batch, format) • Compile and check error messages • Insert/Delete breakpoints and execute with Go command • Input in the console window • Inspect in Watch and Memory windows • Modify values in Watch window

```c
#include<stdio.h>
int main(void)
{
    int i,k,tmp;
    char str[];
    printf("input a string:");
    i=0;
    while((str[i]=getchar())!='\n')
        i++;
    str[i]='\0';
    k=i-1;
    for(i=0;i<k;i++)
    {  tmp=str[i];
        str[i]=str[k];
        str[k]=tmp;
        k++;
    }
    for(i=0;str[i]!='\0';i++)
        putchar(str[i]);
}
```
Debug

Figure 7.64: Debugging the reversed string program.

7.6.2.1 Editing the code

After setting up the project, Mr. Brown copied the file attached in the email to the editor in IDE, only to find many "?" in the code as shown in Figure 7.65. Why was this the case? It turned that programs copied from other files might contain other characters due to using a different encoding. These characters should be eliminated before compilation. To remove all "?" in this example in batch, we can use the Replace command in the menu Edit.

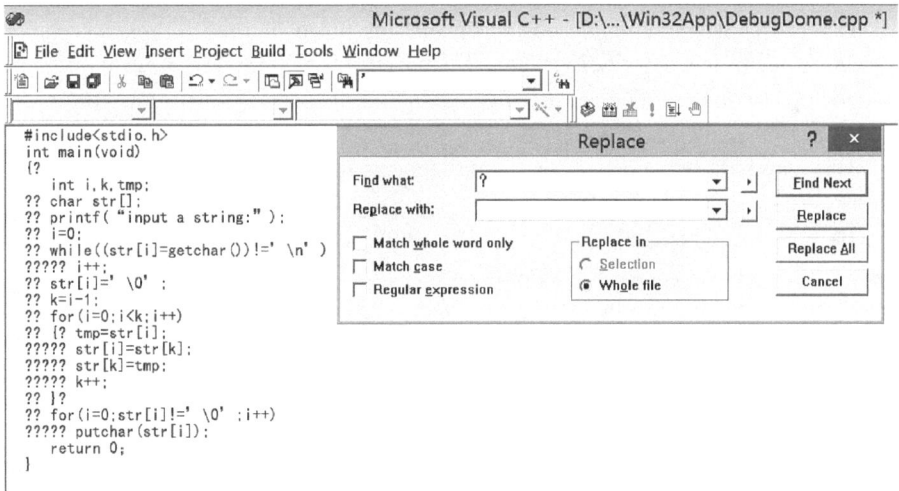

Figure 7.65: Copied program.

After removing abnormal characters, Mr. Brown noticed that the code was not aligned as required in the coding style. Because adjusting the alignment line by line was tedious, it would be helpful if there were a "One-click align" command. Fortunately, VC6 provides the "Format Selection" command, whose hotkey is Alt + F8. As shown in Figure 7.66, the format of the code can be adjusted with a single click using this command.

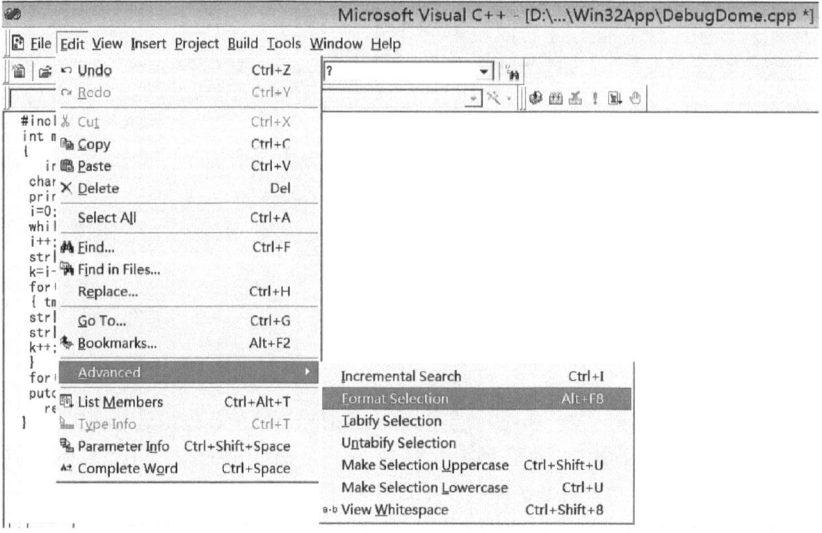

Figure 7.66: Format selected text.

7.6.2.2 Compilation

Mr. Brown clicked the compile button on the left end of Build MiniBar to compile the source code currently being opened. Upon the first compilation, a dialog box popped up as shown in Figure 7.67, asking whether he would like to create a project workspace. He selected "Yes."

The compilation result was shown in the message panel as shown in Figure 7.68. There were 27 errors in total.

He double-clicked the first error, a tiny blue arrow appeared in the editor as shown in Figure 7.69. The error message stated that the size of array str was unknown, which is indeed a bug as the size of an array was necessary.

However, errors still existed after he changed the array size to 8 so he double-clicked on the first error again and the screen was as shown in Figure 7.70. The error message was "unknown character." However, the print statement seemed correct at a glance. Having no idea what was wrong, the professor had to type in this line again. After careful observation, he noticed that the double quotation marks were incorrect. Both half-width characters and full-width characters were

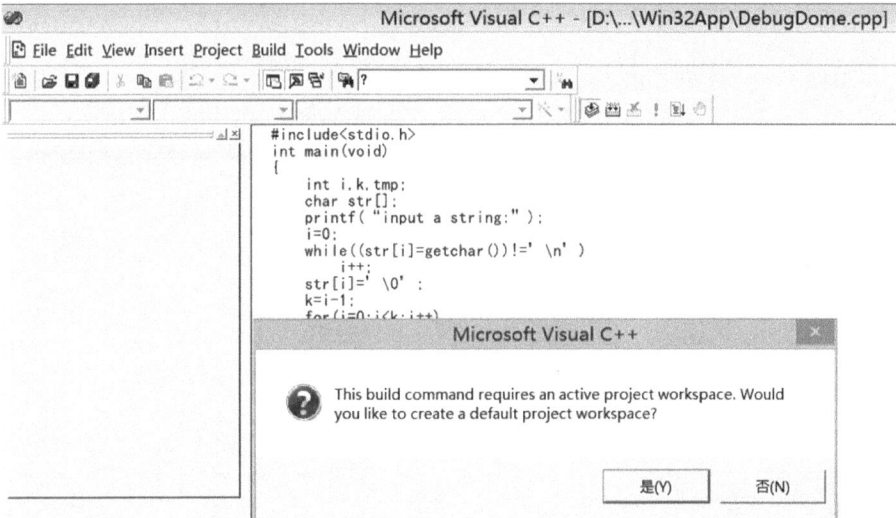

Figure 7.67: First compilation.

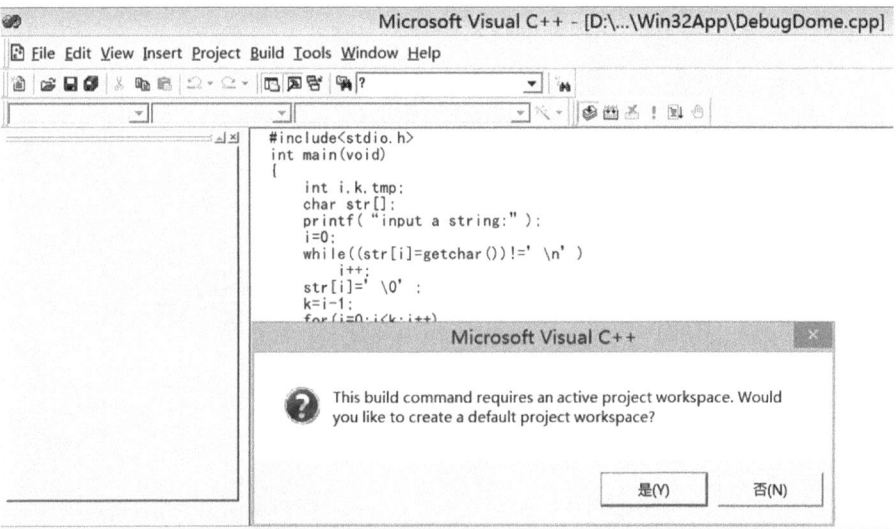

Figure 7.68: 27 errors in compilation result.

acceptable in the editor, but only half-width punctuation marks were correct in programs.

Mr. Brown recompiled the program after fixing the problem. There were 19 errors this time as shown in Figure 7.71. The number had decreased a lot compared with the initial one.

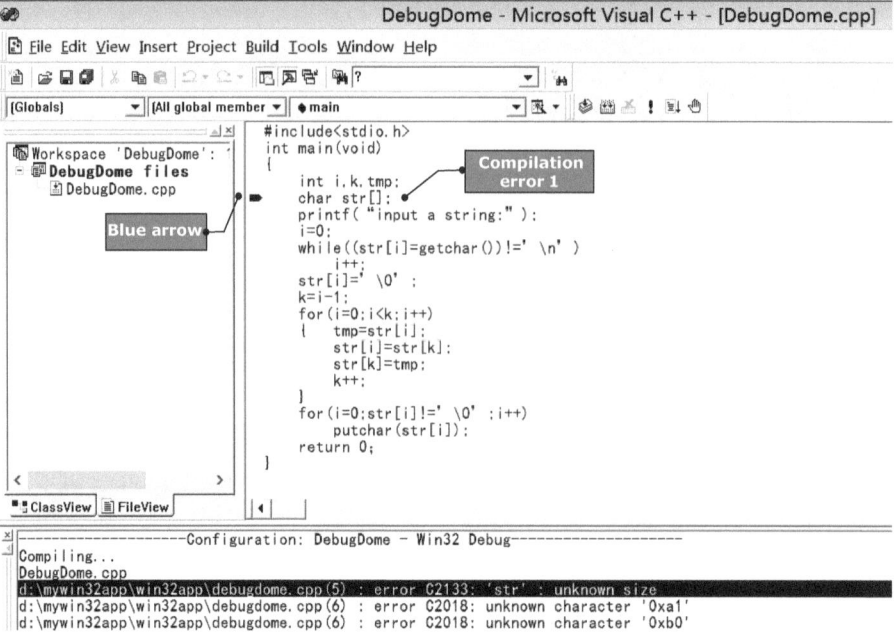

Figure 7.69: Compilation error 1.

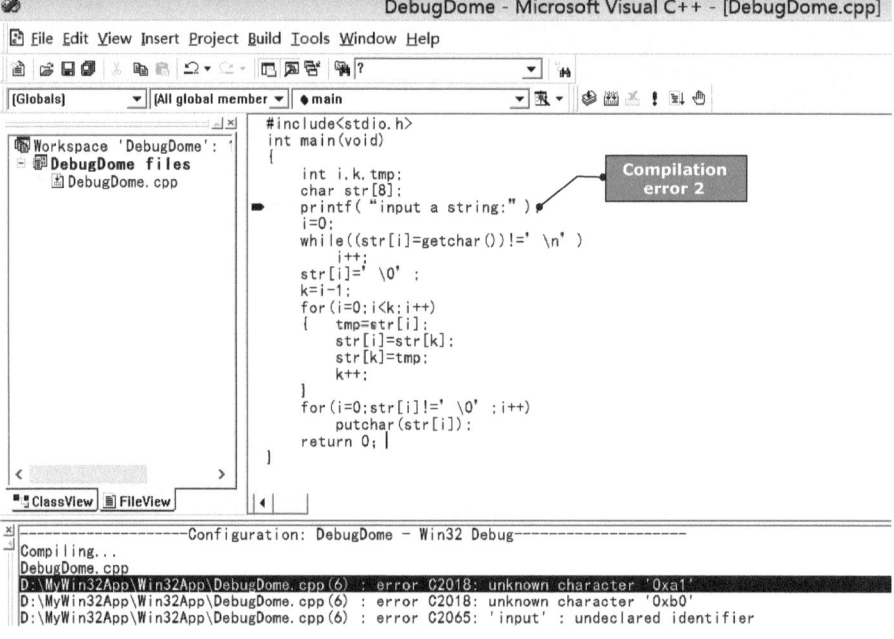

Figure 7.70: Compilation error 2.

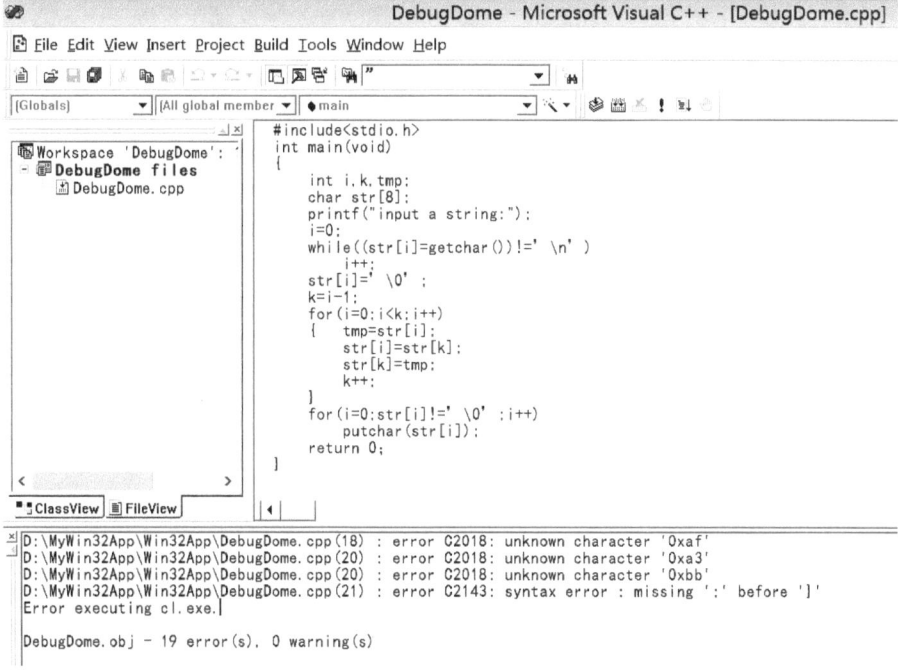

Figure 7.71: 19 errors in compilation result.

Double-clicking on the first error again, he found that the error indication arrow was on the line of while statement as shown in Figure 7.72. Having seen the full-width quotation mark error above, Mr. Brown immediately figured out that the single quotation marks were the error. There were multiple occurrences of this error in the program so he fixed them all at once. As shown in Figure 7.73, the number of errors decreased to three after compilation again, which was reduced a lot.

After double-clicking on the first error, the error indication arrow was on the line of the return statement as shown in Figure 7.74. Mr. Brown quickly noticed that it was the full-width semicolon that led to the error. Compilation succeeded after this error was fixed as shown in Figure 7.75.

We can conclude from the above compilation process that the number of errors is often reduced a lot if we recompile after fixing one error. This demonstrates that an error can cause subsequent errors. Hence, we only examine the first error and fix errors one by one when inspecting the compilation result.

7.6.2.3 Linking
Mr. Brown clicked the second button in Build MiniBar to link files after compilation succeeded. DebugDemo.exe was generated after linking succeeded as shown in Figure 7.76.

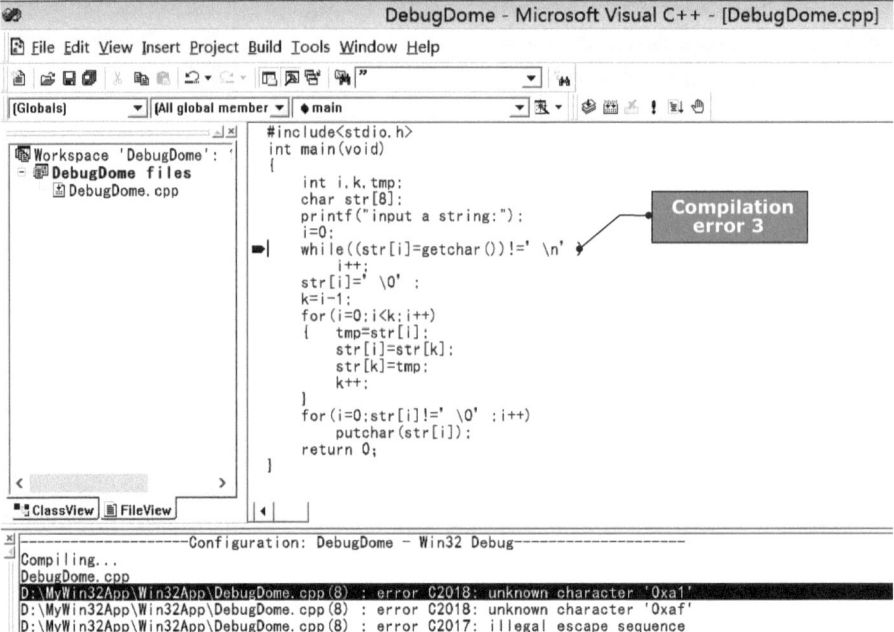

Figure 7.72: Compilation error 3.

Figure 7.73: 3 errors in compilation result.

Figure 7.74: Compilation error 4.

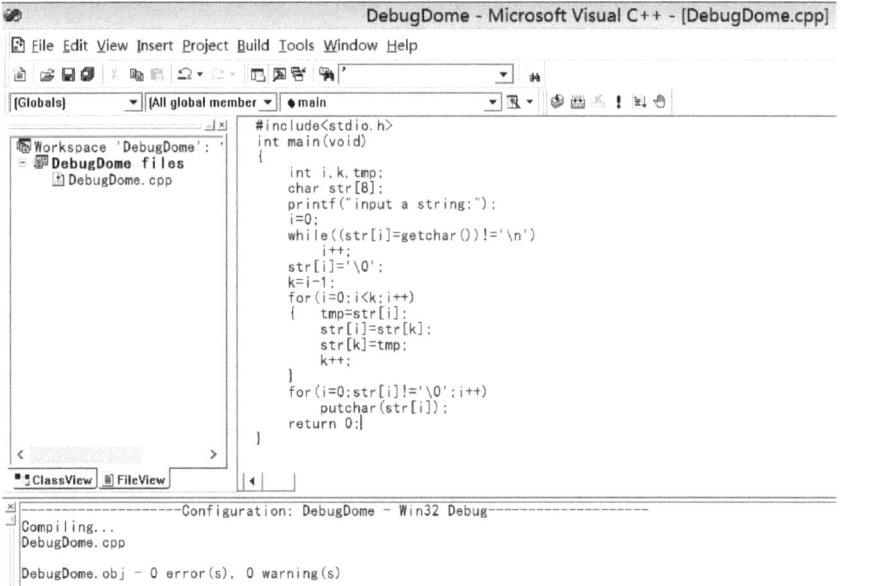

Figure 7.75: Compilation succeeded.

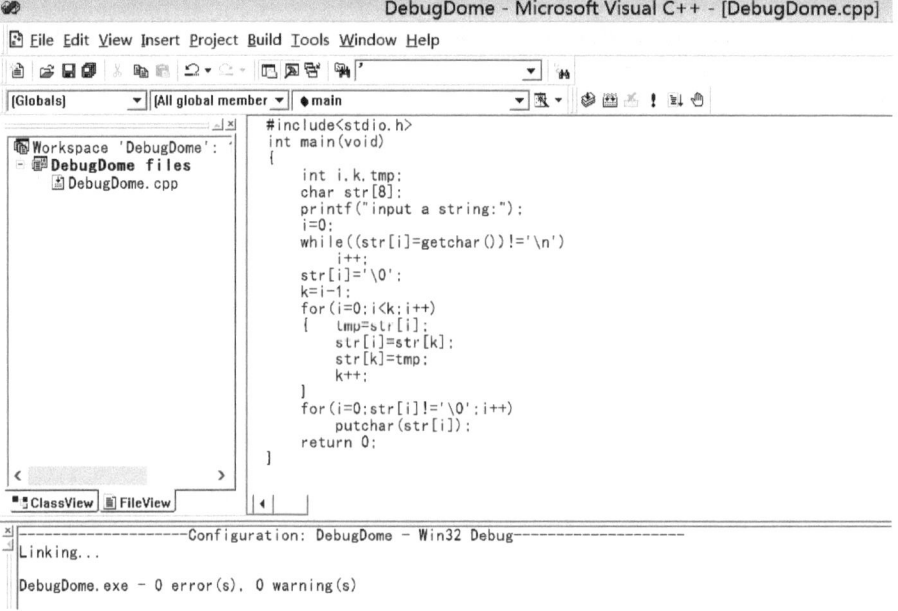

Figure 7.76: Generate exe file by linking.

7.6.2.4 Execution

Mr. Brown clicked the exclamation mark button in the Build MiniBar, and the result of the program was shown in the console. As shown in Figure 7.77, given the input "hello," only character "o" was output, which was incorrect.

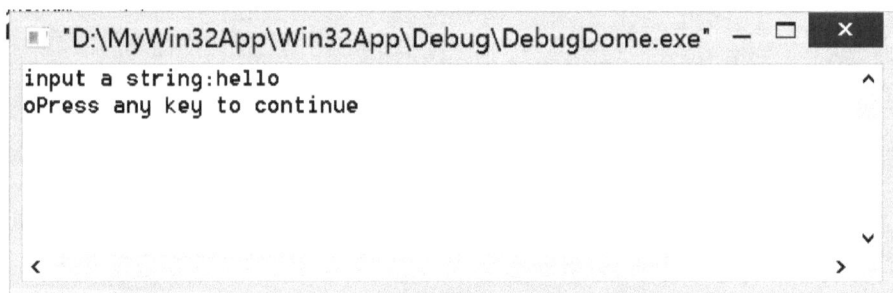

Figure 7.77: Wrong execution result.

7.6.2.5 Debugging

7.6.2.5.1 Insert breakpoint

Mr. Brown noticed that the input was read into array str character by character using a while loop, thus he decided to verify whether the input was correctly read first.

He inserted a breakpoint by left-clicking on the line after while loop and clicking the hand shape button in Build MiniBar as shown in Figure 7.78.

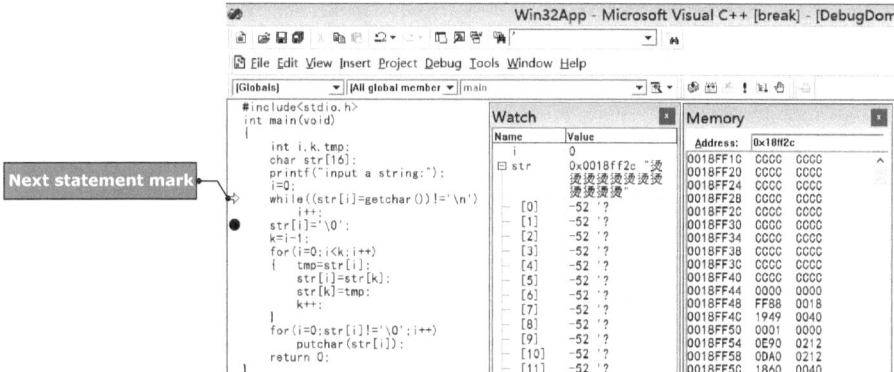

Figure 7.78: Insert a breakpoint.

After entering stepwise tracing by pressing F10, the debugger will add a yellow arrow on the left of program lines, indicating the next statement to be executed. Upon one F10 click, one line of code will be executed (if there are multiple statements in a line, all of them will be executed).

Before while loop was executed, there was no input. The value of i was 0. The address of array str was 0x18ff2c and the elements in it were all -52, which corresponded to the Chinese character "烫" and hexadecimal number CC. To be able to type in more characters, Mr. Brown changed the size of the array to 16.

Good programming habit
Write one statement in a line for easier debugging.

7.6.2.5.2 Inspect input

After clicking the Go button (next to the exclamation mark button) in Build MiniBar, the console window popped up with the message "input a string:" Mr. Brown typed in "hello," the debugger returned to the main screen of IDE, where the program was paused at the breakpoint as shown in Figure 7.79. Meanwhile, in the Watch and Memory window, the value of i was changed to 5 and str[5] was ASCII value 10, which corresponded to a newline. In conclusion, the input was read correctly.

As the input was correct, he proceeded to check whether data processing was correct.

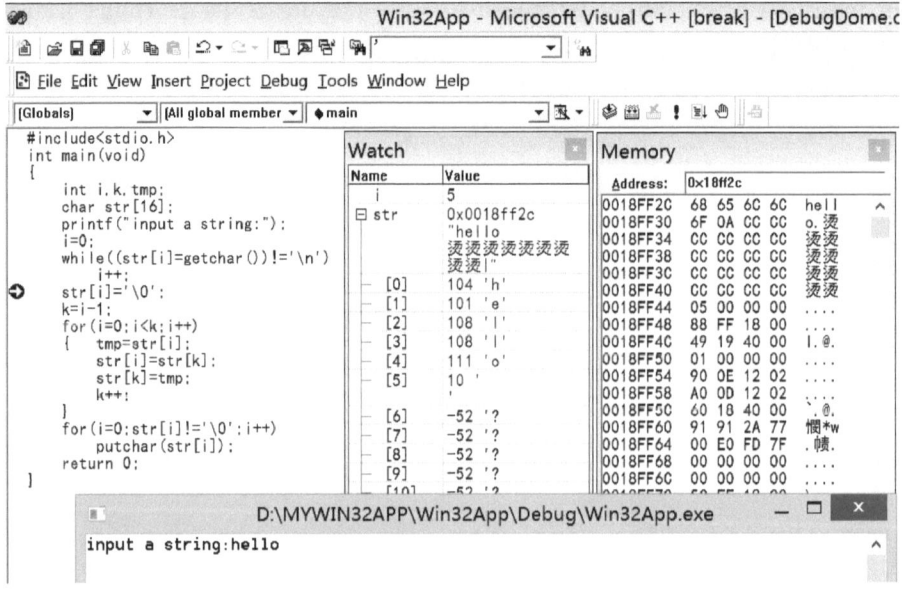

Figure 7.79: Input data.

7.6.2.5.3 Trace

There were two for loops in the processing part. The first one was used to reverse the string, whereas the second one was used for output. Mr. Brown started with the first loop. In each iteration, values of str[i] and str[k] were swapped; therefore, variables he needed to watch were i, k, str[i] and str[k], which were listed in the Watch window. As shown in Figure 7.80, i = 0 and k = 4 before the swap, which were the 0th and last index in the input string. Values of str[i] and str[k] were str[i] = 'h' and str[k] = 'o', which would be swapped in the loop body.

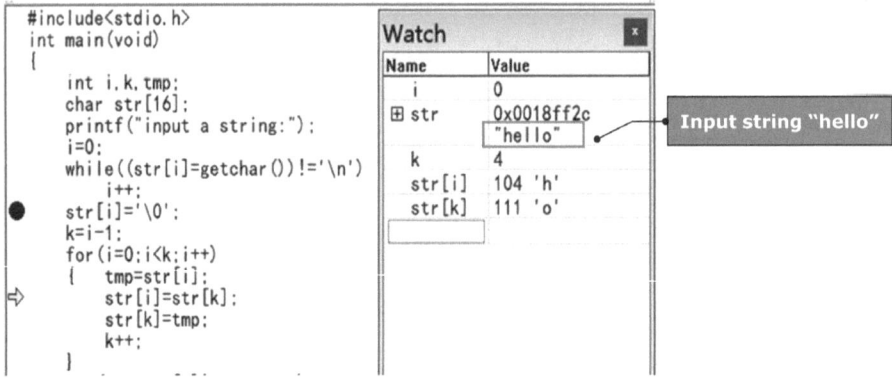

Figure 7.80: Process tracing 1.

In Figure 7.81, the swap in the loop body had been executed and the result indicated that the swap was successful.

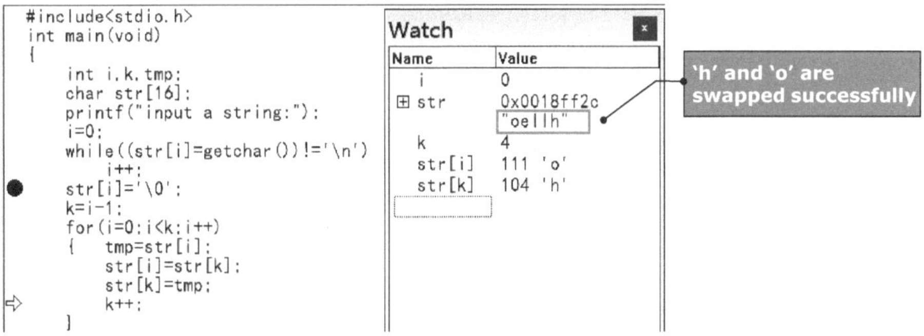

Figure 7.81: Process tracing 2.

In Figure 7.82, i = 1 and the program had entered the next iteration. The characters to be swapped this time were 'e' and 'l.' Although str[i] = 'e' in the Watch window, the value of str[k] was 0. Professor found the problem after careful inspection: logically, the value of k should decrease but not increase, so k++ was the root cause.

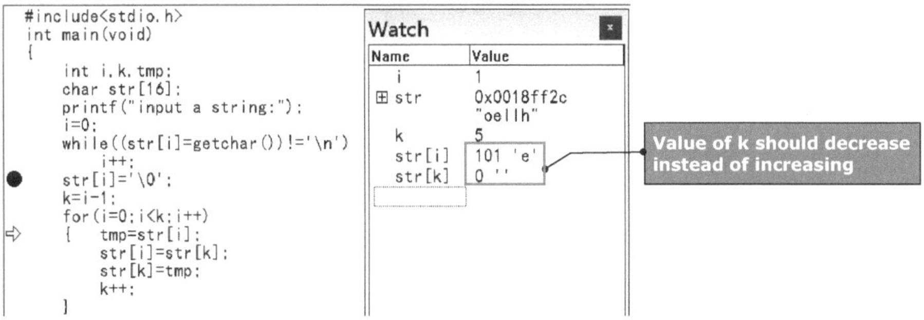

Figure 7.82: Process tracing 3.

7.6.2.5.4 Fix the error

Mr. Brown did not exit debugging to modify the code after finding the error. Instead, he changed the value of k in the Watch window to 3 and continued tracing as shown in Figure 7.83.

He then repeated this process by modifying the value of k in each iteration until the array str was completely reversed as shown in Figure 7.84. The second for loop should be correct as all it did was outputting the contents of array str. Professor inserted another breakpoint at the return statement.

After clicking the Go button, the correct output "olleh" was displayed in the console as shown in Figure 7.85.

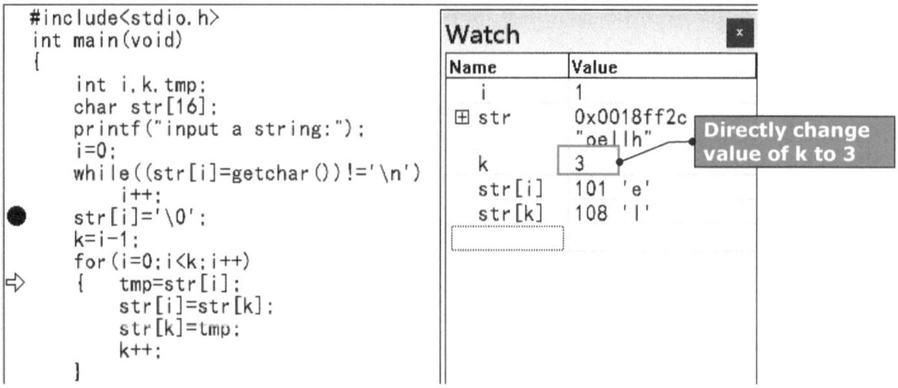

```
#include<stdio.h>
int main(void)
{
    int i,k,tmp;
    char str[16];
    printf("input a string:");
    i=0;
    while((str[i]=getchar())!='\n')
        i++;
    str[i]='\0';
    k=i-1;
    for(i=0;i<k;i++)
    {   tmp=str[i];
        str[i]=str[k];
        str[k]=tmp;
        k++;
    }
```

Watch		x
Name	**Value**	
i	1	
⊞ str	0x0018ff2c "oellh"	
k	3	
str[i]	101 'e'	
str[k]	108 'l'	

Directly change value of k to 3

Figure 7.83: Process tracing 4.

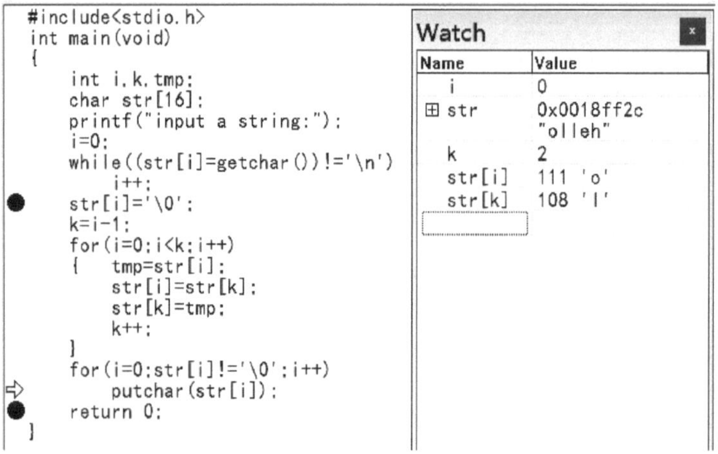

```
#include<stdio.h>
int main(void)
{
    int i,k,tmp;
    char str[16];
    printf("input a string:");
    i=0;
    while((str[i]=getchar())!='\n')
        i++;
    str[i]='\0';
    k=i-1;
    for(i=0;i<k;i++)
    {   tmp=str[i];
        str[i]=str[k];
        str[k]=tmp;
        k++;
    }
    for(i=0;str[i]!='\0';i++)
        putchar(str[i]);
    return 0;
}
```

Watch		x
Name	**Value**	
i	0	
⊞ str	0x0018ff2c "olleh"	
k	2	
str[i]	111 'o'	
str[k]	108 'l'	

Figure 7.84: Process tracing 5.

As debugging was done, Mr. Brown added captions to the screenshots of the debugging process and sent them back to the student. When reviewing them later, he noticed that there was a typo in the file name and multiple screenshots were affected. He really hoped the compiler could do a spell check, but that would be another story.

7.6.3 Example of using the call stack

We mentioned in Section 7.4.4 that the execution path of a program is automatically recorded in a stack. Doing so enables us to quickly locate the function in which errors occur by inspecting the top of the stack upon a program crash. In particular, this stack refers to the Call Stack in IDE, which we can see in Debug windows as shown in Figure 7.86.

```c
#include<stdio.h>
int main(void)
{
    int i,k,tmp;
    char str[16];
    printf("input a string:");
    i=0;
    while((str[i]=getchar())!='\n')
        i++;
    str[i]='\0';
    k=i-1;
    for(i=0;i<k;i++)
    {   tmp=str[i];
        str[i]=str[k];
        str[k]=tmp;
        k++;
    }
    for(i=0;str[i]!='\0';i++)
        putchar(str[i]);
    return 0;
```

Watch		
Name	**Value**	
i	5	
⊞ str	0x0018ff2c "olleh"	
k	2	
str[i]	0 ' '	
str[k]	108 'l'	

D:\MYWIN32APP\Win32App\Debug\Win32App.exe

```
input a string:hello
olleh
```

Figure 7.85: Process tracing 6.

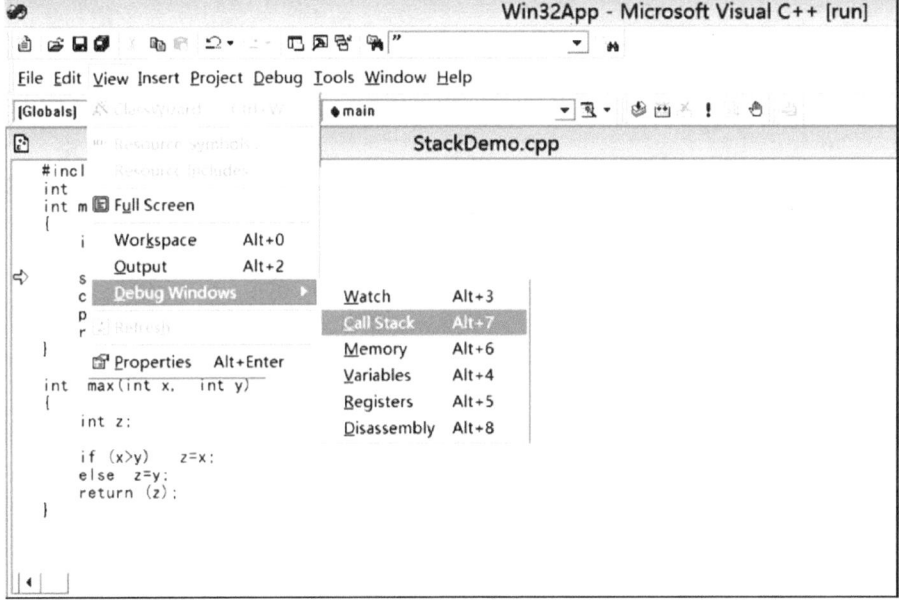

Figure 7.86: Call stack.

We shall use an actual program as shown in Figure 7.87, to demonstrate how this stack works. Upon execution, we enter the main function first and the stepwise tracing arrow points to the function to be called, namely max(). In the Call Stack window shown in Figure 7.88, we can see that the current function arrow points main, line 8. We count line numbers starting from the line of include (empty lines also count). The function call happens exactly on line 8.

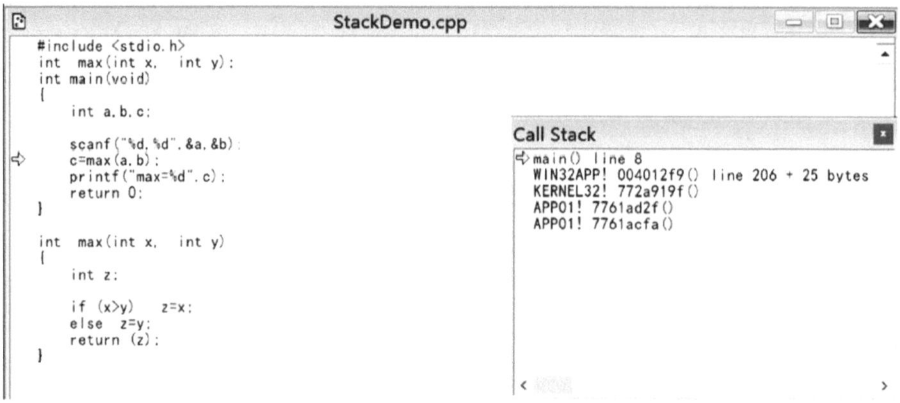

Figure 7.87: Call stack inspection 1.

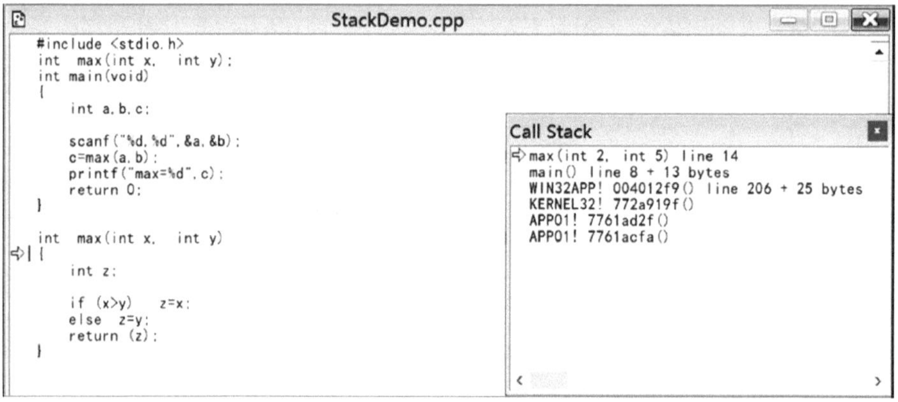

Figure 7.88: Call stack inspection 2.

Observing the Call Stack window after the program jumps to child function max, we notice that max becomes the current function and the program is currently at line 14. Arguments of the function are also displayed, which are consistent with the arguments shown in the Watch window as shown in Figure 7.89.

As shown in Figures 7.90 and 7.91, the value of line in Call Stack increases as stepwise tracing continues. Hence, we do not have to trace a program step by step

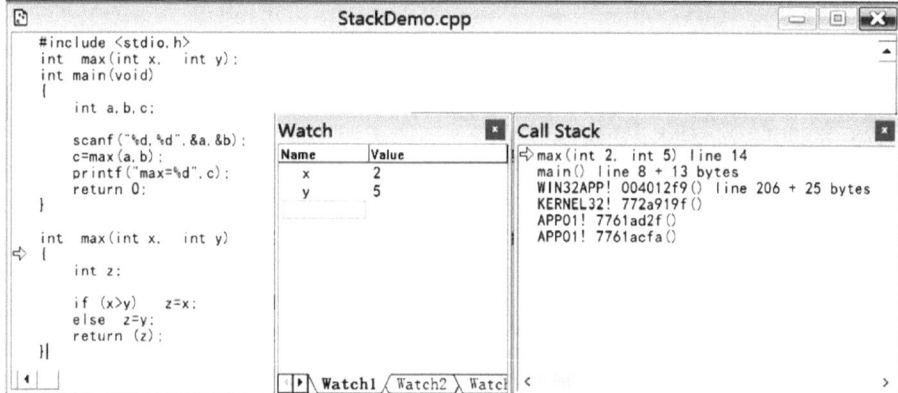

Figure 7.89: Call stack inspection 3.

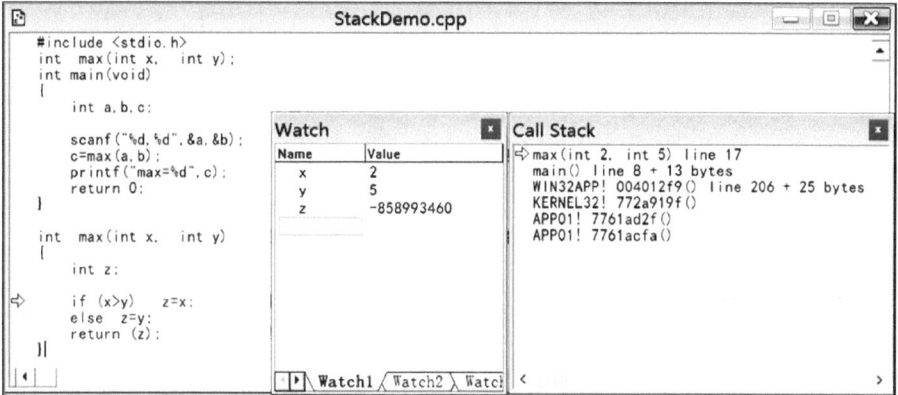

Figure 7.90: Call stack inspection 4.

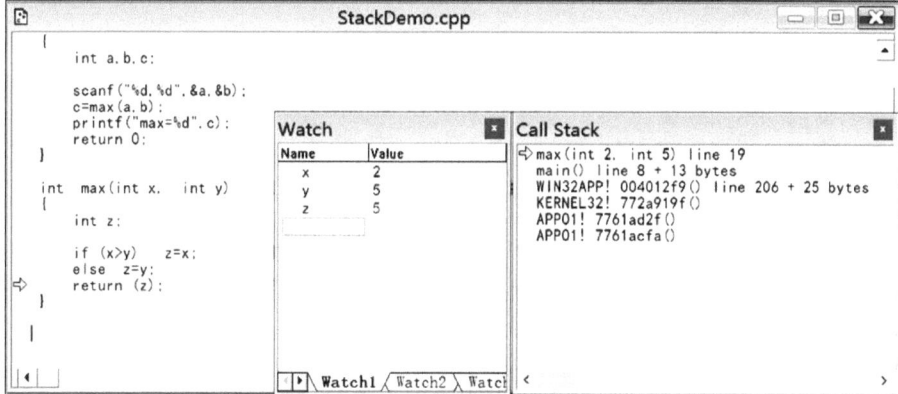

Figure 7.91: Call stack inspection 5.

when some statement causes system failure. As long as the failure is reproducible, we can find the error statement by inspecting the top of the call stack.

7.6.4 Example of using data breakpoint

We have seen examples of location breakpoints and now we are going to cover how to set and use data breakpoints.

7.6.4.1 Source code and execution result

The source code and execution result are shown in Figure 7.92.

```
#include "stdio.h"
#include"string.h"

int main(void)
{
    char str1[12]="hello world";
    char str2[4];
    int i=0;

    printf("str1 : %s\n", str1);
    printf("putchar str2: 12345678\n");
    do
    {
        str2[i]=getchar();
    }
    while (str2[i++]!='\n');
//  str2[i]='\0';

    printf("str1 : %s\n", str1);
    printf("str2 : %s\n", str2);
    return 0;
}
```

```
str1 : hello world
putchar str2:  12345678
12345678
str1 : 5678
world
str2 : 12345678
world
```

Figure 7.92: Example program of debugging using data breakpoints.

The program is straightforward. There are two character arrays, str1 and str2. The value of str1 is determined during initialization, whereas that of str2 comes from keyboard input. At the end of the program, characters in these arrays are output.

The execution result is problematic, in any case. The output of str1 was correct before str2 was input. However, it was changed after str2 was input.

7.6.4.2 Debugging plan

According to the erroneous result being displayed, the first half of str is changed while the second half is not. Values of str1[0] and str1[1] are 5 and 6, respectively.

Because 5 is entered before 6, the error must have happened when str1[0] was changed. By pausing the program at this point, we can investigate the root cause of the error. Our debugging steps are as follows:

- Execute the program step by step, observe str1 and str2 before str2 is input.
- Insert a data breakpoint to watch str1[0].
- Type in all input data at once in the console in tracing mode.
- Execute Go command and wait for the moment data breakpoint is encountered, which is also the moment error occurs.

7.6.4.3 Tracing and debugging

7.6.4.3.1 Inspect contents of array str1 and str2 before str2 is assigned a value

As shown in Figure 7.93, the content of str1 is its initial value. However, the content of str2 is weird. This is because the IDE stops the display of character array upon seeing the terminating character. The length of str2 is 4, which corresponds to two Chinese characters, "烫烫" (recall that a Chinese character takes up 2 bytes).

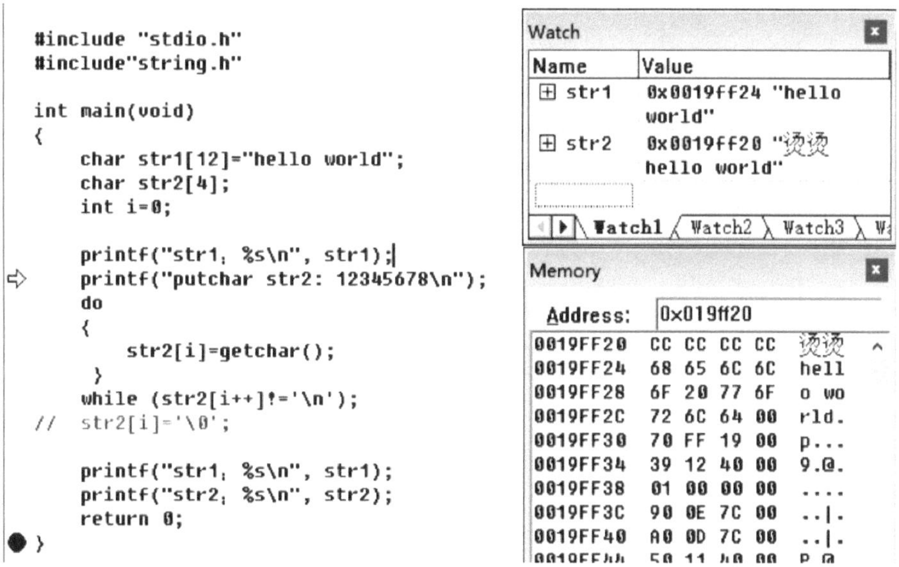

Figure 7.93: Debugging using data breakpoint step 1.

7.6.4.3.2 Insert data breakpoint

As shown in Figure 7.94, we select the breakpoints sub-menu in the menu "Edit" and choose the "Data" tab. After that, we enter the expression to be watched, str1[0], in the text area, "Enter the expression to be evaluated." The IDE then automatically completes other configurations and adds a breakpoint at the end of the program.

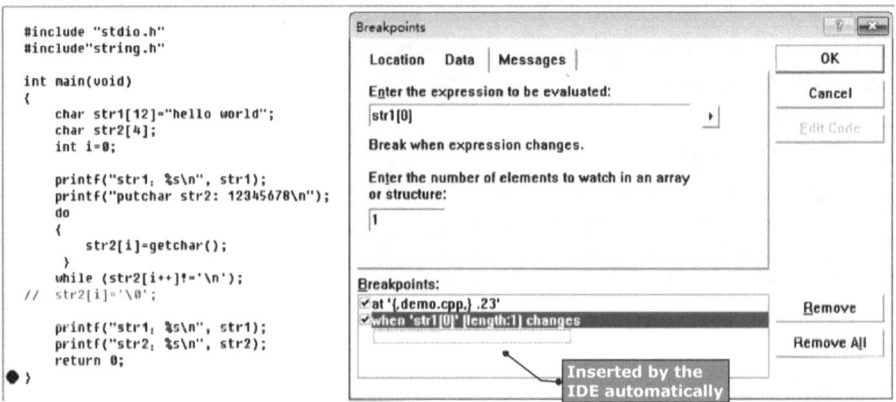

Figure 7.94: Debugging using data breakpoint step 2.

Returning to stepwise tracing, we enter "12345678" in one go when encountering getchar() in the do-while loop as shown in Figure 7.95. Then we continue the execution by pressing the "Go" button in Build MiniBar. A window pops up, indicating that a break has happened and the value of str1[0] has been modified as shown in Figure 7.96.

```
        printf("str1: %s\n", str1);
        printf("putchar str2: 12345678\n");
        do
        {
            str2[i]=getchar();
        }
  ⇨     while (str2[i++]!='\n');
    //  str2[i]='\0';

        printf("str1: %s\n", str1);
        printf("str2: %s\n", str2);
        return 0;
```

```
▣ D:\MYWIN32APP\Win32App\

str1: hello world
putchar str2: 12345678
12345678
▪
```

Figure 7.95: Debugging using data breakpoint step 3.

In the Watch window shown in Figure 7.97, we see that str1[0] has been changed to 5. The value of i is 4, so the change happened when assigning values to elements of str2 in the do-while loop. The root cause of this error is that str1 is right after str2 in memory. The size of array str2 is 4 bytes, thus a string longer than 4 bytes will override elements of str1. It is not hard to find the error in this case. One could figure out the error if he noticed the addresses of two arrays. However, the lesson we can learn from this example is that with proper use of breakpoints, we can capture the moment an error occurs without tracing the program step by step.

```
        printf("str1: %s\n", str1);
        printf("putchar str2: 12345678\n");
        do
        {
            str2[i]=getchar();
        }
⇨       while (str2[i++]!='\n');
    //  str2[i]='\0';

        printf("str1: %s\n", str1);
        printf("str2: %s\n", str2);
        return 0;
●   }
```

```
Microsoft Visual C++                          ×

   ⓘ  Break when 'str1[0]' (length:1) changes

                                         OK
```

Figure 7.96: Debugging using data breakpoint step 4.

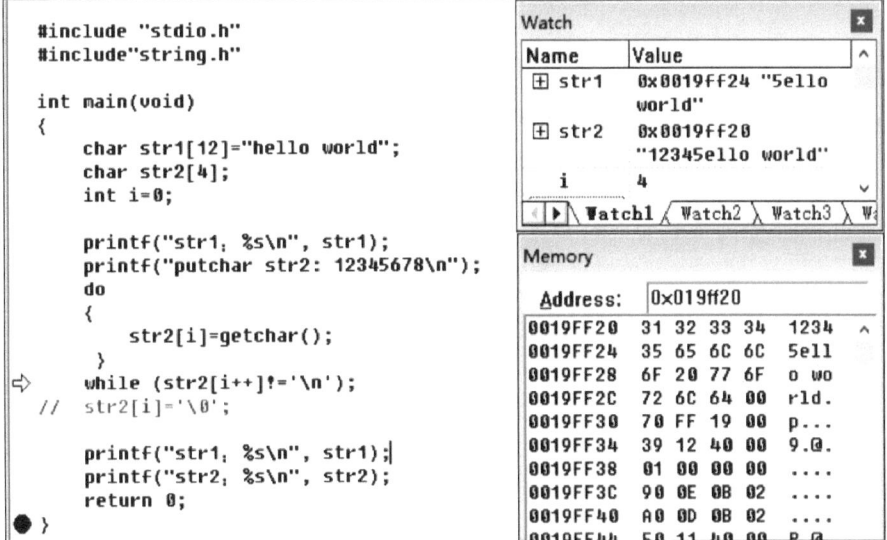

Figure 7.97: Debugging using data breakpoint step 5.

7.7 Summary

The main concepts in this chapter and their relations are shown in Figure 7.98.

We need to design test cases carefully before debugging.

Input and expected output should be determined.

Normal, exceptional, and edge cases should be considered.

We must be detail-oriented to be perfect

Do not panic when errors occur in the compilation.

Read the error message carefully to find what is wrong.

Errors may be caused by other errors.

So we should fix them one at a time.

We should compare execution result with the expected result,

And review the program logic if they are inconsistent.

Setting breakpoints, tracing step by step, watching variables, inspecting memory are all debugging techniques.

We find bugs by thinking and analyzing carefully, and eventually, we get the result right.

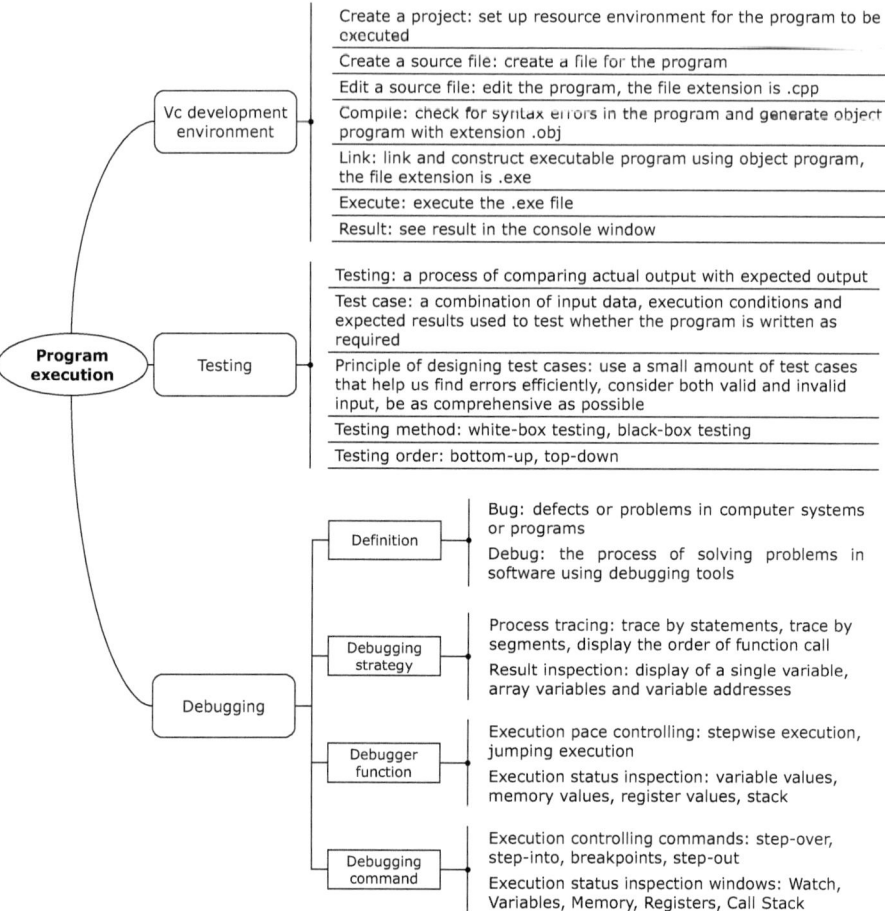

Figure 7.98: Concepts related to program execution and their relations.

7.8 Exercises

7.8.1 Multiple-choice questions

(1) [Program errors]

Which of the following errors is not detectable by computers during debugging? ( )

A) Compilation error

B) Runtime error

C) Logic error

D) All errors are detectable

(2) [Debugging and testing]

Which of the following statements is wrong? ( )

A) The purpose of testing is finding and correcting errors.

B) Locating the error is a necessary step in debugging.

C) We call the process of finding bugs "debug."

D) We should follow the testing plan in testing to eradicate randomness.

(3) [Software testing]

When using the white-box testing method, we should design test cases based on

( )

A) The internal logic of the program

B) Complex structure of the program

C) The functionality of the program

D) Manual

7.8.2 Debugging exercises

Note: In each problem, we show the lines with bugs. Please correct them without adding or deleting any line.

(1) Functionality

Given an integer n, the program outputs the sum of its digits (e.g., if n = 1308, the program outputs 12. If n = −3204, the program outputs 9.)

There is an error on line 6 and one on line 9.

```
01  #include <stdio.h>
02  int main(void)
03  {
04      int n,s=0;
```

```
05    scanf("%d",&n);
06    while (n<0)
07    {
08        s=s+n%10;
09        n=n%10;
10    }
11    printf("%d\n",s);
12    return 0;
13 }
```

(2) Functionality

Given a string input, the program stores its characters into a string t, reverses the string and appends it to t*. For example, given input "ABCD", the string t should be "ABCDDCBA."

There is an error on line 10 and one on line 11.

```
01 #include <stdio.h>
02 #include <string.h>
03 void fun(char *s,char *t)
04 {
05    int i,sl;
06    sl=strlen(s);
07    for (i=0; i<sl; i++)
08        t[i]=s[i];
09    for (i=0; i<sl; i++)
10        t[sl+i]=s[sl-i];
11    t[sl]="\0";
12 }
13
14 int main(void)
15 {
16    char s[100],t[100];
17    scanf("%s",s);
18    fun(s,t);
19    printf("%s",t);
20    return 0;
21 }
```

(3) Functionality

Given input a = 3 and n = 6, the program outputs the value of the following expression: 3 + 33 + 333 + 3333 + 33333 + 333333

There are two errors on lines 10 and 11.

```
01  #include <stdio.h>
02  int main(void)
03  {
04    int i,a,n;
05    long t=0,s=0;
06    scanf("%d%d",&a,&n);
07    t=a;
08    for (i=1; i<=n; i++)
09    {
10      t=t*10+i;
11      s=s+t;
12    }
13    printf("%ld\n",s);
14    return 0;
15  }
```

(4) Functionality

Given input *n*, the program outputs all prime factors of it (e.g., if n = 13860, the program outputs 2, 2, 3, 3, 5, 7, 11)

There is an error on line 7 and one on line 15.

```
01  #include <stdio.h>
02  int main(void)
03  {
04    int n,i;
05    scanf("%d",&n);
06
07    i=1;
08    while (n>1)
09      if (n%i==0)
10      {
11        printf("%d\t",i);
12        n/=i;
```

```
13        }
14        else
15           n++;
16        return 0;
17  }
```

Appendix A Precedence and associativity of operators

In the C language, we call the number of operands of an operator its "arity." A unary operator has only one operand. For example, +i, −j and x+\+ are unary operations. A binary operator has two operands. x+y and p%q are both binary operations.

When we use the same operator multiple times in a statement, some operators are evaluated from left to right, whereas some are evaluated from right to left. This attribute is called the associativity of operators in C (see Table A.1).

Table A.1: Operators in C.

Precedence	Operator	Meaning	Type	Associativity
1	() [ ] −> ,	Parentheses Index operator Structure pointer member operator Structure member operator	Unary	Left-to-right
2	! ~ ++ −− (type) + − * & sizeof	Negation operator Bitwise not operator Increment and decrement operators Forced-type conversion Positive and negative operators Dereference operator Address-of operator Size operator	Unary	Right-to-left
3	* / %	Multiplication, division, and remainder operators	Binary	Left-to-right
4	+ −	Addition and subtraction operator	Binary	Left-to-right
5	<< >>	Left-shift operator Right-shift operator	Binary	Left-to-right
6	< <= > >=	Less than, less than or equal to, greater than, greater than or equal to	Relational	Left-to-right
7	== !=	Equal to, not equal to	Relational	Left-to-right
8	&	Bitwise and operator	Bitwise	Left-to-right
9	^	Bitwise xor operator	Bitwise	Left-to-right
10	|	Bitwise or operator	Bitwise	Left-to-right

https://doi.org/10.1515/9783110692327-008

Table A.1 (continued)

Precedence	Operator	Meaning	Type	Associativity
11	&&	Conjunction operator	Bitwise	Left-to-right
12	||	Disjunction operator	Bitwise	Left-to-right
13	? :	Conditional operator	Ternary	Right-to-left
14	= += −= *= /= %= <<= >>= &= ^= |=	Assignment operator	Binary	Right-to-left
15	,	Comma operator	Sequential	Left-to-right

Appendix B ASCII table

ASCII value	Character	ASCII value	Character	ASCII value	Character	ASCII value	Character
0	NUT	32	(Space)	64	@	96	`
1	SOH	33	!	65	A	97	a
2	STX	34	"	66	B	98	b
3	ETX	35	#	67	C	99	c
4	EOT	36	$	68	D	100	d
5	ENQ	37	%	69	E	101	e
6	ACK	38	&	70	F	102	f
7	BEL	39	'	71	G	103	g
8	BS	40	(	72	H	104	h
9	HT	41	)	73	I	105	i
10	LF	42	*	74	J	106	j
11	VT	43	+	75	K	107	k
12	FF	44	,	76	L	108	l
13	CR	45	–	77	M	109	m
14	SO	46	.	78	N	110	n
15	SI	47	/	79	O	111	o
16	DLE	48	0	80	P	112	p
17	DCI	49	1	81	Q	113	q
18	DC2	50	2	82	R	114	r
19	DC3	51	3	83	S	115	s
20	DC4	52	4	84	T	116	t
21	NAK	53	5	85	U	117	u
22	SYN	54	6	86	V	118	v
23	TB	55	7	87	W	119	w
24	CAN	56	8	88	X	120	x
25	EM	57	9	89	Y	121	y
26	SUB	58	:	90	Z	122	z

https://doi.org/10.1515/9783110692327-009

(continued)

ASCII value	Character	ASCII value	Character	ASCII value	Character	ASCII value	Character
27	ESC	59	;	91	[	123	{
28	FS	60	<	92	\	124	|
29	GS	61	=	93	]	125	}
30	RS	62	>	94	^	126	~
31	US	63	?	95	_	127	DEL

Appendix C Common library functions of C

Library functions are not part of the C language. They are programs provided by compilers based on users' needs. Each C compiler has a collection of library functions. Each collection has a different number of functions. Functions in each compiler have different names and functionality. The American National Standards Institute (ANSI) C standard proposes a set of standard library functions for compilers to provide. This set includes library functions in most C compilers. However, these are still some functions in the set that have never been implemented in some compilers. Concerning generality, this appendix lists standard library functions proposed by ANSI C.

There are many types of library functions (e.g., screen and graphical functions, date/time functions, and system functions). Each of these types contains a series of functions that have different functionality. We cannot introduce all of them due to space limitations. Hence, we only list those needed in classes. When writing C programs, readers may refer to function manuals of the compiler they use.

1 Mathematical functions

To use mathematical functions (see Table C.1), we should use the following preprocessing command in our source file.

```
#include <math.h> or #include "math.h"
```

Table C.1: Mathematical functions.

Name	Prototype	Functionality	Return value
acos	double acos (double x);	Compute the value of arccos x, where −1≤x≤1.	Computation result
asin	double asin (double x);	Compute the value of arcsin x, where −1≤x≤1.	Computation result
atan	double atan (double x);	Compute the value of arctan x.	Computation result
atan2	double atan2(double x, double y);	Compute the value of arctan x/y.	Computation result
cos	double cos (double x);	Compute the value of cos x, where x is measured in radians.	Computation result

https://doi.org/10.1515/9783110692327-010

Table C.1 (continued)

Name	Prototype	Functionality	Return value
cosh	double cosh (double x);	Compute the value of cosh x (hyperbolic cosine).	Computation result
exp	double exp (double x);	Compute the value of e^x.	Computation result
fabs	double fabs (double x);	Compute the absolute value of x.	Computation result
floor	double floor (double x);	Compute the greatest integer less than or equal to x.	Double representation of the integer
fmod	double fmod (double x, double y);	Compute the floating point remainder of x/y.	Double representation of the remainder
frexp	Double frexp (double val, int *eptr);	Break double number val into its binary significand and an exponent of 2, namely val=x*2^n. n is stored in a variable pointed to by eptr.	Significand x, where $0.5 \le x < 1$
log	double log (double x);	Compute the value of lnx.	Computation result
log10	double log10(double x);	Compute the value of $\log_{10}x$.	Computation result
modf	double modf (double val, int *iptr);	Break double number val into its integer part and fraction part. The integer part is stored in a variable pointed to by iptr.	Fraction part of val
pow	double pow (double x, double y);	Compute the value of x^y.	Computation result
sin	double sin (double x);	Compute the value of sin x, where x is measured in radians.	Computation result
sinh	double sinh (double x);	Compute the value of sinh x (hyperbolic sine).	Computation result
sqrt	double sqrt (double x);	Compute the square root of x, where x≥0.	Computation result
tan	double tan (double x);	Compute the value of tan x, where x is measured in radians.	Computation result
tanh	double tanh (double x);	Compute the value of tanh x (hyperbolic tangent).	Computation result

2 Character functions

To use character functions (see Table C.2), we should use the following preprocessing command in our source file.

```
#include <ctype.h> or #include "ctype.h"
```

Table C.2: Character functions.

Name	Prototype	Functionality	Return value
isalnum	int isalnum (int ch);	Check if ch is a letter or a number.	1 if ch is either a number or a letter, 0 otherwise
isalpha	int isalpha (int ch);	Check if ch is a letter.	1 if ch is a letter, 0 otherwise
iscntrl	int iscntrl (int ch);	Check if ch is a control character (ASCII value between 0 and 0x1F, both inclusive).	1 if ch is a control character, 0 otherwise
isdigit	int isdigit (int ch);	Check if ch is a digit.	1 if ch is a digit, 0 otherwise
isgraph	int isgraph (int ch);	Check if ch has a graphical representation (ASCII value between 0x21 and 0x7e, both inclusive).	1 if ch has a graphical representation, 0 otherwise
islower	int islower (int ch);	Check if ch is a lowercase letter.	1 if ch is a lowercase letter, 0 otherwise
isprint	int isprint (int ch);	Check if ch is a printable character (ASCII value between 0x20 and 0x7e, both inclusive).	1 if ch is a printable character, 0 otherwise
ispunct	int ispunct (int ch);	Check if ch is a punctuation character (printable characters except letters, digits, and space).	1 if ch is a punctuation character, 0 otherwise
sspace	int isspace (int ch);	Check if ch is a white-space (space, tab, or newline).	1 if ch is a white-space, 0 otherwise
isupper	int isupper (int ch);	Check if ch is an uppercase letter.	1 if ch is an uppercase letter, 0 otherwise
isxdigit	int isxdigit (int ch);	Check if ch is a hexadecimal digit (0–9, A-F, a-f).	1 if ch is a hexadecimal digit, 0 otherwise
tolower	int tolower (int ch);	Convert ch into lowercase.	Lowercase letter of ch
toupper	int toupper (int ch);	Convert ch into uppercase.	Uppercase letter of ch

3 String functions

To use string functions (see Table C.3), we should use the following preprocessing command in our source file.

```
#include <string.h> or #include "string.h"
```

Table C.3: String functions.

Name	Prototype	Functionality	Return value
memchr	void memchr(void *buf, char ch, unsigned count);	Locate the first occurrence of ch in the first count characters in memory block buf.	A pointer to the first occurrence of ch in the block of memory pointed by buf. If ch is not found, the function returns NULL.
memcmp	int memcmp(void *buf1, void*buf2, unsigned count)	Compare the first count bytes of the block of memory pointed by buf1 to the first num bytes pointed by buf2.	buf1<buf2, return a negative number buf1=buf2, return 0 buf1>buf2, return a positive number
memcpy	void *memcpy (void *to, void*from, unsigned count);	Copy the values of count bytes from the location pointed to by from directly to the memory block pointed to by to. The arrays should not overlap.	to
memove	void *memove (void *to, void*from, unsigned count);	Copy the values of count bytes from the location pointed to by from directly to the memory block pointed to by to. The arrays may overlap.	to
memset	void *memset (void *buf, char ch, unsigned count);	Set the first count bytes of the block of memory pointed by buf to the specified character ch.	buf
strcat	char *strcat(char *str1, char *str2);	Append a copy of string str2 to str1. The first character of str2 overwrites the terminating null character '\0' in str1.	str1
strchr	char *strchr(char *str, int ch);	Find the first occurrence of ch in the string str.	A pointer to the first occurrence of ch in str. If ch is not found, the function returns NULL.

Table C.3 (continued)

Name	Prototype	Functionality	Return value
strcmp	int *strcmp(char *str1, char *str2);	Compare the string str1 to the string str2.	str1<str2, return a negative number str1=str2, return 0 str1>str2, return a positive number
strcpy	char *strcpy(char *str1, char *str2);	Copies the string pointed by str2 into the array pointed by str1, including the terminating null character.	str1
strlen	unsigned int strlen(char *str);	Return the length of the string str (without including '\0').	The length of string
strncat	char*strncat (char*str1, char*str2, unsigned count);	Append the first count characters of str2 to str1, plus a terminating null-character.	str1
strncmp	int strncmp(char *str1, *str2, unsigned count);	Compare up to count characters of the C string str1 to those of the C string str2.	str1<str2, return a negative number str1=str2, return 0 str1>str2, return a positive number
strncpy	char*strncpy (char*str1, *str2, unsigned count);	Copy the first count characters of str2 to str1.	str1
strnset	void *setnset (char *buf, char ch, unsigned count);	Set the first count characters of string buf to character ch.	buf
strset	void *setset(void *buf, char ch);	Set all characters of string buf to character ch.	buf
strstr	char *strstr(char *str1, *str2);	Find first occurrence of str2 in str1.	A pointer to the first occurrence in str1 of the entire sequence of characters specified in str2, or a null pointer if the sequence is not present in str1.

4 Input/output functions

To use input/output functions (see Table C.4), we should use the following preprocessing command in our source file.

```
#include <stdio.h> or #include "stdio.h"
```

Table C.4: Input/output functions.

Name	Prototype	Functionality	Return value
clearerr	void clearer (FILE*fp);	Reset both the error and the eof indicators of the file stream fp.	None
eof	int eof(int fp);	Check if the file pointed by fp has reached end.	1 if the file ends, 0 otherwise
fclose	int fclose(FILE *fp);	Close the file associated with fp, and release the buffer.	0 if successfully closed, nonzero value otherwise
feof	int feof(FILE *fp);	Check whether the eof indicator associated with fp is set.	A nonzero value if set, 0 otherwise
ferror	int ferror(FILE *fp);	Check if the error indicator associated with fp is set.	A nonzero value if set, 0 otherwise
fflush	int fflush(FILE *fp);	Save all control information and data of the file pointed by fp.	0 if successfully saved, nonzero value otherwise
fgets	char *fgets(char *buf, int n, FILE *fp);	Read characters from fp and stores them as a C string into buf until (n-1) characters have been read.	buf on success. EOF if end-of-file reached or error occurs.
fgetc	int fgetc(FILE *fp);	Return the character currently pointed by the internal file position indicator of the specified fp.	The character read is returned. If error occurs, EOF is returned.
fopen	FILE*fopen (char*filename, char *mode);	Open the file whose name is filename in the specified mode.	A pointer to the file on success, 0 otherwise
fprintf	int fprintf(FILE *fp, char *format, args, . . .);	Write the C string pointed by format to fp.	Total number of characters written
fputc	int fputc(char ch, FILE *fp);	Write a character ch to fp.	The character written is returned on success, EOF is returned if an error occurs

Table C.4 (continued)

Name	Prototype	Functionality	Return value
fputs	int fputs(char str, FILE *fp);	Writes the C string pointed by str to fp.	0 on success, EOF on error
fread	int fread(char*pt, unsigned size, unsigned n, FILE *fp);	Read an array of n elements, each one with a size of size bytes, from fp and store them in the block of memory specified by pt.	The total number of elements successful read is returned. If an error occurs or the end-of-file is reached, 0 is returned.
fscanf	int fscanf(FILE *fp, char *format, args, . . .);	Read data from fp and store them according to the parameter format into the locations pointed by args.	Number of items successfully filled
fseek	int fseek(FILE *fp, long offset, int base);	Set the position indicator associated with fp to a new position (base + offset).	The current position is returned on success, −1 is returned otherwise
ftell	long ftell(FILE *fp);	Return the current value of the position indicator of fp.	The current position is returned on success, 0 otherwise
fwrite	int fwrite(char *ptr, unsigned size, unsigned n, FILE *fp);	Write an array of n elements, each one with a size of size bytes, from the block of memory pointed by ptr to the current position in fp.	The total number of elements successfully written is returned
getc	int getc(FILE *fp);	Return the character currently pointed by the internal file position indicator of fp.	The character read is returned on success, −1 is returned if end-of-file is reached or an error occurs.
getchar	int getchar();	Return the next character from the standard input.	The character read is returned on success, −1 is returned if end-of-file is reached or an error occurs.
gets	char *gets(char *str);	Reads characters from the standard input and stores them as a C string into str.	str on success, NULL otherwise [Note: C11 standard introduces a safer function, gets_s(), to substitute gets()].
printf	int printf(char *format, args, . . .);	Write the C string pointed by format to the standard output. The additional arguments following format are formatted and inserted in the resulting string replacing their respective specifiers.	The total number of characters written is returned on success, a negative number is returned on error.
putc	int prtc(int ch, FILE *fp);	Write a character ch to fp and advance the position indicator.	The character written on success, EOF on error
putchar	int putchar (char ch);	Write a character ch to the standard output.	The character written on success, EOF on error

Table C.4 (continued)

Name	Prototype	Functionality	Return value
puts	int puts(char *str);	Write the C string pointed by str to the standard output and appends a newline character ('\n').	A non-negative value on success, EOF on error
read	int read(int fd, char *buf, unsigned count);	Read up to count bytes from file descriptor fd into the buffer starting at buf.	The number of bytes read on success, −1 on error
remove	int remove(char *fname);	Delete the file whose name is specified in fname.	0 on success, −1 on error
rename	int remove(char *oname, char *nname);	Change the name of the file or directory specified by oname to nname.	0 on success, −1 on error
rewind	void rewind (FILE *fp);	Set the position indicator associated with fp to the beginning of the file, clear the end-of-file and error internal indicators.	None
scanf	int scanf (char *format, args, . . .);	Read data from stdin and stores them according to the parameter format into the locations pointed by the additional arguments.	The number of items of the argument list successfully filled. EOF if end-of-file is reached. 0 on error.
sscanf	nt sscanf (const char *str, const char * format, ...);	Read data from str and stores them according to parameter format into the locations given by the additional arguments, as if scanf was used, but reading from s instead of the standard input.	The number of items in the argument list successfully filled. −1 on error, and the error reason is stored in errno.
write	int write(int fd, char *buf, unsigned count);	Write up to count bytes from the buffer starting at buf to the file referred to by the file descriptor fd.	The number of bytes written on success, −1 on error.

5 Dynamic storage allocation functions

To use dynamic storage allocation functions (see Table C.5), we should use the following preprocessing command in our source file.

```
#include <stdlib.h> or #include "stdlib.h"
```

Table C.5: Dynamic storage allocation functions.

Name	Prototype	Functionality	Return value
callloc	void*calloc (unsigned n, unsigned size);	Allocate a block of memory for an array of *n* elements, each of them size bytes long, and initialize all its bits to zero.	A pointer to the beginning of the memory block allocated by the function on success, 0 otherwise.
free	void free(void *p);	A block of memory p previously allocated by a call to malloc, calloc or realloc is deallocated.	None
malloc	void*malloc (unsigned size);	Allocate a block of size bytes of memory.	A pointer to the beginning of the memory block allocated by the function on success, 0 if not enough memory.
realloc	void*realloc (void *p, unsigned size);	Change the size of the memory block pointed to by p to size. size can be either larger or smaller than the original size.	A pointer to the reallocated memory block on success, NULL on failure.

6 Other functions

Table C.6 lists functions that are not in the above categories. We should use the following preprocessing command in our source file to use these functions.

```
#include <stdlib.h> or #include "stdlib.h"
```

Table C.6: Other functions.

Name	Prototype	Functionality	Return value
abs	int abs(int num);	Compute the absolute value of num.	Computation result
atof	double atof(char *str);	Parse the C string str, interpreting its content as a floating point number.	Computation result in double precision
atoi	int atoi(char *str);	Parse the C-string str, interpreting its content as an integral number of type int.	Conversion result
atol	long atol(char *str);	Parse the C-string str, interpreting its content as an integral number of type long.	Conversion result
exit	void exit(int status);	Terminate the program, return status to the caller.	None

Table C.6 (continued)

Name	Prototype	Functionality	Return value
itoa	char *itoa(int n, char *str, int radix);	Convert an integer n to a null-terminated string using base radix and store the result in the array given by str.	A pointer to str
labs	long labs(long num);	Compute the absolute value of long integer num.	Computation result
ltoa	char *ltoa(long n, char *str, int radix);	Convert a long integer n to a null-terminated string using base radix and store the result in the array given by str.	A pointer to str
rand	int rand();	Return a pseudo-random integral number in the range between 0 and RAND_MAX. RAND_MAX is defined in the header file.	A pseudo-random integer
random	int random(int num);	Generate a random integer between 0 and num.	A random integer

Appendix D Common escape characters

Escape characters are character sequences starting with "\". They are special character constants in C. Table D.1 lists common escape characters.

Table D.1: Common escape characters.

Character	Meaning	ASCII value
\0	Null character	0
\n	Newline, which moves the cursor to the beginning of the next line	10
\t	Horizontal tab, which moves the cursor to the next output field (each field has eight columns)	9
\v	Vertical tab	
\b	Backspace, which moves the cursor back by one column	8
\r	Carriage return, which moves the cursor to the beginning of the current line	13
\f	Form feed, which moves the cursor to the beginning of the next page	12
\a	Alarm	
\\	Backslash	92
\'	Single quote	39
\"	Double quote	34
\?	Question mark	63
\ddd	Octal number with three digits	
\xhh	Hexadecimal number with two digits	

https://doi.org/10.1515/9783110692327-011

Appendix E Bitwise operations

1 Bitwise AND (&)

Bitwise AND (&) is mainly used for two purposes:

(1) Zeroing out

For example, we have a number x = 0010 1011. Then we can use y = 1101 0100 or y = 0000 0000 to zero out x so that x&y = 0.

(2) Extracting the specified bit

For example, suppose we have a number a = 0010 1100 1010 1100, which takes up two bytes. To obtain its lower byte, we can use a number y = 0000 0000 1111 1111 and do

$$a\&y = 0000\ 0000\ 1010\ 1100$$

Suppose we have a number a = 0101 0100, and we want to preserve the 3rd, 4th, 5th, 7th, and 8th bits of it (counting from the left). We can use a number b = 0011 1011 and do

$$c = a\&b = 0001\ 0000$$

2 Bitwise OR (|)

Suppose a = 0011 0000, b = 0000 1111, then a|b = 0011 1111.

Usage: We use bitwise OR to change specified bits of a binary number to 1, without knowing what those bits were.

Mask: The specified bits of the mask are 1's, whereas the remaining bits are 0's.

For example, suppose we have int a = 055555 and we want to change the highest bit to 1. Then we can use a mask b = 0x8000.

a: 0101 1011 0110 1101
b: 1000 0000 0000 0000
a|b: 1101 1011 0110 1101

3 Bitwise XOR (^)

The bitwise XOR operation takes two binary operands of equal length. The result in each position is 1 if and only if the two bits in this position are different. Bitwise XOR is also known as bitwise addition (corresponding bits are added, and the carry is discarded). It is used for the following purposes:

https://doi.org/10.1515/9783110692327-012

(1) Flipping specified bits (1 changed to 0, 0 changed to 1)

Mask: The specified bits of the mask are 1's, while the remaining bits are 0's.

For example, suppose we have a = 0x0F 0000 0000 0000 1111. Then we can use number b = 0x18 0000 0000 0001 1000 and

$$a^b = 0000000000010111$$

(2) XOR with 0 to preserve the value
(3) Swapping values of two variables without any intermediate variable

The method is as follow:

$$a = a\char`\^b;\ \ b = b\char`\^a;\ \ a = a\char`\^b;$$

Proof: Based on the second equation, we have

$$b = b\char`\^a = b\char`\^(a\char`\^b) = b\char`\^a\char`\^b = a\char`\^b\char`\^b = a\char`\^0 = a$$

Based on the third equation, we have

$$a = a\char`\^b = (a\char`\^b)(b\char`\^(a\char`\^b)) = a\char`\^b\char`\^b(a\char`\^b) = a\char`\^0\char`\^a\char`\^b = a\char`\^a\char`\^b = 0\char`\^b = b$$

4 Bitwise NOT (~)

~ is a unary operator for bitwise NOT.

For example, ~025, namely ~0000 0000 0001 0101, is 1111 1111 1110 1010.

Note:

(1) ~025 is not −025.
(2) Performing bitwise NOT on a number twice yields the original number.
(3) We often use bitwise NOT together with bitwise AND, bitwise OR or shift operations to complete specific tasks.

For example, the expression x&~077 extracts the bits in front of the lower 6 bits of x and zeroes out the lower 6 bits.

5 Shift operations (>>, <<)

Shift operations are in the form m<<n and m>>n, in which m is the number to be shifted and n is the number of bits of the shift. m and n are both integer expressions. The type of the result depends on the type of m.

In a << operation, the higher bits of operand m, which are shifted out of the left end, are discarded. The remaining bits are padded with 0 on the right.

In a >> operations, the lower bits of operand m, which are shifted out of the right end, are discarded. If m is an unsigned number, the remaining bits are padded with 0 on the left. If m is signed, a sign is added in an arithmetic shift, and 0 are padded in a logical shift.

Associativity: << and >> are left-associated. A left shift is equivalent to multiplying a power of 2 and a right shift is equivalent to dividing the operand with a power of 2.

Table E.1 shows examples of multiplication and division operations using shift operations.

Table E.1: Multiplication and division with shift operations.

Character x	x after shift	Value of x
x = 7	00000111	7
x≪1	00001110	14
x≪3	01110000	112
x≪2	11000000	192
x≫1	01100000	96
x≫2	00011000	24

Bitwise operators can be used with assignment operators to obtain extended assignment operators such as &= , |=, >>=, <<=, and ^=.

a&= b is equivalent to a = a&b

a<<= 2 is equivalent to a = a<<2

For example, the expression x>>p+1-n&~(~0<<n) does the following: it extracts n bits of x starting from the pth position (counting from the right end, which is the 0th position) and stores these bits in the lower bits. Suppose p =4 and n = 3. Then the result is bits of x between the second bit and the fourth bit.

Index

https://doi.org/10.1515/9783110692327-013